Lecture Notes in Educational Technology

Lecture Notes in Educational Technology

The series *Lecture Notes in Educational Technology* (LNET), has established itself as a medium for the publication of new developments in the research and practice of educational policy, pedagogy, learning science, learning environment, learning resources etc. in information and knowledge age, – quickly, informally, and at a high level.

More information about this series at http://www.springer.com/series/11777

Daniel Churchill · Jie Lu
Thomas K.F. Chiu · Bob Fox
Editors

Mobile Learning Design

Theories and Application

Springer

Editors
Daniel Churchill
Faculty of Education
The University of Hong Kong
Hong Kong
Hong Kong

Thomas K.F. Chiu
Faculty of Education
The University of Hong Kong
Hong Kong
Hong Kong

Jie Lu
Educational Resources Centre, BGI College, Genomics
Shenzhen
China

Bob Fox
Faculty of Arts and Social Sciences
University of New South Wales
Sydney
Australia

ISSN 2196-4963 ISSN 2196-4971 (electronic)
Lecture Notes in Educational Technology
ISBN 978-981-10-0025-6 ISBN 978-981-10-0027-0 (eBook)
DOI 10.1007/978-981-10-0027-0

Library of Congress Control Number: 2015955878

Springer Singapore Heidelberg New York Dordrecht London

Printed on acid-free paper

Springer Science+Business Media Singapore Pte Ltd. is part of Springer Science+Business Media (www.springer.com)

Foreword

What Will They Think of Next?

This book arrives at an exciting time. The technology stories that circulate in the media talk about access through WIFI anywhere and diminishing (or at least increasing variability of) device size. So it is not unexpected that a group of scholars should start discussing the possibilities and opportunities of such developments in how we can learn and collaborate in such a digital world that has broken away from traditional classrooms. Daniel Churchill, his co-editors and the contributors have created a text that summarises what designers and researchers believe are the range of influences that this emerging field is facing. In the first part of six sections, the chapters deal with the definitional and emerging nuances being identified with the field of study. The key concepts are not unexpectedly: mobility, interactivity and collaboration, and augmentation. But as the later chapters explore it is also about how the world can be represented, accessed and overlaid with digital support. Overall, the opportunities of mobile learning and the barriers it breaks by social and other forms of communication and collaboration promise more that the rather limited and still largely didactic e-learning models available in many educational ecosystems.

In the second part, the focus shifts to the current adoption of mobile learning and how students perceive its value. Importantly, trends that have been noted here have been the possibilities of working on real-world contexts with overlays of digital structures and mentoring. The third part explores the combination of technology, pedagogy and context improving the flexibility of the new mobile learning contexts to provide increasing student personalization and to collect data of individual learning styles and strategies. This long hoped for adaptive learning system approach has been a goal of learning technologies for many years but it has largely eluded many designers; the different chapters illustrate how mobile learning contexts support collaboration and sharing in ways that have not been designed into most standard eLearning contexts.

The next two sections of this book focus on how mobility and the combination of technologies can "fit" in learning broadly and in specific discipline domains. The writers have written about particular strategies and the "fit" with each discipline. This emphasis is important, early approaches to learning science did not identify the importance of domain knowledge and how it could be could supported and enhanced with the combination of elements—technology, learning approach and pedagogical context.

In the last section, the one chapter seeks to explore how future options might influence how mobile approaches might effectively support learning in a digital age. In this summarization, the possibilities of mobility and smart devices support a learner to explore their world by providing an organising lens to display the evidence and to aggregate it in ways that support the learner's meaning-making. I believe that the chapter raises both the plus and minus sides of this new learning ecology; this combination enables convenience of access to ideas and tools to support creation of many ways and modalities of representing them, and also to the increasing loss of personal privacy as the learner collates and makes sense of the phenomena they are studying. Overall, this book pulls together all of the elements that have been used as a solution to student motivation, increasing impact, supporting diversity and enabling personalization of the tools that support mLearning.

John G. Hedberg
Macquarie University

Preface

This book has been written and published at the time of growing interest in and a need for mobile learning in education at all levels. The chapters in this book are primarily concerned with theories and practices related to the adoption of mobile and emerging technologies in education. These chapters are collected from three sources. The first source comprises a pool of papers directly submitted for consideration for inclusion in this book. The second source includes papers from a small number of invited authors. The third source comprises a small number of rigorously selected papers from the pool of papers presented at the International Mobile Learning Festival (IMLF). The IMLF conference is a regular international gathering of scholars and educational practitioners interested in mobile and emerging learning design. The conference features evidence-based developments surrounding mobile and emerging learning design for the twenty-first century learning.

Educational usages of e-books, streaming videos, podcasts, social networking, cloud computing, blogs, multimedia and video editing and many other mobile applications have been adopted by innovative educators and institutions around the world. To scale-up these innovative practices mediated by mobile technologies, there is a pressing need to harness research studies with a solid theoretical underpinning, and empirically validated practical recommendations to inform research, practices and policies. The purpose of this book, therefore, is to update contemporary developments surrounding theories and applications of mobile technologies in education at all levels. In particular, attention is given to emerging learning design models as well as exemplary cases of adoption of mobile technologies.

It can be suggested that mobile technology today offers a spectrum of tools for teachers, educational opportunities as well as new options for student–technology partnerships in learning. Empowered with interactive multimedia presentational capabilities, handheld technology permits the delivery of a range of multimedia material such as video, audio, graphics and integrated media. When appropriately designed for the context, educationally useful digital resources for learning can be effectively delivered via mobile technologies to students at any time, inside and outside of classrooms. The powerful technical features of mobile technologies, and available mobile applications powered with social media and cloud computing enable new forms of learning platforms which can serve contemporary pedagogies across a variety of educational contexts (see Churchill and Churchill 2008; Evans 2008; Kaleebu et al. 2013, Lai et al. 2007).

Relevant studies report a variety of issues in relation to use of mobile technologies in education. Examples of issues reported include use of mobile technology during classes, enabling teachers and students to share files; allowing students to ask anonymous questions, answer polls and give teachers feedback (e.g., Ratto et al. 2003); deliver an intelligent tutoring systems and quizzes (e.g., Segal et al. 2005); the dissemination information, the collection data during field trips and the support of students' inquiries (e.g., Churchill et al. 2010; Jong and Tsai, in press); supporting computer collaborative learning (e.g., Roschelle and Pea 2002; Zurita and Nussbaum 2004); using mobile instant messenger to support second language learning (Lai, in press); the improvement of literacy and numeracy for disadvantaged young adults (Attewell 2005); as a personal technology for lifelong learning (Sharples 2000); as personalized learning environments (e.g., Song and Fox 2008); as instructional tools and a replacement to laptops (e.g., Shen et al. 2009); as a tool for learning on the move (e.g., Wong et al. 2010); as a mediating tool for ubiquitous, seamless, authentic and situated learning experiences, (e.g., Hedberg 2014; Looi et al. 2010; Wong and Looi 2011), teacher use of iPads as a transformative strategy (Churchill and Wang 2014), and so on. Liu et al. (2014), who conducted a comprehensive analysis of the literature on mobile learning from 2007 to the present, argue that the most contemporary studies explore issues from four distinct perspectives, which include comparison studies (e.g., studies of learning outcomes), non-comparison studies (e.g., studies of communication and collaboration with mobile technology), mobilized learning studies (e.g., studies of learning outside of classrooms) and academic content studies (e.g., studies of mobile technology in natural science education).

For Liu et al. (2014), the key problem with the research and practice on mobile learning is a weak connection and even complete absence of any connection to learning theories. This connection is essential if the new theoretical frontiers and affordances of mobile technology are to be explored. Therefore, for the effective integration of mobile technology in education, an appropriate learning design that builds on sound learning-theoretical foundation is essential. From the literature, it has been suggested that mobile learning has been designed according to three paradigms, including (see Churchill et al. 2014): "learning with mobile technologies" (e.g. Anderson and Blackwood 2004; Churchill and Churchill 2008; Song and Fox 2008), "learners on the move" (e.g., Gu et al. 2011; Seppälä et al. 2003; Wong et al. 2010), and "dynamic, seamless and ubiquitous learning experiences" (e.g., Wong and Looi 2011; Kearney 2014; Song 2014; Ting 2013). For Churchill, Lu and Chiu (2014), the most critical aspect of effective mobile learning today is integration of mobile technology, social media and a learning design. A learning design should serve as a powerful intervention strategy to transform teacher thinking in a productive direction (e.g., Churchill et al. 2013; Churchill, Fox and

King, Chapter 1 in this book). This book proposes the RASE learning design framework, which emphasizes four core components to a mobile-enabled learning environment, namely resources, activity, support and evaluation.

This book comprises 24 chapters written by authors and co-authors from across the world. The book is sorted into the main six parts as follows:

- Mobile Learning Design—explores learning design frameworks and approaches for integration of mobile and emerging technologies in education, including the RASE (Churchill, Fox and King), authentic learning approaches (Burden and Kearney), social media and collaboration (Cochrane and Narayan) and Activity-theoretical perspective (Rozario, Ortlieb and Rennie). An additional chapter by Notari and Hielscher provides a useful classification/ontology of educational Apps. Understanding of this ontology might contribute to a more effective integration of Apps into learning designs. The final chapter by Kidd and Crompton explores augmented reality, its affordance and possibilities for application via mobile learning technologies.
- Mobile Learning Adoption and Student Perception—attention is given to the issues of acceptance, adoption and student perception related to educational integration of mobile learning technology. The issues addressed include adoption factors (Balakrishnan and Lay), student conception of mobile learning (Khan, Abdou and Clement), student concerns and attitudes (Putnik), and student usage and perception (Hu et al.). The chapters in this part provide unique perspectives on some specific applications of mobile technology, such as in interactive lectures, and integration with a learning management system.
- Mobile Learning Analytics—examines the important and increasingly emerging issue of learning analytics, and explores how mobile technology might be adopted to provide more systematic understanding of student engagements. Tam, Yi, Xu and Lam explore learning analytics in the context of application of a cloud-based technology platform, while Wong provides a unique perspective on "flipped classrooms", and how mobile technology might assist the process of examining student learning.
- Mobile Learning Across the Curriculum—explores the integration of mobile technology across the curriculum and educational entities. This part explores integration into K-12 education (Turner; Wang), early childhood education (Tavernier), out-of-the-class learning (Hayes and Weibelzahl) and workplace learning (Gu). Though there is no specific focus on higher education in this part, the concepts and ideas introduced are highly applicable and useful to this sector.
- Mobile Learning in Subject Domains—provides more specific perspectives on the integration of mobile technology in specific curriculum areas and topics, including Geometry (Crompton), Healthcare (Cook and Santos), college English education (Wang and Cui), English vocabulary learning (Sytwu and Wang) and Mathematics (Khoo; Chiu). This part highlights the need for further research and documentation of practices in the development of emerging literacies related to mobile learning. For now, however, the reports on English and Mathematics education appear to dominate the discussion.

- Future Development—a single chapter by Pegrum is included in this part. The chapter offers an outstanding conclusion to the ideas presented in this book and sets the context for further development, underlining various aspects and factors surrounding effective adoption of technologies in education.

In summary, contemporary mobile technologies offer a set of tools and affordances for the advancement of teaching and learning. Furthermore, research and practice should incorporate not just mobile technology, such as smartphones and tablets, but the need to follow developments with emerging technologies, such as a variety of wearable devices (e.g., glasses and watches), "internet-of-things" and other emerging technological innovations that introduce and make possible educationally useful affordances at new levels. A stronger connection between mobile technology integration and learning-theoretical frameworks is essential to guide research, practice and policy. Rather than focusing on technology, a key proposition of this book is to lead education integration of mobile and emerging technologies through an appropriate evidence-based learning design framework. Equally important is the achievement of curriculum specified outcomes; the development of new literacies; learner satisfaction; relevance of educational activities given the work practices of young individuals; and more effective work management, change and performance by teachers. The potential of intellectual partnerships with mobile and emerging learning technologies is promising, however, without empirical research input, a learning design framework and relevant policy, success will be hard to realise. Further studies are required to investigate aspects of such methodologies, framework and policies. In addition, research needs to pay attention to aspects of the design of mobile learning Apps across various categories ranging from multimedia content, communication, digital storytelling to social networking and cloud computing.

In conclusion, on behalf of the editorial team, I wish to give special thanks to the authors and reviewers of the papers, and others who assisted in the development of this project. Working with more than 35 authors and co-authors from across the world has been a challenging but rewarding experience for the editorial team. Special thanks to the Mobile Learning Faculty Research Theme of the Faculty of Education, The University of Hong Kong, Consultants International for Human Development and the International Mobile Learning Festival for the support invested in this project. In the future we intend to expand this collection through further involvement with scholars and practitioners and their participation in forums such as the IMLF conference. I am sure that this book will contribute to the advancement of knowledge and practice in the implementation of mobile and emerging learning technologies.

Daniel Churchill

References

Anderson, P., & Blackwood, A. (2004). Mobile and PDA technologies and their future use in education. *JISC Technology and Standards Watch*, *4*(3), 3–33.

Attewell, J. (2005). *Mobile technologies for learning*. London, UK: Learning and Skills Development Agency.

Churchill, D., & Churchill, N. (2008). Educational affordances of PDAs: A study of a teacher's exploration of this technology. *Computers & Education, 50*(4), 1439–1450.

Churchill, D., Kennedy, D. M., Flint, D., & Cotton, N. (2010). Using handhelds to support students' outdoor educational activities. *International Journal of Continuing Engineering Education and Life-Long Learning, 20*(1), 54–72.

Churchill, D., King, M., & Fox, B. (2013). Learning Design for Science Education in the 21st Century. *Journal of the Institute for Educational Research, 45*(2), 404–421.

Churchill, D., Lu, J, & Chiu, K. F. T. (2014). Integrating mobile technologies, social media and learning design. *Educational Media International, 51*(3), 163–165.

Churchill, D., & Wang, T. (2014). Teacher's use of iPads in higher education. *Educational Media International, 51*(3), 214–225.

Evans, C. (2008). The effectiveness of m-learning in the form of podcast revision lectures in higher education. *Computers & Education, 50*(2), 49–498.

Gu, X., Gu, F., & Laffey, J. M. (2011). Designing a mobile system for lifelong learning on the move. *Journal of Computer Assisted Learning*, *27*(3), 204–215.

Hedberg, J. (2014). Extending the pedagogy of mobility. *Educational Media International, 51*(3), 237–253.

Jong, S. M., & Tsai, C. C. (In press). Understanding the concerns of teachers about harnessing mobile technology to facilitate outdoor social inquiry learning: the EduVenture experience. *Interactive Learning Environments.*

Kaleebu, N., Gee, A., Jones, R., & Watson, A.H.A. (2013). SMS story impact assessment report. *VSO, Papua New Guinea.* Retrieved from http://www.vsointernational.org/sites/vso_international/files/sms-story-impact-assessment-report_tcm76-41038_0.pdf

Kearney, M. (2014). Investigating teachers' adoption of signature mobile pedagogies. *Computers & Education.*

Lai, A. (In press). Mobile immersion: an experiment using mobile instant messenger to support second language learning. *Interactive learning Environments.*

Lai, C.-H., Yang, J. C., Chen, F. C., Ho, C. W., & Chan, T. W. (2007). Affordances of Mobile Technologies for Experiential Learning: The Interplay of Technology and Pedagogical Practices. *Journal of Computer Assisted Learning, 23*(4), 326.

Liu, M., Scordino, R., Geurtz, R., Navarrete, C., Ko, Y., & Lim, M. (2014). A look at research on mobile learning in K-12 education from 2007 to the present. *Journal of Research on Technology in Education, 46*(4), 325–372.

Looi, C. K., Seow, P., Zhang, B., So, H. J., Chen, W-L., & Wong, L. H. (2010). Leveraging mobile technology for sustainable seamless learning. *British Journal of Educational Technology*, *41*(2). 154–169.

Ratto, M., Shapiro, R. B, Truong, T. M., & Griswold, W. G. (2003). *The active class project: experiments in encouraging classroom participation.* Retrieved from http://www-cse.ucsd.edu/~wgg/Abstracts/activeclass-cscl03.pdf

Roschelle, J., & Pea, R. (2002). A walk on the WILD side: How wireless handhelds may change CSCL. *International Journal of Cognition and Technology, 1*(1), 145–168.

Segall, N., Toni, L., Doolen, J., and Porter, D. 2005. A usability comparison of PDA-based quizzes and paper-and-pencil quizzes. *Computers & Education*, *45*(4),417–432.

Seppälä, P., & Alamäki, H. (2003). Mobile learning in teacher training. *Journal of computer assisted learning*, *19*(3), 330–335.

Sharples, M. (2000). The design of personal mobile technologies for lifelong learning. *Computers & Education, 34*(3-4), 177–193.

Shen, R., Wang, M., Gao, W., Novak, D., & Tang, L. (2009). Mobile learning in a large blended computer science classroom: system function, pedagogies, and their impact on learning. *IEEE Transaction on Education*, *52*(4), 538–546.

Song, Y. (2014). "Bring Your Own Device (BYOD)" for seamless science inquiry in a primary school. *Computers & Education*, *74*, 50–60.

Song, Y., & Fox, R. (2008). Using PDA for undergraduate student incidental vocabulary testing. *ReCALL*, *20*(3), 290–314.

Ting, Y. L. (2013). Using mobile technologies to create interwoven learning interactions: An intuitive design and its evaluation. *Computers & Education*, *60*(1), 1–13.

Wong, L. H., Chin, C. K., Tan, C. L., & Liu, M. (2010). Students' Personal and Social Meaning Making in a Chinese Idiom Mobile Learning Environment. *Educational Technology & Society*, *13*(4), 15–26.

Wong, L. H & Looi, C. K. (2011). What seams do we remove in mobile-assisted seamless learning? A critical review of the literature. *Computers & Education*, *57*(4), 2364–2381.

Zurita, G., & Nussbaum, M. (2004). Computer supported collaborative learning using wirelessly interconnected handheld computers. *Computers & Education, 42*, 289–314.

Contents

About the Editors

Daniel Churchill is an Associate Professor at the Faculty of Education, The University of Hong Kong. He has been teaching for more that two decades at universities around the world, including Europe, Australia and Asia. He specialises in educational technology, education system development, and multimedia design. Over the years, he has been active in researching topics including learning 2.0, mobile learning, e-learning instructional design, learning objects, and teacher professional development. He is a recipient of a prestigious "HKU Outstanding Teacher Award," and "the Macromedia's Award for the Best Multimedia on the Internet." Also, Dr. Churchill is active as a novelist, and his latest novel "The Refugees" has been published and distributed globally in 2015.

Jie Lu currently is director of educational resources centre, BGI College. She received her Ph.D. degree from the University of Hong Kong in 2012 and worked as postdoctoral fellow from 2013 to 2014 at the same university. Her research interests include learning engagement, mobile learning and educational application of social media.

Thomas K.F. Chiu Thomas Chiu, an experienced educator, is a lecturer in Faculty of Education at the University of Hong Kong. He published papers that specialize in mobile learning, multimedia learning and technology adoption in reputed journals and has been serving as editors and reviewers of repute. With solid software technology knowledge and teaching experience, he developed large scale computer applications, and also designed and delivered IT training courses. He has won several awards including teaching, and design of digital educational materials and learning systems. His research areas include learning design, ICT educational policies and big data in education.

Bob Fox Deputy Director, Learning and Teaching Unit at The University of New South Wales (UNSW). He holds concurrent appointments as Honorary Professor, Faculty of Education, University of Hong Kong (HKU) and Adjunct Professor, School of Education, Faculty of Arts and Social Sciences, UNSW. He was awarded

HKU's University Teaching Fellowship for excellence in teaching. His research interests focus on technological practice and change in higher education; building institutional capacity and staff capabilities; teacher professional development; innovative use of new technology; blended learning, BOOCs and MOOCs; learning environments and the impact of technology on physical learning spaces.

Part I
Mobile Learning Design

Chapter 1
Framework for Designing Mobile Learning Environments

Daniel Churchill, Bob Fox and Mark King

Abstract In this chapter the RASE learning design framework is proposed as a key strategy for utilizing multiple affordances of mobile learning technology. This learning design framework is based on the premise that an effective learning environment must include and integrate at least four core components, namely: Resources, Activity, Support and Evaluation. The activity component is the most important, requiring students to engage with intellectual and knowledge-based developments. Mobile technology offers a number of affordances that support learning, including: Resources, Connectivity, Collaboration, Capture, Representation, Analytical and Administration tools. Effective use of mobile technology includes deployment of these affordances in the learning design in a way that supports different components of the RASE framework and achievement of set learning outcomes. This chapter presents and discusses concepts, arguments, and a discussion of an example of an app that integrates multiple affordances, supported by all components of the RASE learning design framework.

1.1 Introduction

Mobile devices such as tablets, mobile phones and iPods are being increasingly used in education around the world. Since 2011, the annual Horizon Report has emphasized the importance of mobile technology, and coupled with cloud computing, these technologies will continue to have a major impact on education

D. Churchill (✉)
The University of Hong Kong, Hong Kong, SAR China
e-mail: dchurch@hku.hk

B. Fox · M. King
The University of New South Wales, Sydney, NSW, Australia
e-mail: bobfox@unsw.edu.au

M. King
e-mail: mark.king@unsw.edu.au

D. Churchill et al. (eds.), *Mobile Learning Design*, Lecture Notes in Educational Technology, DOI 10.1007/978-981-10-0027-0_1

(see New Media Consortium 2011). Educational uses of, for example, e-books, digital videos, podcasts, social networking, cloud computing, and many other mobile apps have been adopted by different groups of innovative educators and institutions around the world.

Mobile technology offers a spectrum of tools for teachers, twenty-first century educational opportunities and new options for student-technology partnerships in learning. Empowered with interactive multimedia presentational capabilities, mobile technology enables the delivery of a range of multimedia material such as video, audio, graphics and integrated media. If appropriately designed for the context, educationally useful digital resources for learning can be effectively delivered via mobile technologies to students at any time, inside and outside classrooms. Furthermore, features of mobile technologies, and available mobile applications powered with social media and cloud computing enable new forms of learning platforms that can serve in a variety of educational contexts (see Churchill and Churchill 2008; Evans 2008; Lai et al. 2007). However, for Liaw et al. (2010) although mobile technologies have power to improve education, there is a lack of recommendations for educators, as the current research and practical recommendations are still in an embryonic stage. For Churchill et al. (2014), mobile learning has been designed following three paradigms, including: 'learning with mobile technologies' (e.g. Anderson and Blackwood 2004; Churchill and Churchill 2008; Song and Fox 2008), 'learners on the move' (e.g. Gu et al. 2011; Seppälä and Alamäki 2003; Wong et al. 2010), and 'dynamic, seamless and ubiquitous learning experience' (e.g. Wong and Looi 2011; Kearney 2014; Song 2014; Ting 2013). However, we find these paradigms to be incomplete, and that a more comprehensive and applicable framework for learning design is needed to provide teachers, educational policy-makers and researchers with a representation of how affordances of emerging technologies can be utilized in the context of teaching and learning. In this chapter, we explore the RASE (Resources-Activity-Support-Evaluation) learning design framework (see Churchill et al. 2013), and discuss how it can be utilized to integrate affordances of mobile technologies in a learning environment.

1.2 RASE Learning Design

The central idea behind the RASE learning design framework is that *Resources* are not sufficient for full achievement of learning outcomes. In addition to resources, teachers need to consider the following:

- *Activity* for students to engage in using resources and working on tasks such as experiments and problem solving leading through active experience towards achievement of learning outcomes.
- *Support* to ensure that students are provided assistance, and where possible with tools to independently or in collaboration with other students, solve emerging difficulties.

- *Evaluation* to inform both students and teachers about progress and to serve as a tool for understanding what else needs to be done in order to ensure learning outcomes are achieved.

Figure 1.1 is a visual representation and summary of the RASE learning design.

The RASE learning design framework builds upon important theoretical work and concepts described below.

- *Constructivist learning environment* (Jonassen 1999). In this view, learning should be arranged around activities, and occur in an environment that supports knowledge construction, as opposed to knowledge transmission. Knowledge construction is a process where students individually construct their understanding of the content of the curriculum based on exploration, social engagement, testing of understandings and consideration of multiple perspectives.
- *Activity Theory* (Engeström 1987). Activity Theory specifies the components that are part of a human activity system. To understand what is learning, it is important to understand the specifics of this activities, as well as tools used in the process, the rules and the division of labor, community involved in the process, parallel and vertically related activities, interactivity, and contradictions.
- *Problem solving* (Jonassen 2000). For Jonassen, learning is most effective in the context of the tasks in which students engage to solve ill-structured, authentic,

Fig. 1.1 The RASE learning design

complex and dynamic problems. These kinds of problems are significantly different from logical, well-structured problems with one solution. These types of problems are dilemmas, case studies, strategic decision-making and planning, all of which require students to actively engage in deep reflection, consideration of theoretical perspectives, the use of tools, creation of artefacts and analysis of various possible solutions. Students learn through complex problems, and not through the absorption of ready-made knowledge, rules and procedures.

- *Problem-based learning (PBL)* (Savery and Duffy 1995). Savery and Duffy propose PBL as an optimal design model for student centred learning. Similar to those approaches above, the PBL builds upon constructivist philosophy and contends that learning is a process of knowledge construction and social co-construction. A key feature of the PBL is that students actively work on authentic tasks, and construct knowledge in contexts that reassemble those in which they would use that knowledge. Creativity, critical thinking, metacognition, social negotiation, and collaboration are all perceived as a critical component of a PBL process. One of the key characteristics of PBL is that teachers should not primarily be concerned with the knowledge students' construct, but should focus, more attention to metacognitive processes.
- *Rich environments for active learning* (Grabinger and Dunlap 1997). Similar to Savery and Duffy, Grabinger and Dunlap propose PBL as a highly effective educational intervention. However, in their approach, further attention is given to the context of the environment in which PBL occurs, considering components and complexities that such an activity requires. In particular, emphasis is placed upon making students more responsible, willing to provide initiatives, reflective and collaborative in the context of dynamic, authentic and generative learning. This approach also emphasizes the importance of the development of lifelong learning skills.
- *Technology-based learning environments and conceptual change* (Vosniadou et al. 1995). In this view, the central role of technology is to support students' conceptual changes and concept learning rather than simple knowledge transfer. Students construct mental models and other internal representations via attempts to explain the external world. Students often bring prior misconceptions to learning situations.
- *Interactive learning environments* (Harper and Hedberg 1997; Oliver 1999). Oliver proposes that a learning module must contain resources, tasks and support to serve the complexity required for learning. For learning to take place, a task must engage students to make purpose-specific uses of resources. The teacher's role is to support learning. These three integrated components will lead to interactivity essential for learning.
- *Collaborative knowledge building* (Bereiter and Scardamalia 2003). Knowledge building is a theoretical construct developed by Bereiter and Scardamalia to provide interpretation of what is required in the context of collaborative learning activity. Personal knowledge is seen as an internal, unobservable phenomenon, and the only way to support learning and understand what is taking place is to deal with so-called public knowledge (which represent what a community of

learners know). This public knowledge is available to students to work on, expand and modify through discourse, negotiation, and collective synthesis of ideas.

- *Situated learning* (Brown et al. 1989). Brown and colleagues build upon the Activity Theory perspective to emphasize the central role of an activity in learning. An activity is where conceptual knowledge is developed and used. It is argued that this situation produces learning and cognition. Thus, activity, tools and learning should not be considered as separate. Learning is a process of enculturation where students become familiarized with uses of cognitive tools in the context of working on an authentic activity. Both activity and how these tools are used are specific to a culture of practice. Concepts are not only situated in an activity, but also are progressively developed through it, shaped by emerging meaning, culture and social engagement. Brown and colleagues strongly suggest that activity, concept and culture are interdependent, in that "the culture and the use of a tool determine the way practitioners see the world, and the way the world appears to them determines their cultural understanding of the world and of the tools… To learn to use tools as practitioners use them, a student, like an apprentice, must enter that community and its culture" (p. 33). Hence learning is a process of enculturation, where students learn to use a domain's conceptual tools in an authentic activity.

What can be observed from these theoretical approaches is that an Activity is central to learning. Learning is an experience where learners construct and use knowledge. Furthermore, each of the four components of the RASE is discussed is more details.

1.2.1 Resources

Resources include (a) content (e.g. digital media, textbooks and a lecture by a teacher), (b) material (e.g. chemicals for an experiment, paint and canvas) and (c) tools that students use when working on their activity (e.g. laboratory tools, brushes, calculators, rulers, statistical analysis software and word processing software). When integrating technology resources in teaching, it ought to be done in a way that leads students to learn with, rather than just learn from these resources. In this way, students can develop elements of their overall new literacies. There are various software tools that students can use in learning (e.g. Mind Mapping tool such as MindMeister, image/video editing tool such as iMovie, professional tools such as AutoCAD and Mathematica, and model building and experimentation tools such as Interactive Physics and Stella).

1.2.2 Activity

An activity is a critical component for the achievement of learning outcomes. It provides learners with an experience where learning occurs in context of emerging understanding, testing ideas, generalizing and use of knowledge. The following are two key characteristics of an effective activity. An activity must be 'learning-centred'

- It must focus on what learners will do to learn, and how their rather conceptual changes will develop, rather than on what students will remember to reproduce at examinations.
- Resources are tools in students' hands, which assist them to complete tasks.
- Teachers are facilitators who participate in the learning process as partners and critical friends.
- Learners produce artefacts that demonstrate their learning process, not just outcomes.
- Learners learn about the process by actively experimenting with approaches and reflecting on effective strategies (metacognition).
- Learners develop new literacies required for twenty-first century learning, working and living.

Furthermore, an activity must be 'authentic'. This means that:

- It should contain real-life scenarios and ill-structured problems.
- It should reassemble professional practice and thinking.
- It should use tools specific to professional practice.
- It should result in artefacts that demonstrate professional performance (intellectual and practical knowledge use), not only knowledge.

The following are examples of what an activity may be:

- A design project (e.g. design an experiment to test scientific hypothesis).
- Case study (e.g. a case of how a scientist identified new physics regularity).
- A problem solving learning task (e.g. minimizing friction in a design of ski).
- Develop a documentary movie on a specific issue of interest (e.g. GM food pros and cons).
- A poster to promote a controversial scientific issue (e.g. Nuclear energy).
- Planning a history day in your school (e.g. create a model and display to inform about ancient Egyptian culture).
- Develop as software to control mechanical transfer of power (e.g. use Scratch to design a digital model to assist an analytical task)
- Role-play (e.g. defending science experiment with small animals).

1.2.3 Evaluation

An activity engages learners in working on tasks, and developing artefacts that evidence their learning. This evidence of student learning enables the teacher to monitor student progress and provide further formative guides to help improve students' learning achievement. Outcomes of an activity can be a conceptual artefact (e.g. an idea or a concept presented in a written report), a hard artefact (e.g., a model of an electric circuit), or a soft artefact (e.g. a computer-based creation). Artefacts produced by students must undergo peer and expert review and a revision before final submission. This process may also involve learner/group presentations and peer/expert feedback. Also, students need to record their progress, so they too can monitor own learning and the improvements they make. Rubrics can be provided to enable students to conduct self-evaluation as well. The produced artefacts ought to be evaluated in ways that students can reflect upon feedback and take further action towards a more coherent achievement of the learning outcomes. Evaluation of learning is an essential part of effective learning-centred experiences. It needs to be formative in order to enable students to constantly improve their learning, and provide feedback on progress.

1.2.4 Support

The purpose of support is to provide students with essential scaffolding while enabling the development of learning skills and independence. Support might anticipate students' difficulty, such as understanding an activity, using tool or working in groups. In addition, teachers must track and record ongoing difficulties and issues that need to be addressed during learning, and share these with students. Four modes of support are possible: teacher-student, student-student, student-artefact (additional resources) and student-community (seeking assistance from other people and sources). Support can take place in a classroom and in-online environments such as through forums, Wikis, Blogs and social networking spaces.

Also, support can be seen as anticipatory of student needs. Depending on the course, proactive support structures such as FAQs can be planned and implemented in light of such needs. The objective of anticipatory support is to ensure students have access to a body of resources when they need help, rather than being dependent on asking teachers for help. Here are some specific strategies:

- Build a body of resources and materials which form a FAQ Page
- Create a "How Do I?" or "Help Me" Forum
- Create a Glossary of course-related terms
- Use checklists and rubrics for activities
- Use other social networking platforms and synchronous tools such as chat and Skype.

Overall, support should aim to lead students to become more independent learners. For example, before a student can ask a teacher for help, they must first ask their classmates through one of the forums and/or search the Internet for solutions to their problem(s). In this way, students are expected to take responsibility for their learning and to support other students in their cohort.

Designing learning environments based on the RASE, whether for online, blended or classroom based learning, should have integration of all four components. Learning design usually begins by articulation of an activity for learners to engage. Planning evaluation would then take place, followed by provision of resources and support.

Furthermore in this chapter, possibilities of how mobile learning might support the RASE will be examined through consideration of affordances of mobile technology.

1.3 Affordances of Mobile Learning Technologies and the RASE Learning Design

Studies report a variety of possibilities in relation to use of mobile technologies in education. Examples of issues reported include: use of mobile technology during classes, enabling teachers and students to share files; allowing students to ask anonymous questions, answer polls, and give teachers feedback (e.g. Ratto et al. 2003); delivering an intelligent tutoring system and quizzes (e.g. Segal et al. 2005); disseminating information, collecting data during field trips and supporting students' inquiries (e.g. Churchill et al. 2010); supporting computer collaborative learning (e.g. Roschelle and Pea 2002; Zurita and Nussbaum 2004); improving literacy and numeracy for disadvantaged young adults (Attewell 2005); as a personal technology for lifelong learning (Sharples 2000); as a personalized learning environments (e.g. Song and Fox 2008); as instructional tool and a replacement to laptops (e.g. Shen et al. 2009); as a tool for learning on the move (e.g. Wong et al. 2010); as a mediating tool for ubiquitous, seamless and situated learning experiences, (e.g. Looi et al. 2010; Wong and Looi 2011), and so on.

An affordance is a useful concept that can be applied to interpret how teachers engage technology in their practice. Norman (1988) defines affordances as "the perceived and actual properties of the thing, primarily those fundamental properties that determine just how the thing could possibly be used" (p. 9). For Barnes (2000), a teacher's use of new technology in teaching and learning is carried out with a belief that this technology will afford learning in some way. Affordances can include actual uses, and those uses that emerge in teachers practice. Therefore, how mobile technology will be used in education depends largely on teachers' understandings of affordances of this technology.

What do we know about educational affordances of mobile technology at this stage? The literature related to early adoption of mobile technology suggests that it might assist students to learn anytime, anywhere, by empowering them "to access internet resources and run experiments in the field, capture, store and manage everyday events as images and sounds, and communicate and share the material with colleagues and experts throughout the world" (Sharples et al. 2002, p. 222). For Luchini, Quintana and Soloway (2004), the key benefit of such mobile technology is that powerful personal devices can "provide access to tools and information within the context of learning activities" (p.135). For Hsieh, Jang, Hwang and Chen (2011) mobile technology has potential to support students' reflection leading to improved learning achievement when there is an appropriate match between a teacher's teaching style and students' learning style.

Klopfer and Squire (2005) describe five potential educational affordances of PDAs: (1) portability, as mobile technology can be taken to different locations; (2) social interactivity, as mobile technology can be used to collaborate with other people; (3) context sensitivity, as mobile technology can be used to gather real or simulated data; (4) connectivity, as mobile technology enables connection to data collection devices, other handhelds, and to a network; and (5) individuality, as mobile technology can provide scaffolding to the learners. Patten, Sánches and Tangney (2006) present a framework that consists of the following affordances of PDA technology: administration, referential, interactive, microworld, data collection, location awareness and collaboration. Liaw, Hatala and Huang (2010) suggest five affordances of mobile technology for education: (a) educational content and knowledge delivery, (b) adaptive learning applications, (c) interactive applications, (d) individual applications and (e) collaborative applications. Churchill and Churchill (2008) expanded upon these studies and examined a teacher's use of PDA technology. Their study articulated a number of affordances of PDA technology including as a multimedia access, connectivity, capture, representational and analytical tool. In our own study of teacher use of iPads in higher education (see Churchill and Wang 2014), we explicated a set of categories of apps utilized by educators. These include: (a) productivity, (b) teaching administration, (c) note taking, (d) communication, (e) cloud management, (f) social content creation and (g) content accessing tools.

These affordances from these reports are sorted through our analysis into emerging groups of affordances that include (see Table 1.1): (a) resources tool, (b) connectivity tool, (c) collaboration tool, (d) capture tool, (e) analytic tool, (f) representation tool, and (g) administration tool. These groups are used as an analytical framework for understanding affordances that emerge in this study.

The following is a brief description of these key affordances of mobile technology:

- *Resources* A variety of multimedia resources can be delivered using this technology, such as e-books, web pages, presentations, interactive resources, audio files and video segments. These resources can be accessed at anytime, anywhere, by connecting to the Internet mobile network or wireless network connections, from the memory of the device or storage card if the resources were previously downloaded, or through synchronization of the device with a

Table 1.1 Affordances of mobile learning technology

Klopfer and Squire (2005)	Patten et al. (2006)	Churchill and Churchill (2008)	Liaw et al. (2010)	Churchill and Wang (2014)	Summary of affordances emerging from across these studies
Portability	Administration (7)	Multimedia access (1)	Educational content and knowledge delivery (1)	Productivity tools (3, 6)	1. Resources
Social interactivity (3[a])	Referential (1)	Connectivity tool (2)	Adaptive learning applications (1)	Teaching Administration tools (2, 7)	2. Connectivity
Context sensitivity (4)	Interactive resource (1)	Capture tool (4)	Interactive applications (3)	Note taking tools (4, 6,)	3. Collaboration
Connectivity (2, 3)	Microworld environment (1)	Representational tool (6)	Individual applications (6)	Communication tools (2, 3)	4. Capture
Individuality (1)	Data collection (4)	Analytical tool (5)	Collaborative applications (3)	Cloud management tools (3, 4, 7)	5. Analytical
	Location awareness (4)			Social content creation tools (2, 3, 4, 6)	6. Representation
	Collaboration (3)			Content accessing tools (1, 5)	7. Administration

[a]Corresponds to an affordance listed in the summary (final column)

computer. A variety of apps available for mobile devices support delivery and access to resources such as e-books, multimedia material and video content, as for example, iBooks, Kindle, YouTube, Perfect Reader, iTunes and iTunesU.

- *Connectivity tool* Mobile technology empowers students to connect to each other, facilitators and experts in the field, exchange ideas and files, socially construct and negotiate meanings, manage activities and negotiate roles in their projects, etc. Connection might be established synchronously and asynchronously over mobile telephony and wireless networks that support voice and multimedia data transmission. These include tools that support communication and social networking, such as for example apps as Facebook, Skype, Google Hangouts, WhatsApp, WeChat, Viber, FaceTime, Facebook and MyPad.
- *Collaboration tool* This affordances enables student to co-design artefacts that demonstrate their learning, collaborate on projects a problem-based task, and share roles and responsibilities. Also, these include tools that allow connectivity to the Cloud, network drives and a computer and co-development of resources. Examples of apps include Air Shawing, FileBrowser, Dropbox, ZumoDrive, Air Drive, AirDisk. Goodle Drive and Office2HD.
- *Capture tool* Mobile technology is equipped with capture capabilities that include capture of video, audio and still photographs. Students might, for example, photograph and videotape machines and people during their industry visits, or photograph diagrams from a book or catalogue (e.g. by using apps such as Genius Scan, Cam Scaner, Dragon, ProCapture or Movie Pro). The capture affordance also includes audio capture (e.g. Smart Voice Recorder App). For example, students might interview experts and capture their own audio notes, or capture characteristic sounds of a faulty engine. There is a possibility for specially designed extensions and consoles to be attached to a mobile device and used to capture, store and process other kinds of data such as, for example, recording global positioning of certain air pollution sources.
- *Analytical tool* A mobile device might be used as an analytical tool to aid students' tasks. For example, these might include standard, scientific and graphic calculators such as Algeo Graphing Calculator App, or specially designed analytical tools created by teachers and designers to allow students to analyse certain data.
- *Representation tool* Mobile technology might be used by students and teachers to create representations which demonstrate their thinking and knowledge. These might be, for example, mind maps, captured and edited images, audio and videos. Apps such as iMovie, HansOn, Bamboo Paper, Penultimate, AudioNote, Draw Free, iPocketDraw. Blogsy and Wordpress enable content creating and editing via mobile devices directly to blogs and websites.
- *Administration* These include mobile tools that support classroom teaching, such as those that support connection to a projector, mark-book, presentation tools and classroom management tools. Examples of apps used are Moodle, Clicker School, TeacherPal, Prezi Viewer, Slides Shark, LanSchool Teacher. A variety of productivity apps can support a spectrum of administrative

activities of teachers, for example apps such as Mail, iAnnotate, Docs2PDF, Neu.Annotate, PDF Notes and Dragon.

We can use this set of affordances to assist us in understanding how mobile technology supports the RASE learning design and its component. Some of these possibilities are presented in the Table 1.2:

Table 1.2 Affordances of mobile learning technology and how these might support components of the RASE learning design

RASE Component	Affordances supporting RASE elements
• Resources	• Resources—Variety of multimedia resources and tools can be delivered via mobile technology (e.g. e-books, Youtube videos, PDF documents, SlideShare presentations, lecture recording, and learning object). These can serve to support students learning activities occurring in a variety of other contexts and platforms (inside and outside of a classroom, with or without use of other technologies and resources, etc.) • Connectivity—Enables access to resources via Cloud, or resources can be distributed to learners via these tool
• Activity	• Representation—numerous apps are available which, for example, support content editing and creation of multimedia (e.g. iMovie), building models (e.g. Autodesc ThinkerPlay) and mind maps (e.g. SimpleMind) • Connectivity—apps such as Skype, WeChat, WhatsApp, Adobe Connect, etc. all support connectivity and synchronous engagement and file sharing • Collaborative—apps supporting Cloud-based applications such as Google Docs/Sheets/Slides, can be used to support, for example collaborative project report development, financial analysis spreadsheet, or and an online presentation of a proposal • Analytical—apps such "DataAnalysis by Data Evaluation Systems" can be used to assist analysis of data. Use of Cloud-based applications, such as Google Sheets, can enable collaboration on analytical tasks. Some conceptual representation resources might be effectively utilized for analytical purposes (see Churchill 2013) • Capture—this affordance enables learners to capture data in multimedia forms and use in learning activities such as, digital storytelling, project documentation, or development of presentations
• Support	• Connectivity—This affordance empowers students and teachers to remain connected continuously, seek and provide support and monitor emerging learning difficulties • Resources—apps with multimedia resources can be used as additional information to be provided to students based on emerging difficulties
• Evaluation	• Administration—A spectrum of apps available can be used to assist teachers' administration. However, in context of RASE, apps such as "Easy Assessment - Rubric Creation & Assessment Tool for Teachers" can be used to create and administer rubrics for evaluation. Google Cloud apps might be used to create rubrics and collect evaluation data, annotating PDF apps can support effective feedback, while EdModo or similar learning system supporting mobile application, might be used to create and administer surveys • Connectivity—This affordances can effectively be utilized to communicate with students and provide feedback on their progress and completion of learning artefacts • Collaborative—Cloud-based application supporting collaborative artefact development would provide a teacher with and app for easy access and features for providing feedback and initiating improvements

1.4 A Case for an App that Integrates Multiple Affordances of Mobile Technology and Supports the RASE Learning Design

Mobile technology and its affordances might serve a single or multiple components of a learning design (or even part of a component). Different strategies might be used to achieve other requirements of a learning design that mobile technology cannot meet. However, mobile technology alone, in rare cases, might be able to support all the four components and integrate multiple affordances in a single learning environment. In this part of the paper, the authors will describe a case for an app designed with an aim to bring together multiple affordances and support multiple components of the RASE learning design.

In 2009, one of the authors met an exceptional teacher of Chinese as a Foreign Language at the University of Hong Kong. The teacher was dedicated and keen to solve some on fundamental problems in teaching Chinese as a foreign language she understood exist in context of her practice. Over the years, she experienced challenges in teaching of Chinese characters to non-Chinese students. Specifically, the teacher understood that it was difficult for her students to learn 'order of strokes'. In her current practice, she would write a character on the white board, explaining stroke order and students would follow by completing exercised prescribed in their book.

To assist this teacher with content of 'Chinese Character Writing', we designed 'Mobilese' App that allows a learner to examine characters from a set of 200 Chinese simplified characters via a mobile device. The 200 characters were carefully selected and included in the Mobilese by the participating Chinese language teacher. At the home interface the learner could select one of the displayed characters, or narrow down their choices by selecting one of the categories that classify characters according to a number of strokes required to write it. Alternatively, a learner could search for a character according to so-called 'pinyin' (pronunciation written in alphabetical characters).

Once a character is selected, it will be displayed, and a learner will be able to preview and practice writing it by recreating the correct stroke order. Unique to the approach is that a learner cannot write a character in any incorrect order, that is, rather than receiving a feedback for being incorrect, he or she must discover a proper sequence in his or her own way. Another unique feature of this resource is that of 'similar characters'. The teacher, based on her almost 40 years of teaching experience, understood that her students often make mistakes and mix-up characters that are somehow similar. Such characters have been identified and the feature created to allow students to preview similar characters, and focus on specific aspects of these characters that cause mix-up. An additional functionality allows a learner to examine history of a specific character, listen to pronunciation in Putonghua and Cantonese versions, and explore examples of words in Putonghua and Cantonese that use that character. The interface of the Mobilese App is presented in the Fig. 1.2.

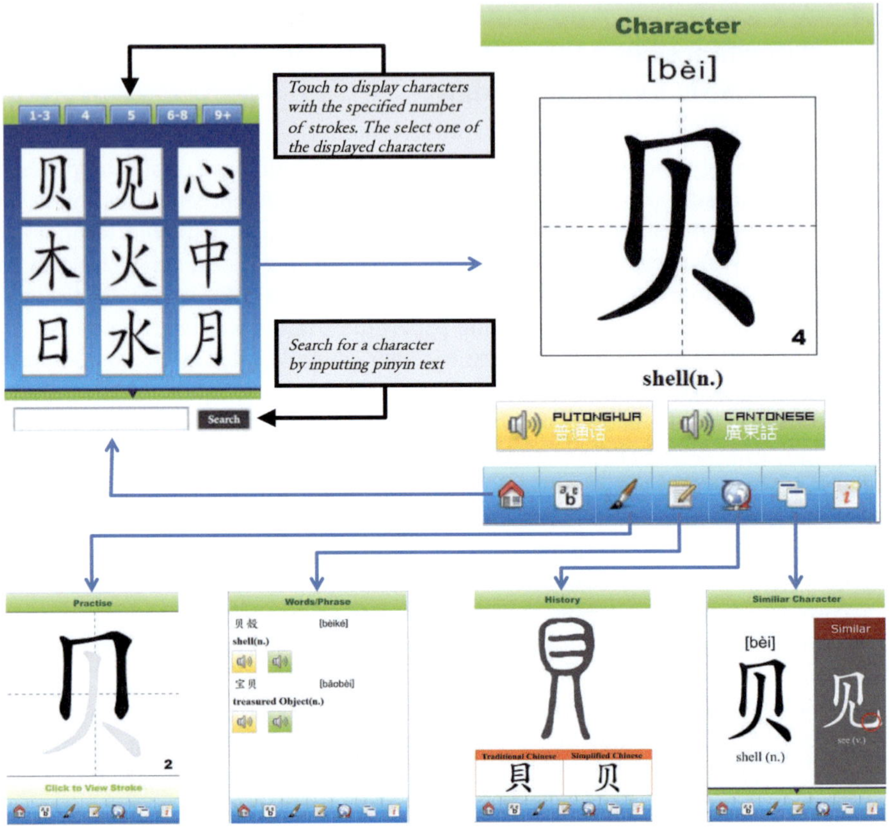

Fig. 1.2 Interface of the Mobilese app

This first design an app for Chinese character learning utilized on 'Resources' affordance of mobile technology. Therefore, this app was capable only of contributing to the 'Resources' component of the RASE learning design (and possibly 'Support' component). It had not utilized any other mobile affordance. Our further effort concentrated on exploring utilization of other affordances as a mean of creating possibilities for support of other components of the RASE.

As public interest in the Mobiliese grew, soon there were a number of schools interested to use this digital resource in their context. One of international schools from Hong Kong become in particular interested in Mobilese, however, they required certain modifications, and wanted it to be available for implementation via iPod Touch mobile devices. We saw this as an opportunity to further develop our idea of integrating affordances of mobile technology, and supporting components across the RASE. We have begun redevelopment of the Mobilese with integration of new features, and reorganization of content according to the school's needs, and in the way that is suitable for their early primary learners and the Chinese language curriculum. The school's principal, Chinese language curriculum leaders and technology coordinators favoured this proposal, and soon the school was ready to provide substantial funding and enter in the collaborative project agreement for redevelopment of this digital resource for learning via iOS devices.

A considerable redesign was made from the initial version of the Mobilese. Although the fundamental idea of character learning through visuals and interactivity remains the same in the both digital resources for learning via mobile technology, the redesigned digital resources focused on traditional rather than simplified characters according to the specific curriculum of Chinese language at the school. Additional functionalities and arrangements were build in the resources in a way that reflect the recommendations from the school's teachers and the curriculum leader, advices of experts from the team, and a number of affordances of mobile technology. The new version of the app was named 'iMobilese.'

In the iMobilese, similar as in the previous design of the Mobilese, a learner was able to select a character from the list of 200 included in the app. The menu for selecting characters has taken a new approach. Characters on the menu are sorted by default according to a number of strokes in ascending order. However, a learner was able to rearrange these in alphabetical order according to pinyin, or according to units predefined flexibly by teachers according to the school's curriculum programme. A learner could preview how characters are written in stroke-by-stroke order, examine similar characters, and explore words using those characters. However, there were a number of new features. First, a learner is now able to mark a character he or she is exploring as 'favourite', 'need to learn' or 'known' for ones that are learnt already. This information is then made available for easy reference at the main interface where a learner selects characters to explore. A learner could then quickly access a list of character he or she marked as favourite, need to learn or already known. This information was saved, and could be examined by a teacher or parents. In this way certain aspects of 'Connectivity' affordances have been build in the app as a mean of providing for 'Support' component of the RASE. The main interface of the iMobilese is shown in the Fig. 1.3.

An additional feature was integrated allowing a learner to save his or her own attempts of independent writing of characters for later preview, or for emailing to the teacher or other students. Further attempt in the redesign was made to integrate Capture affordances of mobile technology. While viewing specific characters, a learner can record his own voice for purposes of comparing own pronunciation to pronunciation of an expert, which is available in the resource. In fact, by using these features a learner could capture any audio, such as audio notes, or record a teacher or a class friend. Furthermore, a learner could take photographs of objects, people, book pages and scenes that they associate with a specific character. These audio recordings, photographs taken and images of characters written by a learner were saved on the device used. The learner could later use these media files in creating presentations with SonicPics (app favourably suggested by the teachers involved in the project). These were just a small step forward towards utilizing 'Representation' affordances in support of Activity component of the RASE, while supplementing limitations of iMobilese with use of other app available for mobile devices.

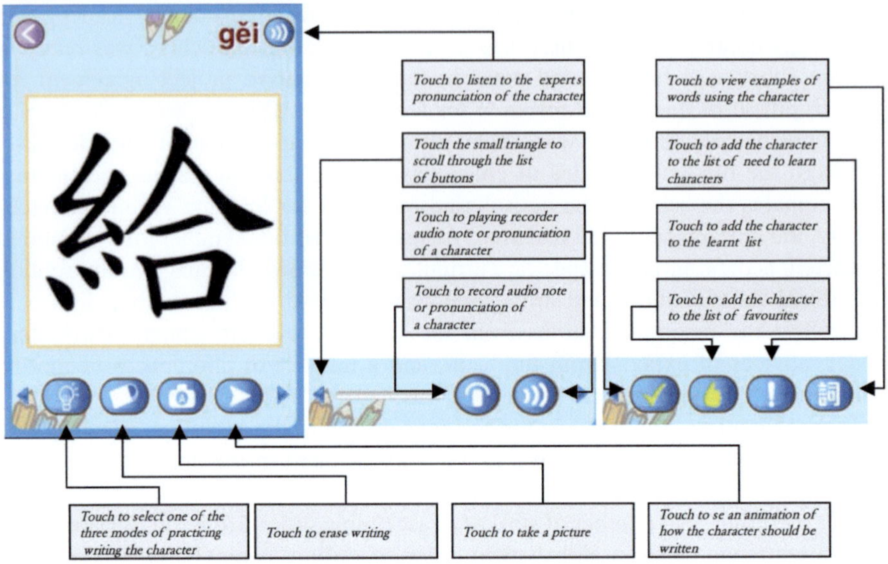

Fig. 1.3 Interface of the iMobiles app

The iMobilese also includes three practice resources designer with intention to serve as support resources: (a) Word Maker, used by learners to match two characters and form a word, (b) Pinyin Match, engages learners in associating characters with correct pinyin, and (c) Hear and Match, learners hear a sound and select a related character. Practices calculate the number of correct matches within a minute. Results of the practices were stored on the device and could be accessed by a teacher. The practices have been developed based on the worksheet exercise that teachers used in their teaching.

Ther iMobilese remains to largely provide 'Resources' affordance and support 'Resources' component of the RASE. The remaining components of the RASE were achieved by other ways. Here is how the iMobilese was used:

- The app was used in combination with workbooks (served as an 'Activity' component). For example, learners used the iMobilese to explore demonstrations and practice writing in correct stroke order first. They then used the workbook to practice handwriting with pencils while writing down a word constructed with the character. They referred to the iMobilese when encountering unknown characters.
- The iMobilese was used by learners to independently explore writing and speaking. Learners in groups received a worksheet with a list of characters they needed to review and practice. They took screenshots of their writing and recorded their speaking of a sentence constructed with the character. In most cases, learners played back their recordings immediately.

- The iMobilese was used to support practice-based group learning for reinforcing pinyin pronunciation and listening skills. Learners were grouped to use '*Pinyin Match'* and '*Hear and Match'* practices under the supervision of a teaching assistant. They showed their scores and competed with one another.
- The app was used to facilitate characters' meaning making by taking photos of real-world objects. Learners drew a picture on a white paper that was related to meaning of a character, and then took photos of the drawing and kept it in the iMobilese. In another example, learners took their iPod Touches home and captured photos of the real-world objects that associate them with characters they were learning.

Although iMobilese was an important step forward, it still fall short of utilizing multiple affordances of mobile technology, and the RASE components. A logical progression of iMobilese was to develop it in the iPad version, which would provide more possibilities. Although both, small devices and tablets, essentially used the same systems, and the same apps can be delivered via both, the large screen size of iPads assured greater design flexibility, functionalities and integration of tools made possible by this technology.

The author met a director of a technology company from Hong Kong who showed interest in iMobilese. An idea emerged to explore development of a commercial iPad app for learning of Chinese characters, that can be used by variety of audiences, such as school learners, independent learners across the Globe, language schools, corporate training, etc. Subsequently, the director made a decision to invest in the development of such resources, and made an offer to the author to lead this project. From the authors perspective, having unconstrained budget to develop and experiment with multiple affordances was an ideal situation, and soon the agreement was formalized, and the design process begun.

Most immediate idea was to redesign iMobilese, given the larger screen area, and expand on its content to include up to 1,000 characters. However, further thinking lead the author to ambitiously pursue other possibilities, including development of this resource in a platform that makes possible variety of activities, such as digital story telling ('Representation' affordance) and social networking ('Connectivity' affordance), and provides a tool for teachers to manage processes of learning in their own way ('Administration' affordance). The new app was labelled 'Chinese Learning Apps System', or CLAS. Screen capture of the prototype of CLAS is presented in the Fig. 1.4.

The new app is more complex, and there are variety tools incorporated by building upon different affordances of mobile technology. The new app is not only a digital resource for learning, but it incorporates features of learning management, social networking and representational tools in a platform that supports learning activities.

In addition to the development of this app for Chinese language, the author is engaged in designing a similar app for mathematics education. This app will contain a collection of conceptual representation resources for mathematics education ('Resources' component of a learning design). Learners will be able to

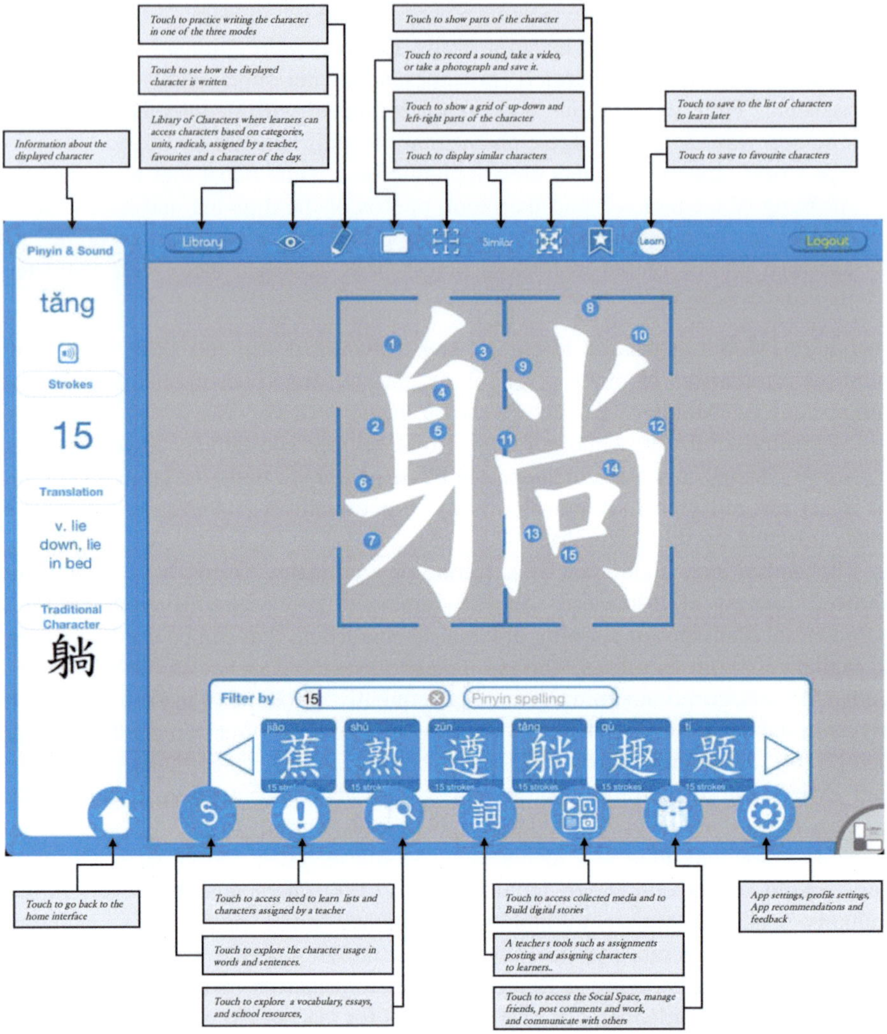

Fig. 1.4 Interface from the iPad version of app for Chinese character learning

explore these resources in various ways, engage with tools supporting 'Activity', such as annotation, digital storytelling, concept mapping, while teachers will have an additional component that allows them to set up activities for students ('Administration' affordance) for individual or group work ('Collaboration' affordance). Build in communication features ('Connectivity' affordance) will assist the 'Support' and 'Evaluation' components of a learning design. This project is in a very early stage; however, some preliminary designs have been put in place.

1.5 Conclusion

Overall, contemporary mobile technology has matured sufficiently to support integration of multiple affordances in a learning environment. A learning environment is articulated based on learning design. In this chapter we proposed that a learning design framework composed of the four components: Resources, Activity, Support and Evaluation, is the most suitable for contemporary learning-centred pedagogy. We call this approach to learning design the 'RASE.' Designers of multimedia resources should consider this recommendation when developing apps, that support the integration of multiple affordances in order to design resources that can support teachers in delivering learning designs based on the RASE, fully benefiting from the use of mobile technology. These key affordances of mobile learning technologies include: Resources, Connectivity, Collaboration, Representation, Capture, Analytical and Administrative tools.

In conclusion, we emphasize a special form of resource suitable for learning design with the use of mobile learning technology; what we call 'Conceptual Representation Resource'. Over the last decade, we have conducted extensive research on the design and educational uses of learning objects (see Churchill 2005, 2007, 2008, 2013, 2014; Churchill and Hedberg 2008; Jonassen and Churchill 2004). The concept is broadly understood as a specific form of cognitive structure that enables a knower to understand new information, and engage in specific disciplinary thinking, problem solving and further learning. The literature underlines the importance of conceptual learning, and refers to evidence that incomplete conceptual knowledge and misconceptions seriously impede learning (see Mayer 2002; Smith et al. 1993; Vosniadou 1994). Models have been described in the literature as effective tools for conceptual learning. Their educational use has been in the areas of model-centred learning and instruction (e.g. Dawson 2004; Gibbons 2008; Johnson and Lesh 2003; Lesh and Doerr 2003; Mayer 1989; Norman 1983; Seel 2003; van Someren et al. 1998). A conceptual resource is designed to represent a specific concept (or a set of related concepts) and its properties, parameters and relationships. A learner can manipulate these properties and parameters with interactive components (e.g.,sliders, buttons, hotspot areas, text input boxes) and observe changes displayed in a variety of modes (e.g. numerical, textual, auditory and visual). These resources require little contact time for maximum learning and conceptual knowledge to be constructed, thus holding potential to be the most appropriate multimedia design for mobile apps for learning. This form of resources can serve as external supplements to learners' conceptual knowledge limitations, thus serving as important resources in learning activities requiring conceptual knowledge to be represented. In addition to serving as resources in learning design, these apps can serve as 'Analytical' tools within an activity requiring the collection and processing of certain data from which conceptual knowledge can be derived. Furthermore, design of these kinds of resources might integrate other affordances of mobile technology, such as capture, posting of content and communication. Mobile technologies enable these resources to take on authentic contexts, moving between classrooms, laboratories and the external real world and used by students independently beyond their schools and whenever needed.

In summary, affordances of contemporary mobile technologies offer a set of tools for effective integration in teaching and learning. For this effectiveness to be achieved, it is becoming essential that an evidence-based learning design framework be in place. We call upon further studies to investigate aspects of the framework proposed in this chapter and we recommend that further research gives attention to aspects of design in mobile learning apps.

References

Anderson, P., & Blackwood, A. (2004). Mobile and PDA technologies and their future use in education. *JISC Technology and Standards Watch, 4*(3), 3–33.

Attewell, J. (2005). *Mobile technologies for learning*. London, UK: Learning and Skills Development Agency.

Barnes, S. (2000). What does electronic conferencing afford distance education? *Distance Education, 21*(2), 236–247.

Bereiter, C., & Scardamalia, M. (2003). Learning to work creatively with knowledge. In E. De Corte, L. Verschaffel, N. Entwistle, & J. van Merriënboer (Eds.), *Powerful learning environments: Unravelling basic components and dimensions* (pp. 55–68). Bingley, UK: Emerald Group Publishing.

Brown, J. S., Collins, A., & Duguid, P. (1989). Situated cognition and the culture of learning. *Educational Research, 18*(1), 32–42.

Churchill, D. (2005). Learning object: An interactive representation and a mediating tool in a learning activity. *Educational Media International, 42*(4), 333–349.

Churchill, D. (2007). Towards a useful classification of learning objects. *Education Technology Research and Development, 55*(5), 479–497.

Churchill, D. (2008). Learning objects for educational applications via PDA technology. *Journal of Interactive Learning Research, 19*(1), 5–20.

Churchill, D. (2013). Conceptual model design and learning uses. *Interactive Learning Environments, 21*(1), 54–67.

Churchill, D. (2014). Presentation design for "conceptual model" learning objects. *British Journal of Education Technology, 45*(1), 136–148.

Churchill, D., & Churchill, N. (2008). Educational affordances of PDAs: A study of a teacher's exploration of this technology. *Computers and Education, 50*(4), 1439–1450.

Churchill, D., & Hedberg, G. (2008). Learning object design considerations for small-screen handheld devices. *Computers and Education, 50*(3), 881–893.

Churchill, D., & Wang, T. (2014). Teacher's use of iPads in higher education. *Educational Media International, 51*(3), 214–225.

Churchill, D., Kennedy, D. M., Flint, D., & Cotton, N. (2010). Using handhelds to support students' outdoor educational activities. *International Journal of Continuing Engineering Education and Life-Long Learning, 20*(1), 54–72.

Churchill, D., King, M., & Fox, B. (2013). Learning design for science education in the 21st century. *Journal of the Institute for Educational Research, 45*(2), 404–421.

Churchill, D., Lu, J., & Chiu, K. F. T. (2014). Integrating mobile technologies, social media and learning design. *Educational Media International, 51*(3), 163–165.

Dawson, M. R. (2004). *Minds and machines: Connectionism and psychological modeling*. Oxford: Blackwell Publishing.

Engeström, Y. (1987). *Learning by expanding*. Helsinki: Orienta-konsultit.

Evans, C. (2008). The effectiveness of m-learning in the form of podcast revision lectures in higher education. *Computers and Education, 50*(2), 49–498.

Gibbons, A. (2008). Model-centered instruction, the design and the designer. In D. Ifenthaler, P. Piarnay-Dummer, & J. M. Spector (Eds.), *Understanding models for learning and instruction* (pp. 161–173). New York, NY: Springer.

Grabinger, R. S., & Dunlap, J. C. (1997). Rich environments for active learning: A definition. *Research in Learning and Teaching, 3*(2), 5–34.

Gu, X., Gu, F., & Laffey, J. M. (2011). Designing a mobile system for lifelong learning on the move. *Journal of Computer Assisted learning, 27*(3), 204–215.

Harper, B., & Hedberg, J (1997). Creating motivating interactive learning environments: A constructivist view. *Paper presented at the ASCILITE 97*, Retrieved from http://www.ascilite.org.au/conferences/perth97/papers/Harper/Harper.html.

Hsieh, S. W., Jang, Y. R., Hwang, G. J., & Chen, N. S. (2011). Effects of teaching and learning styles on students' reflection levels for ubiquitous learning. *Computers and Education, 57*(1), 1194–1201.

Johnson, T., & Lesh, R. (2003). A models and modeling perspective on technology-based representational media. In R. Lesh & H. Doerr (Eds.), *Beyond constructivisim: A models and modeling perspectives on mathematics problem solving, learning and teaching* (pp. 3–34). Mahwah, NJ: Lawrence Earlbaum.

Jonassen, D. (1999). Designing constructivist learning environments. In C. M. Reigeluth (Ed.), *Instructional design theories and models: A new paradigm of instructional theory* (Vol. 2, pp. 215–239). Hillsdale, NJ: Lawrence Erlbaum Associates.

Jonassen, D. (2000). Towards design theory of problem solving. *ETR&D, 48*(4), 63–85.

Jonassen, D., & Churchill, D. (2004). Is there a learning orientation in learning objects? *International Journal on E-Learning, 3*(2), 32–41.

Kearney, M. (2014). Investigating teachers' adoption of signature mobile pedagogies. *Computers and Education.*

Klopfer, E., & Squire, K. (2005). *Environmental detectives: The development of an augmented reality platform for environmental simulations*. Retrieved from http://website.education.wisc.edu/kdsquire/manuscripts/ETRD-handheld-Draft.doc.

Lai, C.-H., Yang, J. C., Chen, F. C., Ho, C. W., & Chan, T. W. (2007). Affordances of mobile technologies for experiential learning: The interplay of technology and pedagogical practices. *Journal of Computer Assisted learning, 23*(4), 326.

Lesh, R., & Doerr, H. (2003). Foundations of a models and modelling perspective on mathematics teaching, learning and problem solving. In R. Lesh & H. Doerr (Eds.), *Beyond constructivisim: A models and modeling perspectives on mathematics problem solving, learning and teaching* (pp. 3–34). Mahwah, NJ: Lawrence Earlbaum.

Liaw, S. S., Hatala, M., & Huang, H. M. (2010). Investigating acceptance toward mobile learning to assist individual knowledge management: Based on activity theory approach. *Computers and Education, 54*(2), 446–454.

Looi, C. K., Seow, P., Zhang, B., So, H. J., Chen, W.-L., & Wong, L. H. (2010). Leveraging mobile technology for sustainable seamless learning. *British Journal of Educational Technology., 41*(2), 154–169.

Luchini, K., Quintana, C., & Soloway, E. (2004). Design guidelines for learner-centered handheld tools. *CHI, 6*(1), 135–141.

Mayer, R. E. (1989). Models for understanding. *Review of Educational Research, 59*(1), 43–64.

Mayer, R. E. (2002). Multimedia learning. *Psychology of Learning and Motivation, 41*, 85–139.
New Media Consortium. (2011). The NMC Horizon Report: 2011K-12 Education.
Norman, D. A. (1988). *The psychology of everyday things*. New York: Basic Books.
Norman, D. A. (1983). Some observation on mental models. In D. Gentner & A. L. Stevens (Eds.), *mental models* (pp. 7–14). Hillsdale, NJ: Erlbaum.
Oliver, R. (1999). Exploring strategies for online teaching and learning. *Distance Education, 20* (2), 240–254.
Patten, B., Sánches, I. A., & Tangney, B. (2006). Designing collaborative, constructivist and contextual applications for handheld devices. *Computers and Education, 46*, 294–308.
Ratto, M., Shapiro, R. B, Truong, T. M., & Griswold, W. G. (2003). *The active class project: Experiments in encouraging classroom participation.* Retrieved from http://www-cse.ucsd.edu/~wgg/Abstracts/activeclass-cscl03.pdf.
Roschelle, J., & Pea, R. (2002). A walk on the WILD side: How wireless handhelds may change CSCL. *International Journal of Cognition and Technology, 1*(1), 145–168.
Savery, J. R., & Duffy, T. M. (1995). Problem based learning: an instructional model and its constructivist framework. *Educational Technology, 35*(5), 31–38.
Seel, N. M. (2003). Model-centered learning and instruction. *Technology Instruction Cognition and Learning, 1*(1), 59–85.
Segall, N., Toni, L., Doolen, J., & Porter, D. (2005). A usability comparison of PDA-based quizzes and paper-and-pencil quizzes. *Computers and Education, 45*(4), 417–432.
Seppälä, P., & Alamäki, H. (2003). Mobile learning in teacher training. *Journal of Computer Assisted learning, 19*, 330–335.
Sharples, M. (2000). The design of personal mobile technologies for lifelong learning. *Computers and Education, 34*(3–4), 177–193.
Sharples, M., Corlett, D., & Westmancott, O. (2002). The design and implementation of a mobile learning resource. *Personal and Ubiquitous Computing, 6*, 220–234.
Shen, R., Wang, M., Gao, W., Novak, D., & Tang, L. (2009). Mobile learning in a large blended computer science classroom: System function, pedagogies, and their impact on learning. *IEEE Transaction on Education, 52*(4), 538–546.
Smith, J. P., diSessa, A. A., & Roschelle, J. (1993). Misconceptions reconceived: A constructivist analysis of knowledge in transition. *Journal of the Learning Sciences, 3*(2), 115–163.
Song, Y. (2014). "Bring Your Own Device (BYOD)" for seamless science inquiry in a primary school. *Computers and Education, 74*, 50–60.
Song, Y., & Fox, R. (2008). Using PDA for undergraduate student incidental vocabulary testing. *ReCALL, 20*(3), 290–314.
Ting, Y. L. (2013). Using mobile technologies to create interwoven learning interactions: An intuitive design and its evaluation. *Computers and Education, 60*(1), 1–13.
van Someren, A., Boshuizen, P. A., de Jong, T., & Reimann, P. (1998). Introduction. In A. van Someren (Ed.), *Learning with multiple representations* (pp. 1–5). Oxford, UK: Elsevier Science.
Vosniadou, S. (1994). Capturing and modeling the process of conceptual change. *Learning and Instruction, 4*(1), 45–69.
Vosniadou, S., De Corte, E., & Mandl, H. (1995). *Technology-based learning environments*. Heidelberg: Springer.
Wong, L. H., Chin, C. K., Tan, C. L., & Liu, M. (2010). Students' personal and social meaning making in a Chinese idiom mobile learning environment. *Educational Technology and Society, 13*(4), 15–26.
Wong, L.-H., & Looi, C.-K. (2011). What seams do we remove in mobile-assisted seamless learning? A critical review of the literature. *Computers and Education, 57*(4), 2364–2381.
Zurita, G., & Nussbaum, M. (2004). Computer supported collaborative learning using wirelessly interconnected handheld computers. *Computers and Education, 42*, 289–314.

Author Biographies

Daniel Churchill is an Associate Professor at the Faculty of Education, The University of Hong Kong. He has been teaching for more that two decades at universities around the world, including Europe, Australia and Asia. He specialises in educational technology, education system development, and multimedia design. Over the years, he has been active in researching topics including learning 2.0, mobile learning, e-learning instructional design, learning objects, and teacher professional development. He is a recipient of a prestigious "HKU Outstanding Teacher Award," and "the Macromedia's Award for the Best Multimedia on the Internet." Also, Dr Churchill is active as a novelist, and his latest novel "The Refugees" has been published and distributed globally in 2015.

Bob Fox Deputy Director, Learning and Teaching Unit at The University of New South Wales (UNSW). He holds concurrent appointments as Honorary Professor, Faculty of Education, University of Hong Kong (HKU) and Adjunct Professor, School of Education, Faculty of Arts and Social Sciences, UNSW. He was awarded HKU's University Teaching Fellowship for excellence in teaching. His research interests focus on technological practice and change in higher education; building institutional capacity and staff capabilities; teacher professional development; innovative use of new technology; blended learning, BOOCs and MOOCs; learning environments and the impact of technology on physical learning spaces.

Mark King is a psychologist and cognitive ethnographer working in the areas of human interactivity and learning. He is currently developing methodologies for studying human learning within distributed cognitive systems, including an eye-tracking methodology for investigating the co-synchronization of language behaviour and visual perception between persons in learning events. He currently serves as the Head of Educational Design and Development at UNSW Australia.

Chapter 2
Conceptualising Authentic Mobile Learning

Kevin Burden and Matthew Kearney

Abstract Conventional accounts of authentic learning focus on contextual factors: tasks, processes, how situated the learning is and the extent to which learners engage in simulated or participative real-world activities. This paper theorises how ubiquitous mobile technologies are fracturing the boundaries that demarcate traditional accounts of authentic learning affording new opportunities to reconceptualise what authenticity means for learners when they use a boundary object such as a mobile device. Whilst some of this has been captured previously with terms like 'seamless', 'contextualised' and 'agile' learning, this paper argues that the concept of authentic mobile learning is a highly fluid construct which will continue to change as the technologies develop and as the pedagogical affordances become better understood by educators and end-users. The paper offers a three-dimensional model of authentic mobile learning and argues that further empirical research is required to understand what is authentic mobile learning from the perception of learners.

2.1 Introduction

> Authenticity remains a concept that is referred to by many, yet poorly defined
>
> (Barab et al. 2000, p. 38)

Contemporary endeavours to understand and define mobile learning (m-learning) draw attention to the situated and seamless nature of activities that are mediated through the affordances of mobile technologies, describing these as

K. Burden (✉)
The University of Hull, Kingston-upon-Hull, UK
e-mail: k.j.burden@hull.ac.uk

M. Kearney
University of Technology Sydney (UTS), Sydney, Australia
e-mail: Matthew.Kearney@uts.edu.au

D. Churchill et al. (eds.), *Mobile Learning Design*, Lecture Notes in Educational Technology, DOI 10.1007/978-981-10-0027-0_2

authentic learning (Herrington and Kervin 2007; Herrington et al. 2008). Learners are considered to be more engaged in contexts which offer high levels of personal significance and cultural relevance. In terms of personal significance, they act as a bridge linking new information and theories to learners' life world outside of formal education and in terms of cultural relevance they enculturate the learner into the practices of the community helping them to think like a member of the discipline (Lombardi 2007; Meyers and Nulty 2009; Stein et al. 2004). Despite considerable research associated with authentic learning (Barab et al. 1989; Brown et al. 1989; CTVG 1990; Petraglia 1998; Radinsky et al. 2001), there are to date relatively few studies which have analysed how mobile technologies support and enhance authentic learning and reciprocally how far authenticity is an inherent characteristic of mobile learning itself (Herrington and Kervin 2007; Herrington et al. 2008; Herrington and Oliver 2000; Kearney et al. 2012; Kearney et al. 2015).

Recent data, collected by the authors from an international survey of educators using mobile technologies in their teaching and learning, highlights one of many confusions associated with the twin concepts of authenticity and mobile learning. Participants consistently ranked the construct of authenticity as 'high', with a mean average of 2.4 on a scale of 1 (low) to 3 (high), when describing a learning scenario where they had used mobile technologies for pedagogical purposes. This high ranking of authenticity by the teachers was despite the fact that 82 % of their self-reported scenarios were situated in formal institutional settings such as schools and universities which might normally be considered rather inauthentic settings (Kearney et al. 2015). This paradox forms the focus for this article which seeks to theorise the concept of authentic learning with mobile technologies. Although authenticity and the learning theories associated with it are often described alongside mobile learning many of the underlying concepts and approaches which have been adopted to enact them as pedagogy are based on a range of assumptions about learning which are rarely articulated or fully explained (Radinsky et al. 2001, p. 406; Selwyn 2014).

The paper is structured as follows. Section 2.2 outlines the background of the paper by exploring why authentic learning is considered important. Section 2.3 seeks to define the term authentic learning identifying two interpretations which are evident in authentic mobile learning. The main body of the paper, (Sect. 2.4), brings together existing research about authentic learning to facilitate and support mobile learning. In so doing, it identifies three distinct and recurring definitions. These are subsequently presented as vectors in a three-dimensional orthogonal model which is offered as an original way to conceptualise authentic mobile learning (Sect. 2.5). In this final section, we discuss the implications of these theorisations and consider the utility of the proposed model for better understanding the phenomenon of mobile learning and authenticity.

2.2 Why Is Authentic Learning Important?

The concept of authentic learning is not new and may have reached its zenith in Europe during the Middle Ages when it functioned as the primary mode of instruction in the craft guilds where apprentices honed their skills vicariously alongside a master craftsman (Lombardi 2007, p. 6). The advent of industrialisation brought about the need to train a mass labour force meaning the apprenticeship model of learning declined and was supplanted by less direct but more cost-efficient institutional systems of mass education (Klopfer et al. 2004). Only in recent years has the interest in more authentic real-world learning resurfaced alongside theories of situated learning (Brown et al. 1989) and cognitive apprenticeships (Collins 1988; Collins et al. 1989). Much of this renewed interest can be traced to economic and technological imperatives which have combined to make authentic learning both economically viable and pedagogically appealing.

The economic drivers stem from the structural shifts in post-Fordist capitalism which have seen the decline in traditional labour-intensive industries and the emergence of new forms of production which are largely ‘immaterial’ in nature, based on the manipulation of networked knowledge and ideas (Lazzarato 1996; Selwyn 2014). These structural shifts demand a new set of skills and dispositions for a largely immaterial workforce which include creativity, networking, cooperation and autonomy (Selwyn 2014).

Technology is also an important driver in the renewed popularity of authentic learning since computers and, more lately, mobile technologies have matured to the point at which previously inefficient models of learning are once again feasible. Mobile technologies are relatively ubiquitous, small and discreet making them ideal for many work-based learning tasks such as capturing images, notes and reflections in situ (Burden et al. 2010). Today’s mobile devices are invariably networked which allow learners to participate in real communities of practice such as Science Citizen projects where they are supported by genuine professionals, akin to the traditional apprenticeship model, although at a greatly reduced cost.

Given this resurgence of interest in models of authentic learning and the world-wide technological shift to post-PC devices (PPD) such as mobile phones and tablet computers, it is timely and important to better understand the assumptions which underpin the concepts of authenticity and mobile learning. Therefore, this article addresses the following research questions:

- What assumptions underpin the concept of authentic learning with mobile technologies?
- What functional value do these conceptualisations serve for educators and the wider academic community seeking to further exploit the potential of mobile technologies?

2.3 Defining Authenticity

The Oxford dictionary definition of the term authentic reveals two etymological strands upon which similar but significantly different interpretations of the phrase have gradually emerged. In its original form, deriving from the Greek term 'authentikos', authentic is defined as meaning of 'undisputed origin', 'not a copy' or 'replica' and this interpretation has been appropriated into the legal lexicon where synonyms like 'genuineness', 'bona fide' and 'veritable' are used to imply the integrity and originality of a person, object or act.

The second etymological derivation, which has become the more commonly used (at least since the eighteenth century) stems from a more representative understanding of the term associated with secondary rather than direct experience. An account of an eye witness is described as authentic if it is accurate in its representation of the facts. Authenticity, in this second sense of the term is a measure of reliability and correspondence between the original artefact (e.g. an accident in the street) and its secondary representation (e.g. by an eye witness). In this secondary interpretation, various proxies such as trustworthiness and authoritative certification replace the certainty afforded by direct sensory first-hand presence (Russell 1959) and in this sense authenticity is a measure of fidelity and correspondence between the primary account and its second-hand representation.

When the term authentic is used in association with learning, both the direct and representative etymological definitions are invoked; but until recently with the emergence of ubiquitous ownership of mobile devices authenticity has most commonly referred to the representative interpretation, whereby students tackle real-world problems and challenges through a simulated, rather than a direct participatory interface. Technology and the affordances of mobile technologies challenge these traditions in ways which will be discussed later in the article.

2.4 Authentic Learning and Mobile Technologies

The term authentic learning is used in various different ways in the field of educational technology and this section explores three different descriptions based on studies of mobile technology use reported in the research literature.

In the first of these authenticity describes the context of the learning activity and the extent to which this is participative or simulated. In these descriptions, authenticity is judged by the extent to which students engage in activities and tasks like those undertaken by professional communities of practice in so-called 'real-world' settings. The second definition relates more to the nature of the tasks and activities undertaken. In these cases, authenticity is a measure of the degree of agency granted to students which is also correlated with the extent to which the learning activity is predefined or emergent, planned or unplanned. The third definition of authenticity is embedded within the student's personal goal structures and emotional engagement with the learning activity. From this perspective, authenticity is a measure of how far learning activities 'engage students' lived experience, enabling students to find meaningful connections with their current views, understandings and experiences' (Stein et al. 2004, p. 240).

2.5 Unpacking Authentic Learning

It is generally agreed that authentic learning ideally requires students to tackle real-world problems located in contexts that mimic the work of professionals and discipline experts (Collins 1988; Herrington et al. 2008; Lombardi 2007; Maina 2004; Renzulli et al. 2004).

> In general, learning environments are considered authentic when there is a similarity between the structured learning activities and some meaningful context for that activity (Barab et al. 2000, p. 38)

In traditional educational paradigms, participative authenticity requires learners to be physically located in the community of practice or professional setting itself as in the apprenticeship model; whereas simulated authenticity allows learners to be located in their normal spaces and contexts where the conditions of the real-world contexts are replicated. Technology blurs these distinctions and mobile technologies are causing them to fracture in ways which are not yet fully understood or appreciated.

2.5.1 Participatory Contexts

In participative authentic contexts, learners participate in genuine real-life communities as 'legitimate peripheral' members (Lave and Wenger 1991) gradually learning the practices, stories and languages of the community or what has been described as "the ordinary practices of th[at] culture" (Brown et al. 1989, p. 34). In effect, learning is a socio-cultural process of identity formation as novices are enculturated into the dominant practices of the community gradually gaining status as experts. Learning is considered to be highly authentic because it is situated in the same context that it will be used making it personally meaningful for the learner.

A practical example using mobile technologies would be use of the sense-it ® app which supports learners in measuring and investigating real-world phenomena. It is based on the principles of Citizen Science whereby members of the public use the app on their mobile device to collaborate with professional scientists, contributing to observation and measurement data such as species identification and air/water pollution monitoring (Henerodotou et al. 2014). A similar participative project using mobile devices was reported by Scanlon et al. (2014) who explain how users of the iSpot application were able to participate in location-based science activities based on the local environment, sharing their findings and data with professional scientists and other activists in an online community of practice.

A simple but highly effective example of participative authenticity is reported by Ebner (2009) who undertook a study of academics using Twitter on their mobile phones as a back channel at an academic conference. Delegates tweeted their responses and impressions of each presentation and these tweets were simultaneously projected on a large screen behind the presenter. In this respect, delegates were physically situated in a highly authentic context (the conference) and were also participating in a genuine community of academic practice, as were those lurkers who could not attend the conference directly but could follow and participate online.

In these examples of participative authenticity, mobile technologies mediate how learners work alongside professionals gradually acquiring the habits and cultural trappings of the community as in a traditional apprenticeship model. However, in many of these examples the learner does not need to be physically located in the actual community since this can now be achieved through virtual participation even from within a formal setting such as a classroom or conference venue. In this sense, mobile technologies are blurring the boundaries or seams between formal and informal learning contexts enabling learners to work in ways which are often described as seamless and unbounded (Looi et al. 2010).

2.5.2 *Simulated Contexts*

Previously, most authentic learning activities have been simulated in a 'practice field' (Brown et al. 1989; Collins et al. 1989) such as the classroom due to the logistical problems associated with direct participation including costs, time and concerns about personal safety. In these benign spaces learners simulate the tasks and processes of real-world contexts. Many apps and tools are now available which mimic the tools and processes used by professionals in the real-world such as measurement tools (e.g. virtual wind tunnels, oscilloscopes and laminators) in science. Where these have been used effectively, such as the 'connected classroom' project (Foley and Reveles 2014), they use real-world online resources to engage students in authentic but simulated science inquiry. In this example, students used handheld devices within the classroom to share data from their own experiments with other students and schools allowing them to compare and analyse across larger data sets and collaboratively identify trends as a community of science learners (Burden and Kearney 2016).

In a similar case study Jones et al. (2013) discussed how their nQuire software tool was used on mobile devices to enable science students to take greater responsibility for their own inquiries without adult help. These inquiries were engaging and personally relevant and allowed students to continue their inquiry seamlessly across different contexts such as an after school club and home. These tools and apps have the potential to support highly authentic forms of simulated learning both in formal and hybrid spaces (see below) but empirical research to date suggests they are often used by teachers for low level, unrealistic tasks which bear few resemblances to authentic practices (Kearney et al. 2012, 2015).

2.5.3 Hybrid Contexts

Current advances in mobile technologies have fractured the traditional boundaries between participative and simulated contexts. In some cases, this has seen students participating virtually from within formal contexts in genuine and real communities such as the nQuire project described above (Jones et al. 2013). In these contexts, learning takes on a hybrid complexion which combines features of both a direct, participative and indirect simulated model of authentic learning, and many of the technology projects which have explored these spaces report that they combine all of the best qualities of simulations with the additional benefits of high ecological validity acquired through participation in a genuine community.

The combination of augmented reality (AR) applications and mobile devices frequently results in hybrid models of authenticity referred to as 'participatory simulations' (Barab and Dede 2007). Wong and Looi (2011) for example, documented a series of games played in a physical environment but augmented by virtual artefacts through the mediation of a mobile device (they called this 'mixed reality learning'). Mobile devices with location-based sensors allowed users in the study to interact with explorations, experiments and challenges for inquiry and game-based learning. Lui et al. (2014) described an immersive, cave-like rainforest simulation (called EvoRoom) and a mobile inquiry platform (called Zyeco) that enabled users to collect and share data. Students were co-located in an immersive and physical digital space, collecting observational data from both the classroom (Evoroom) and out-of-class settings (such as parks or museums), and exploring peers' data using large visualisations displayed at front of room.

2.6 Is Authentic Mobile Learning Predefined or Emergent?

Despite advances in mobile technologies which have afforded learners greater agency in how they access information, where they situate their learning and how they present the outcomes of this as assessment artefacts, some authors have noted the reluctance of educators to cede significant control of learning to students (Kearney et al. 2015). This is reflected in the extent to which learning is predefined or is left more open ended and emergent in design.

Williams et al. (2011) define emergent learning as "learning which arises out of the interaction between a number of people and resources, in which the learners organise and determine both the process and to some extent the learning destinations, both of which are unpredictable" (p. 3). There is an implicit assumption in many of the studies on authenticity that learning is likely to be more unplanned and emergent than predefined or prescribed when students tackle ill-defined problems that defy simplistic or quick solutions. Over prescription and unnecessary intervention by educators is included as one of Herrington et al.'s list of inauthentic strategies for mobile learning (2008).

Some researchers have identified planning related to learner generated contexts as a significant vector in understanding how mobile technologies can make learning more authentic (Toh et al. 2013). These studies show how students spontaneously used their mobile devices to capture and share images or video clips related to a personal interest or hobby (e.g. bird watching) without the direction or prescription of a teacher or adult (Jones et al. 2013). These examples often occur in informal settings outside of institutional control but there is no reason to suppose this kind of incidental learning with mobile technologies, could not, and is not taking place within formal settings in the form of serendipitous learning (e.g. where a learner uses their mobile device to capture an idea or inspirational thought) (Toh et al. 2013; Williams et al. 2011).

One area where emergent learning is more evident is in mobile games-based applications where players can engage in highly realistic simulations and problem-solving exercises that mimic many of the tasks undertaken by real professionals. Gwee et al. (2010) reported one such mobile simulation which featured year 9 social studies students using the game Statecraft X on their iPhones to learn about the concept of governance through role play. What distinguishes the game is the amount of spontaneity and lack of planning. Students worked largely at their own pace without interventions or schedules to regulate them.

These discussions then invite questions as to the extent to which authenticity can or should be designed into the learning experiences of students when they use mobile technologies (Barab et al. 2000; Petraglia 1998). This raises an obvious tension as it is difficult to visualise how instructors can design learning activities that are entirely emergent since the very act itself assumes a degree of deliberate intent. For some researchers, the solution is to 'deny the legitimacy of preauthentication' altogether by which they mean they reject the notion that designers or teachers can construct predefined authentic tasks, even if these have real and practical use to a genuine community of practice (Barab et al. 2000). They argue that these elements of authentic learning cannot be predefined because they do not guarantee 'buy in' from learners. If the learner does not personally perceive the context to be authentic it cannot be 'preauthenticated' or designed by some other person. In this sense authenticity "is manifest in the flow itself, and is not an objective feature of any one component in isolation" (Barab et al. 2000, p. 38).

2.6.1 Personal Commitment of Learners

In considering the nature of authentic learning, it is important to identify for whom the learning will be authentic (Barab and Duffy 2000). Most descriptions of authentic learning describe it from the privileged perspective of the instructor or designer and it difficult to appreciate to what extent learners themselves perceive a learning practice to authentic, or what indeed learners think authentic means. However, ultimately authenticity “lies in the learner perceived relations between the practices they are carrying out and the use value of these practices” (Barab et al. 2000, p. 38).

This is partly a methodological concern and there is an urgent need for researchers to design more authentic methods and tools which will gain access to this largely missing learner perspective. This is a genuine concern since designing realistic, real-world tasks or contexts and processes that mimic or place learners in actual professional communities may count for little if the learner does not perceive these artefacts to have personal significance and meaning in relation to their desired learning objective.

> It is very important to consider what is meant by authenticity and to whom - who is the judge (the educator; the learner or the community upon which they try to emulate?) (Barab and Duffy 2000).

Indeed there is a concern amongst some that what constitute genuine real-world communities of practice for adults may be far from authentic from the perspective of learners who may speak an entirely separate discourse based on the ‘curricular language’ with which they are familiar (Heath and McLaughlin 1994). These critics argue that teachers should attempt to locate authentic learning in what they term ‘institutions of curricular authenticity’ where familiar curricular practices, languages, norms and traditions are the Lingua franca. This position is further supported by Hiebert et al. (1996) who argue that students can be engaged in deeply contextualised and authentic tasks within the curriculum as long as they are personally challenged to engage with the underlying concepts and deep structures of the discipline itself.

These considerations therefore foreground a critical third constituent in authentic learning which is the emotional and extra-rational dimension of learning and the commitment of the learner whilst also highlighting one of the more substantial epistemological challenges in the field of authentic learning: how can we capture and understand the learner’s emotional sense of engagement and commitment?

> This definition of authenticity correlates how well a learning activity matches a student’s personal goal structures (Heath and McLaughlin 1994) or the extent to which learners themselves problematize the elements that make up the context (Stein et al. 2004, p. 240).

In many of the case studies reported in this paper, we can infer that learners were highly motivated and engaged in the mobile-learning activities which are described but meaningfulness is a difficult construct to capture and few of the studies detail to what extent the mobile activity enabled learners to develop personal meanings, or indeed why. One exception is the pilot study for the Ecomobile project

(Kamarainen et al. 2013). This project explored how the use of a mobile AR application (FreshAIR) could be combined with probeware tools and software to enable students to understand the ecosystem of a pond in ways which resembled real scientific practice. Feedback and video evidence from students undertaking the project indicate that it was highly engaging and had considerable personal significance for students working in their local environment. They appear to have engaged with the topic on a highly personal level despite the fact it did not feature a genuine professional community of scientists as such.

2.7 Discussion and Implications

Derived from the above definitions and examples, we propose the following orthogonal model as a means of further conceptualising authentic mobile learning (see Fig. 2.1). We identify *Context* as a critical vector in understanding how and where the learning activity is situated and use the terms 'simulated' and 'participative' as the binaries for this continuum. These are not proposed as normative labels since there is no implication here that either form of authenticity is necessarily more desirable than the other.

The second axis called *Planning Design* measures the extent to which the learning activity is planned or unplanned in a similar way to the model developed by Toh et al. (2013). However, given the emerging affordances of mobile technologies we place greater emphasis on the agency of the learner in co-negotiating

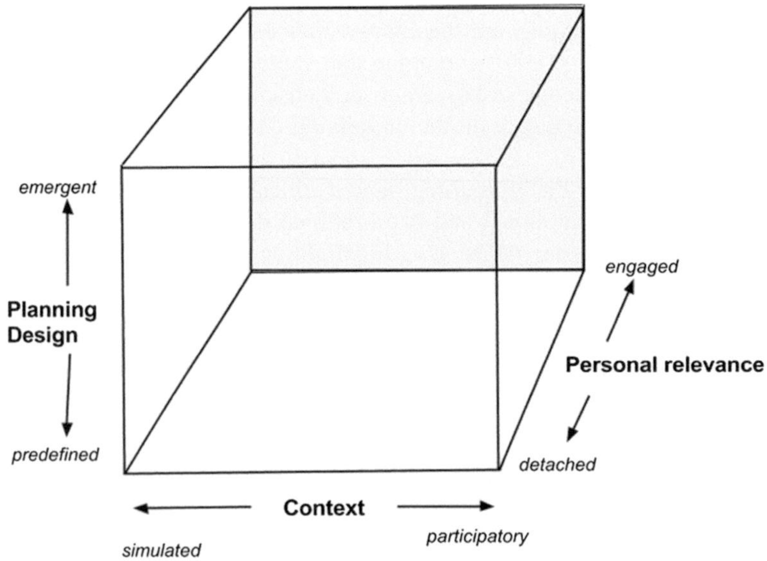

Fig. 2.1 A conceptual model of authentic mobile learning

and designing these contexts. Hence this vector is used to measure both the degree of agency granted to the learner and the extent to which the learning activity as a whole is preplanned or emergent.

Third, we include a vector capturing the *Personal Relevance* and consequent engagement of the learner since this has emerged across many studies as a highly significant but often neglected element of authentic learning. Unlike the other two vectors which are not normative, this vector is more judgemental since it is recognised that learners will elect to disengage from learning which holds little or no personal significance or meaning for them.

2.7.1 How Does the Model Work?

To illustrate how this three-dimensional model might further support the conceptualisation of authentic mobile learning, we have populated it with the three mobile-learning scenarios described earlier in the paper represented by the letters A, B and C (see Fig. 2.2 and Table 2.1).

In terms of the context vector, only the Twitter example (C) was classed as participative since it was set in a genuine real-world context in this professional learning scenario (an academic conference) accessible in both a physical and virtual manner through the mobile device. In the Ecomobile example (A), students participated in real-world tasks and processes using tools in a real-life way and in relevant informal settings but they did not engage with a real community of

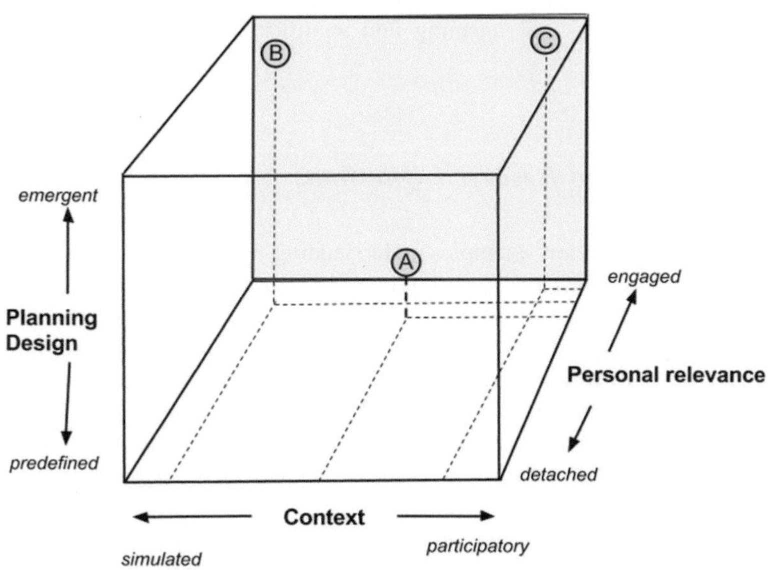

Fig. 2.2 Authentic mobile learning examples

Table 2.1 Characteristics of exemplar authentic mobile learning scenarios

	Exemplar	Context	Planning design	Personal relevance
A	Ecomobile project (Kamarainen et al. 2013)	Hybrid	Predefined	High
B	*Statecraft X* mobile learning game (Gwee et al. 2010)	Simulated	Emergent	High
C	Twitter back channel in an academic conference (Ebner 2009)	Participative	Emergent	High

practice, even though this might have been feasible with the mediation of mobile technology. Therefore, the context was identified as a hybrid. The mobile game example (C) was entirely simulated in terms of context since there was little attempt to involve students in a genuine governance community.

Both examples B and C were classed towards the emergent end of the planning design spectrum since neither was heavily predefined or structured. In the case of the mobile game (B), students were not restricted by fixed schedules and could engage at their own pace. This was also true in the case of the Twitter, for example where participants were left to determine how and when they would structure their responses (if at all). The Ecomobile example (A) was more predefined by the teacher who had devised many of the tasks in advance even though most if it occurred in an informal setting outside of school.

Finally, although students were not directly questioned about their levels of personal engagement in any of these three exemplars, we might infer that motivation and engagement was high judging by the amount of activity which occurred, often unsolicited as in the mobile games example, and this suggests all three examples had high personal meaning and significance from the perspective of learners themselves.

2.7.2 Returning to Research Questions

As shown in even these few examples understanding what is authentic about mobile learning is not straightforward or unproblematic. Therefore this model offers a novel way of conceptualising these issues which rejects simplistic solutions that frame authentic mobile learning in terms of mutually exclusive binaries. Traditionally, this is how authentic learning has been framed epitomised by the dictionary definition duality between first-hand direct experience which equates with the participatory model of authentic learning and, indirect, second-hand experience which equates to the simulated model of authenticity. This article has argued that this traditional duality is no longer valid when students have access to and use mobile devices, blurring the boundaries between simulated and participative forms of real-world learning, between predefined and emergent models of learning and between high or low levels of personal engagement and meaning making.

The concepts of boundary crossing and boundary objects which are inherent features of Activity theory (Engeström et al. 1995) are useful ways of thinking about authenticity and mobile learning because they focus on learning which transcends conventional boundaries such as home/school, formal/informal, physical/virtual using mobile devices as cultural objects which mediate these crossings. Here, "boundaries are understood as a social cultural difference between systems, practices, or social worlds, leading to a discontinuity in action or interaction between these systems" (Snoek 2013, p. 309). In effect, mobile devices fulfil a bridging action since they enable learners to cross-traditional boundaries such as the student who joins an authentic community of scientists on Twitter posting and following tweets as a legitimate member of the community, but from within a formal classroom setting which would traditionally be bounded both physically and culturally in such a manner that this was not feasible. Whilst the mobile device acts as a boundary crossing object in these cases it does so within culturally defined boundaries and practices of the traditional classroom setting. If the teacher, and indeed the institution, prohibit the use of technology across contexts in this seamless fashion (Jones et al. 2013; Wong et al. 2015), or if they attempt to pre authenticate or overly predefine the learning outcomes, it is unlikely these opportunities to cross-boundaries will be ceased upon, or alternatively they become a form of subversive activity undertaken by students looking to escape the rigidity and sterility of classroom learning.

What this chapter has also attempted to highlight is the primacy of affective factors such as perceptions of personal relevance on the part of the learner which is so critical in authentic learning. Research in the pre-mobile era already suggested that authenticity was not a commodity which could be objectified and designed into the context or tasks itself (Barab et al. 2000) but rather it was highly ephemeral and closely associated with the personal perceptions of the individual learner. Current research into authentic mobile learning has identified a significant list of characteristics that are deemed to make learning more authentic (Herrington et al. 2008) but there is little empirical evidence of what these factors mean from the perspective of learners themselves. There is an urgent need, therefore, for the mobile-learning research community to better understand how this kind of data might be elicited and how it would then be used to support in the design of more meaningful and engaging authentic mobile learning scenarios. In this respect, we still face the same epistemological and methodological challenges that were highlighted by researchers investigating the potential of first generation computers to enhance authentic learning: "A major challenge for instructional designers is to develop learning environments that incorporate authentic tasks in realistic contexts" (Barab et al. 2000, p. 60).

2.8 Conclusion

At the beginning of this paper, we identified a conundrum which questioned why educators associate mobile learning so closely with authenticity if most of their learning tasks are situated in formal settings such as schools and universities? The paper has posited that no single criteria or characteristic makes a learning activity authentic (Banas and York 2014) and it has also argued that traditional definitions of authenticity are in need of revision and upgrade to better reflect the boundary crossing potential mediated by mobile devices. Although formal settings such as schools and universities might once have been considered contrived contexts for learning compared to genuine real-world settings such as work placements or apprenticeship this definition is rooted in pre-mobile notions of space and time (Traxler 2009) which are no longer as applicable as they were previously. The conceptual model proposed in this chapter (see Fig. 2.1) has a practical orientation for learning design in mobile environments since it highlights three critical vectors that need to be considered carefully in order to maximise the authenticity of any mobile learning experience. Further research is also required to investigate to what extent educators and learners are reconceptualising their thinking about authentic learning when mobile devices are used seamlessly across the traditional boundaries between formal and informal contexts, virtual and physical worlds and planned and emergent spaces. This paper offers a model to initiate and support this process.

References

Banas, J., & York, C. (2014). Authentic learning exercises as a means to influence preservice teachers' technology integration self-efficacy and intentions to integrate technology. *Australasian Journal of Educational Technology*, *30*(6).

Barab, S. A., & Dede, C. (2007). Games and immersive participatory simulations for science education: An emerging type of curricula. *Journal of Science Education and Technology, 16* (1), 1–3. doi:10.1007/s10956-007-9043-9.

Barab, S. A. & Duffy, T. M. (2000). From practice fields to communities of practice. In D. H. Jonassen & S.M. Land (Eds.), *Theoretical foundations of learning environments* (pp. 25–55). Mahwah, NJ: Erlbaum Associates.

Barab, S. A., Squire, K. D., & Dueber, W. (2000). A co-evolutionary model for supporting the emergence of authenticity. *Educational Technology Research and Development, 48*(2), 37–62.

Brown, J. S., Collins, A., & Duguid, P. (1989). Situated cognition and the culture of learning. *Educational Researcher, 18*(1), 32–42.

Burden, K., Aubusson, P & Schuck, S. (2010). Ethical professional mobile learning for teaching and nursing workplaces, Chapter 12. In N. Pachler, C. Pimmer & J. Seipold (Eds.), *Work-based mobile learning: Concepts and cases. A handbook for academics and practitioners* (pp. 277–305). Oxford: Peter Lang.

Burden, K., & Kearney, M. (in print). Future scenarios for mobile science learning. *Research in Science Education*, *44*(3).

Collins, A. (1988). *Cognitive apprenticeship and instructional technology* (Technical Report No. 6899). Cambridge, MA: BBN Labs Inc.

Collins, A., Brown, J. S., & Newman, S. E. (1989). Cognitive apprenticeship: Teaching the crafts of reading, writing and mathematics. In L. B. Resnick (Ed.), *Knowing, learning and instruction: Essays in Honor of Robert Glaser* (pp. 453–494). Hillsdale: Erlbaum.

CTGV (Cognition and Technology Group at Vanderbilt). (1990). Technology and the design of generative learning environments. *Educational Technology, 31*(5), 34–40.

Ebner, M. (2009). Introducing live microblogging: How single presentations can be enhanced by the mass. *Journal of Research in Innovative Teaching, 2*(1), 91–100.

Engeström, Y., Engeström, R., & Kärkkäinen, M. (1995). Polycontextuality and boundary crossing in expert cognition: Learning and problem solving in complex work activities. *Learning and Instruction, 5*(4), 319–336.

Foley, B. J., & Reveles, J. M. (2014). Pedagogy for the connected science classroom: Computer supported collaborative science and the next generation science standards. *Contemporary Issues in Technology and Teacher Education, 14*(4), 401–418.

Gwee, S., Chee, Y. S., & Tan, E. M. (2010). Game play-time and learning outcomes of boys and girls in a social studies mobile game-based learning curriculum. In M. Montebello, V. Camilleri & A. Dingli (Eds.), *Proceedings of the 9th International Conference on Mobile Learning* (pp. 16–23). Valletta, Malta: University of Malta.

Heath, S. B., & Mclaughlin, M. W. (1994). Learning for anything everyday. *Journal of Curriculum Studies, 26*(5), 471–489.

Herrington, J., & Kervin, L. (2007). Authentic learning supported by technology: Ten suggestions and cases of integration in classrooms. *Educational Media International, 44*(3), 219–236.

Herrington, J., Mantei, J., Herrington, A,. Olney I., & Ferry, B. (2008). New technologies, new pedagogies: Mobile technologies and new ways of teaching and learning. In *Hello! Where are you in the landscape of educational technology? Proceedings ascilite Melbourne 2008.* Retrieved from http://www.ascilite.org.au/conferences/melbourne08/procs/herrington-j.pdf.

Herrington, J., & Oliver, R. (2000). An instructional design framework for authentic learning environments. *Educational Technology Research and Development, 48*(3), 23–48.

Hiebert, J., Carpenter, T. P., Fennema, E., Fuson, K., Human, P., Murray, H., et al. (1996). Problem solving as a basis for reform in curriculum and instruction: the case of mathematics. *Educational Researcher, 25*(4), 12–21.

Herodotou, C., Villasclaras-Fernández, E., & Sharples, M. (2014). The design and evaluation of a sensor-based mobile application for citizen inquiry science investigations. In *Open learning and teaching in educational communities* (pp. 434–439). Springer International Publishing.

Jones, A. C., Scanlon, E., & Clough, G. (2013). Mobile learning: Two case studies of supporting inquiry learning in informal and semiformal settings. *Computers and Education, 61*, 21–32.

Kamarainen, A. M., Metcalf, S., Grotzer, T., Browne, A., Mazzuca, D., Tutwiler, M. S., & Dede, C. (2013). EcoMOBILE: Integrating augmented reality and probeware with environmental education field trips. *Computers and Education, 68*, 545–556.

Kearney, M., Burden, K., & Rai, T. (2015). Investigating teachers' adoption of signature mobile pedagogies. *Computers and Education, 80*, 48–57.

Kearney, M., Schuck, S., Burden, K., & Aubusson, P. (2012). Viewing mobile learning from a pedagogical perspective. *Research in Learning Technology, 20,* 14406. doi:10.3402/rlt.v20i0/14406.

Klopfer, E., Yoon, S., & Rivas, L. (2004). Comparative analysis of Palm and wearable computers for participatory simulations. *Journal of Computer Assisted Learning, 20*, 347–359. doi:10.1111/j.1365-2729.2004.00094.x.

Lave, J., & Wenger, E. (1991*). Situated learning: Legitimate peripheral participation*. Cambridge University Press: Cambridge.

Lazzarato, M. (1996). Immaterial Labour. In M. Hardt & P. Virno (Eds.), *Radical thought in Italy: A potential politics* (pp. 133–147). Minneapolis: University of Minnesota Press.

Lombardi, M. M. (2007). Authentic learning for the 21st Century: An overview. *Educause Learning Initiative Report No. 1,* Boulder, CO, EDUCAUSE Learning Initiative. Retrieved from http://www.educause.edu/ir/library/pdf/ELI3009.pdf.

Looi, C.-K., Seow, P., Zhang, B., So, H.-J., Chen, W., & Wong, L.-H. (2010). Leveraging mobile technology for sustainable seamless learning: A research agenda. *British Journal of Educational Technology, 41*, 154–169. doi:10.1111/j.1467-8535.2008.00912.x.

Lui, M., Kuhn, A., Acosta, A., Niño-Soto, M. I., Quintana, C., & Slotta, J. D. (2014). Using mobile tools in immersive environments to support science inquiry. In *CHI'14 Extended Abstracts on Human Factors in Computing Systems* (pp. 403–406). ACM.

Maina, F. W. (2004). Authentic learning: Perspectives from contemporary educators. *Journal of Authentic Learning, 1*(1), 1–8.

Meyers, N., & Nulty, D. (2009). How to use (five) curriculum design principles to align authentic learning environments, assessment, students' approaches to thinking and learning outcomes. *Assessment and Evaluation in Higher Education, 34*(5), 565–577. doi:10.1080/02602930802226502.
Petraglia, J. (1998). *Reality by design: The rhetoric and technology of authenticity in education.* Mahwah, NJ: Lawrence Erlbaum Associates.
Radinsky, J., Bouillion, L., Lento, E., & Gomez, L. (2001). Mutual benefit partnership: A curricular design for authenticity. *Journal of Curriculum Studies, 33*(4), 405–430.
Renzulli, J. S., Gentry, M., & Reis, S. M. (2004). A time and a place for authentic learning. *Educational Leadership, 62*(1), 73–77.
Russell, B. (1959). *The problems of philosophy* (New ed.). London: Oxford University Press.
Scanlon, E., Woods, W., & Clow, D. (2014). Informal participation in science in the UK: Identification, location and mobility with iSpot. *Journal of Educational Technology and Society, 17*(2), 58–71.
Selwyn, N. (2014). *Distrusting educational technology: Critical questions for changing times.* New York: Routledge.
Snoek, M. (2013). From splendid isolation to crossed boundaries? The future of teacher education in the light of activity theory. *Teacher Development, 17*(3), 307–321.
Stein, S. J., Isaacs, G., & Andrews, T. (2004). Incorporating authentic learning experiences within a university course. *Studies in Higher Education, 29*(2), 239–258.
Toh, Y., So, H. J., Seow, P., Chen, W., & Looi, C. K. (2013). Seamless learning in the mobile age: A theoretical and methodological discussion on using cooperative inquiry to study digital kids on-the-move. *Learning, Media and Technology, 38*(3), 301–318.
Traxler, J. (2009). Learning in a mobile age. *International Journal of Mobile and Blended Learning, 1*(1), 1–12.
Williams, R., Karousou, R., & Mackness, J. (2011). Emergent learning and learning ecologies in Web 2.0. *International Review of Research in Open and Distance Learning, 12*(3) np.
Wong, L. H., & Looi, C. K. (2011). What seams do we remove in mobile assisted seamless learning? A critical review of the literature. *Computers and Education, 57*(4), 2364–2381.
Wong, L. H., Milrad, M., & Specht, M. (Eds.). (2015). *Seamless learning in the age of mobile connectivity.* Singapore: Springer.

Author Biographies

Kevin Burden is a Reader at the University of Hull (UK) in the Faculty of Education where his research focuses on the interface between learning and technology. He is the author of over fifty research and conference papers/chapters and has produced many major reports for governments and industry.

Matthew Kearney is a staff member from School of Education, University of Technology, Sydney, Australia. His main research interests are in the area of technology-enhanced learning in K-12 and teacher education contexts. He has written conference papers for prestigious international and national conferences and produced major research reports for Australian industry and government partners.

Chapter 3
Mobile Social Media: Redefining Professional Development and Collaborative Scholarship

Thomas Cochrane and Vickel Narayan

Abstract In this chapter, we explore a developing model for scaffolding and supporting professional development for higher education lecturers via mobile social media. The framework involves the establishment of communities of practice comprised of lecturers and mobile learning researchers exploring the potential of mobile social media to enable new pedagogical strategies. We identify three key elements of the framework illustrated by two case studies: modelling a community of practice, redefining pedagogy and designing an appropriate technology support infrastructure.

3.1 Introduction

As academic advisors, the authors' of this chapter are primarily interested in challenging and enabling lecturers to explore reflective pedagogical practice that leads to the design and facilitation of engaging and authentic learning experiences for students. A common entry point into discussions around new pedagogical practices is typically around strategies for managing students' use of mobile social media (msm) during class time. A typical enquiry from a frustrated lecturer is: "How can I stop students updating their Facebook status on their phones while in class?" A standard strategy is to require students to turn off their phones while in class, or ban the use of phones during class time. The disruptive nature of mobile devices has been thoroughly discussed within the literature (Anderson and McGreal 2012; Sharples 2001). However, in a world where cellphones will soon out number the number of people on the planet (International Telecommunication Union 2014), and students are increasingly expected to bring their own device (BYOD), a more

T. Cochrane (✉) · V. Narayan
Auckland University of Technology, Auckland, New Zealand
e-mail: thomas.cochrane@aut.ac.nz

V. Narayan
e-mail: vickel.narayan@aut.ac.nz

D. Churchill et al. (eds.), *Mobile Learning Design*, Lecture Notes in Educational Technology, DOI 10.1007/978-981-10-0027-0_3

creative response is needed. Our response to the disruptive nature of student-owned devices in classes is to encourage the engagement of these devices within the learning environment, thus leveraging their disruption of teacher-directed paradigms to enable new pedagogical strategies that are more democratic and student-centred. While mobile learning (mlearning) encompasses a wide variety of devices and approaches, we have focused primarily upon the integration of mobile social media (msm) into the curriculum in authentic ways, moving beyond the substitution of mobile devices as classroom response clickers and PODCast players (for example). The potential of mobile social media in education is in leveraging the potential for ubiquitous connectivity, collaboration, contextual awareness and multimedia content production and sharing that these tools afford.

In order to effectively conceptualize the potential of msm, lecturers need to experience and embed the use of msm within their own personal and professional practice. For lecturers to effectively integrate the use of msm into their curriculum, they first need to understand the potential of msm to form the basis of user-generated content, and communication and collaboration for participation within authentic learning communities. We have found that creating a peer support group in collaboration with educational technologists (via the formation of an intentional community of practice) is one way to facilitate a safe environment for exploring curriculum integration of msm (Cochrane and Narayan 2013).

3.1.1 Why Mobile Social Media?

Smartphones and tablets are powerful computing devices with unique affordances that enable learning and collaboration across multiple contexts. These mobile devices facilitate rich-media production and sharing in the form of images, video, audio and geolocation data, and they can be used to rethink collaboration and develop the potential for enhanced engagement and learning outcomes. Large-scale mobile learning research projects in the UK (Attewell et al. 2010) and Europe (Unterfrauner and Marschalek 2010) have demonstrated that mlearning can empower marginalized learners, and Australian research has shown mlearning can be a catalyst for enabling authentic learning (Herrington et al. 2009). Similarly, the UNESCO (2013) mobile learning report identifies a range of unique benefits of mobile learning, including:

- Expanding the reach and equity of education
- Facilitating personalized learning
- Providing immediate feedback and assessment
- Enabling anytime, anywhere learning
- Ensuring the productive use of time spent in classrooms
- Building new communities of learners
- Supporting situated learning
- Enhancing seamless learning
- Bridging formal and informal learning.

We are particularly interested in mlearning as a catalyst for enabling new pedagogies that focus upon student-generated content in a range of different learning contexts.

3.1.2 Communities of Practice and Social Scholarship

Communities of practice is a social learning theory articulated by Lave and Wenger (1991) and developed (Wenger 1998; Wenger et al. 2002, 2009) across a range of applications and contexts. The concept of COPs has found wide acceptance in educational contexts. Key principles of COPs include: the validity of peripheral participation within a community as members move from observation of more experienced peers to active participation within the community (in our case we are concerned with drawing groups of lecturers into our COP beyond the initial enthusiastic early adopter core group), the glue is the domain or shared interest (in our case this is exploring the potential curriculum impact of a framework for msm), and the brokering of the reified activity of the COP to the wider community (in our case this is the transferable pedagogical practice). Wenger et al. (2005) have proposed the benefits of social media for supporting communities of practice and developed strategies around technology stewardship for guiding COPs in their use of social media (Wenger et al. 2009). While social media is not specifically designed for educational use, epistemologically social media aligns with the main values of social constructivism that is at the heart of communities of practice and many new pedagogical frameworks. Social media also provides a set of interconnected collaborative tools for establishing digital identity and collaborative user-generated content beyond the institutional limits of traditional learning management systems (Herrington et al. 2005).

Social media can also be utilized to transform scholarship from a largely solo pursuit of publication in high-impact peer-reviewed and subscription-based journals, books and conference proceedings. Boyer (1990) is one of the authoritative voices on the scholarship of teaching and learning. It can be argued that Boyer's model requires update in light of the development of the open scholarship movement (Garnett and Ecclesfield 2011) and the emergence of social media tools facilitating collaborative research (for example researchgate.net). For example, Greenhow and Gleason (2014) argue that Boyer's four-dimensional framework of scholarship can be re-envisioned via social media as a framework for social scholarship. These four dimensions include: the scholarship of design (SOD), the scholarship of integration (SOI), the scholarship of teaching and learning (SOTL) and the scholarship of application (SOA). Social scholarship is defined by Greenhow and Gleason as "Social scholarship seeks to leverage social media affordances (i.e. promotion of users, their inter-connections and user-generated content) and potential values (i.e. knowledge as decentralized, co-constructed, accessible and connective) to evolve the ways in which scholarship is accomplished in academia (2014, p. 3). However, Greenhow and Gleason note that "we need more examples of social scholarship in action, within each of the four domains and across them" (2014, p. 9)—this is one of the goals of our framework that we explore later in this chapter.

3.1.3 An MSM Framework for New Pedagogies

Our mobile social media framework for lecturer professional development leverages the effectiveness of mobile devices for enhancing learning outcomes in a tertiary setting. It builds on successful collaborative research carried out across two different higher education institutes involving over 60 mlearning projects since 2006. The framework has three key elements that scaffold and support the integration of mobile social media for new pedagogical strategies:

- Establishing lecturer communities of practice to learn about the affordances of mobile devices in relation to new modes of student learning;
- Redesigning the curriculum in response to shifts in conceptions of teaching;
- Collaborating with ICT services to develop the necessary infrastructure to enable mobile learning across the campus (for example: wireless networks, flexible learning spaces and mobile presentation technologies).

The framework is a blend of concepts that builds on the evidence of over 60 projects, and includes aligning theoretical perspectives such as: the concept of the Pedagogy–Andragogy–Heutagogy (PAH) continuum (Luckin et al. 2010), and Puentedura's (2006) SAMR model (Substitution, Augmentation, Modification, Redefinition) of educational technology transformation and Sternberg et al. (2002) conception of three levels of creativity. Using this developing framework, the researchers have demonstrated how mlearning can enable authentic learning (Herrington et al. 2009) through implementing msm for curriculum change and development (Cochrane and Bateman 2013). The PAH continuum is used as a guiding framework to measure a range of teaching and learning strategies from teacher-directed pedagogy, to student-centred andragogy and student-determined heutagogy (Table 3.1).

In this chapter, we illustrate the effectiveness and scalability of this framework across two educational contexts, and examine the effectiveness of this framework for supporting professional development. The establishment of COPs supports the exploration and implementation of a range of practical strategies for lecturers to utilize the unique attributes of mobile devices for designing transformative teaching and learning practice, and empowering learners to become creative self-directed graduates.

Table 3.1 A mobile social media framework based upon the PAH (modified from Luckin et al. 2010)

	Pedagogy	Andragogy	Heutagogy
Locus of control activity types	Teacher	Learner	Learner
	Content delivery	Teacher as guide	Teacher co-learner
	Digital assessment	Digital identity	Digital presence
	Teacher-delivered content	Student-generated content	Student-generated contexts
	Teacher-defined projects	Student negotiated teams	Student negotiated projects
Cognition level SAMR (Puentedura 2006)	Cognitive	Meta-cognitive	Epistemic
	Substitution and Augmentation	Modification	Redefinition
		Reflection as VODCast	In situ reflections
	Portfolio to eportfolio	Prezi on iPad	Presentations as dialogue with source material
	PowerPoint on iPad	New forms of collaboration	
	Focus on productivity	Mobile device as content creation and curation tool	Community building
	Mobile device as personal digital assistant and consumption tool		Mobile device as collaborative tool
Creativity (Sternberg et al. 2002)	Reproduction	Incrementation	Reinitiation
Knowledge production context	Subject understanding: lecturers introduce and model the use of a range of mobile social media tools appropriate to the learning context	Process negotiation: students negotiate a choice of mobile social media tools to establish an eportfolio based upon user-generated content	Context shaping: students create project teams that investigate and critique user-generated content within the context of their discipline. These are then shared, curated and peer-reviewed in an authentic COP
Mobile social media affordances	Enabling induction into a supportive learning community	Enabling user-generated content and active participation within an authentic design COP	Enabling collaboration across user-generated contexts, and active participation within a global COP
Ontological shift	Learning about: Reconceptualising mobile social media: from a social to an educational domain	Reconceptualising the role of the teacher	Learning to become: Reconceptualising the role of the learner

3.2 Methodology

The projects focus upon establishing a collaborative partnership similar to that advocated by Gunn and Steel (2012, p. 11) involving "learning designers, educational researchers, teachers, technologists and academic support staff". The projects are driven by a reconception of pedagogy rather than technological determinism, whereby mobile social media provide the tools for this pedagogical reconception. The establishment of a community of practice around each project provides the pedagogical and technical support structure for a shift in conceptions of learning and teaching. This involves an ontological shift (Chi and Hausmann 2003), reconceptualizing the role of mobile social media from a purely social domain to an educational domain, and reconceptualizing the roles of the teachers and learners, moving away from a teacher-directed content delivery mode towards a student-determined learning paradigm (heutagogy).

3.2.1 Method

In this paper, we illustrate the implementation of our mobile social media framework within the context of two case studies. The data collected was largely qualitative, triangulated with quantitative survey data. Each case study was comprised of a Community of Practice of researchers and lecturers nurtured via a variety of msm tools linked via a Google Plus Community and curated via a project Twitter hashtag. Wordpress blogs are used to collate and share the outcomes of each COP with a wider audience. Data collection included: a survey at the start of each individual lecturer project to establish the participants' previous experience of mobile devices and social media in learning, the collation of participating lecturers and students social media eportfolios—effectively their own learning journals of their experiences throughout the project—and focus group discussions at the middle and end of each project. At the end of each semester, we encourage participants to reflect critically on their journey by collaborative publishing of msm implementation case studies in appropriate conference proceedings and journal articles. These peer-reviewed reflective practice case studies are used to inform the design of the next iteration of projects. All participants are informed of the aims and requirements of each research project, and are asked to sign informed consent forms as approved by the institution's ethics committee. An example timeline of the stages for each project is shown in Table 3.2.

Data analysis tools included discourse analysis of participant blog posts using collated word clouds, and transcription of participant reflective videos that were uploaded to YouTube and embedded in their blogs. Mobile social media (for example: Blog posts, YouTube videos, Twitter, Google Plus, Vine and Instagram videos) is curated via project hashtags allowing the collation of all social media generated as part of the project. Analysis of the curated social media outputs utilizes

Table 3.2 Framework implementation stages

MSM project stages	Timeframe	Process and outcome
Establish weekly COP of academic advisors and practitioners team	Semester 1	Establish a COP of interested lecturers within a Department. Lecturers reflect upon their prior pedagogical beliefs and practice
Establish support requirements and infrastructure needs, including WiFi coverage and production of mobile airplay screens		Lecturers share their current course outlines and assessment strategies for collaborative editing via Google Docs
Completion of an initial survey that explores participants' prior pedagogical beliefs and practice		Lecturers develop competency with mobile devices
Establish lecturer eportfolios		Lecturers explore msm pedagogies
Establish a collaborative research agenda and research questions, and establish ethics consent procedures		Lecturers develop pedagogical msm activities based on social constructivist pedagogies
BYOD mlearning projects with staff and students	Semester 2	Students establish mlearning eportfolios using their own devices
		Increased student engagement
Implementation of the mlearning activities within each course and assessment		Flexible learning
		Facilitating social constructivist pedagogies and bridging learning contexts
Academic advisors collaborate with participating lecturers to publish and present case studies based on project implementation, these then inform the design of the following iteration of the projects	End of Semester 2	Feedback gathered from students via end of project iteration surveys and focus groups, informing collaborative research writing based on prior and redeveloped course outlines and outcomes via Google Docs
		Conference papers, journal publications and symposia

tools such as TAGSExplorer (Hawksey 2011) and tagboard (http://tagboard.com/) for graphic analysis of how mobile social media enables (or not) collaboration and communication. This enables identification of emerging themes, and triangulation against the observation and identification of critical incidents from COP discussions.

Building a COP is based upon establishing relationships and a sense of trust. The authors identify lecturers in various departments of the institution who are seeking to explore innovation in teaching and learning, and then invite them to gather a group of like-minded peers around them to form a collaboration between the lecturers and the academic advisors. While we have worked across many different curriculum contexts over the past 6 years, the participants of the two example COPs in this paper are drawn from three higher education contexts including: Journalism, Law and Communication Studies. Each of the 10 participating lecturers are exploring, designing and integrating mobile social media into their own curriculum.

Fig. 3.1 A typical msm COP meeting

3.2.2 Establishing the COPs

Each COP is designed to become a model for participants to apply to their own teaching practice contexts as they explore the integration of msm into the curriculum they teach. We begin by negotiating and establishing a range of core-supporting mobile social media tools to enhance the weekly face-to-face COP meetings usually held at a local café with free wifi access (Fig. 3.1). We also spend time collaboratively exploring the mobile social media framework and collaboratively discuss ideas for applying this to one another's teaching contexts.

3.2.3 MSM Tools

A Google Plus Community was established as a hub for each of the two example COPs for scheduling events, posting resources and engaging in discussion relating to organizing the COP and implementing the framework. The cross-platform Google Plus mobile App and the Hangouts App enabled participants to connect and collaborate from a variety of locations. The G+ community was made public, but contribution was allowed by invited users only. This was important to allow peripheral participation by interested people around the world, while allowing moderation of content by the core group of participants and wider invited participation for feedback and input. A Twitter hashtag (for example #mojomlaw) was defined for each COP to curate the use of mobile social media associated with the projects, and a regularly auto updating TAGSExplorer spreadsheet for visual analysis of collaboration via Twitter was created. As Fig. 3.1 shows, TAGSExplorer allows identification of core participants as major nodes and linkages to peripheral participants either following the project or having remote input into the project conversations. This was particularly useful in brokering the activity of each COP to a wider audience of peripherally interested peers within their departments, and time-based snapshots of TAGSExplorer Twitter analysis indicates the growth of these peripheral connections and conversations graphically. TAGSExplorer also provides a motivating tool for students as they can graphically see the impact of their Twitter conversations within the community around the hashtag. In the example from the 2013 mobile Journalism course shown in Fig. 3.2, we can see the largest conversational node is the course lecturer, but there are also significant student nodes that developed over the timeframe of the course.

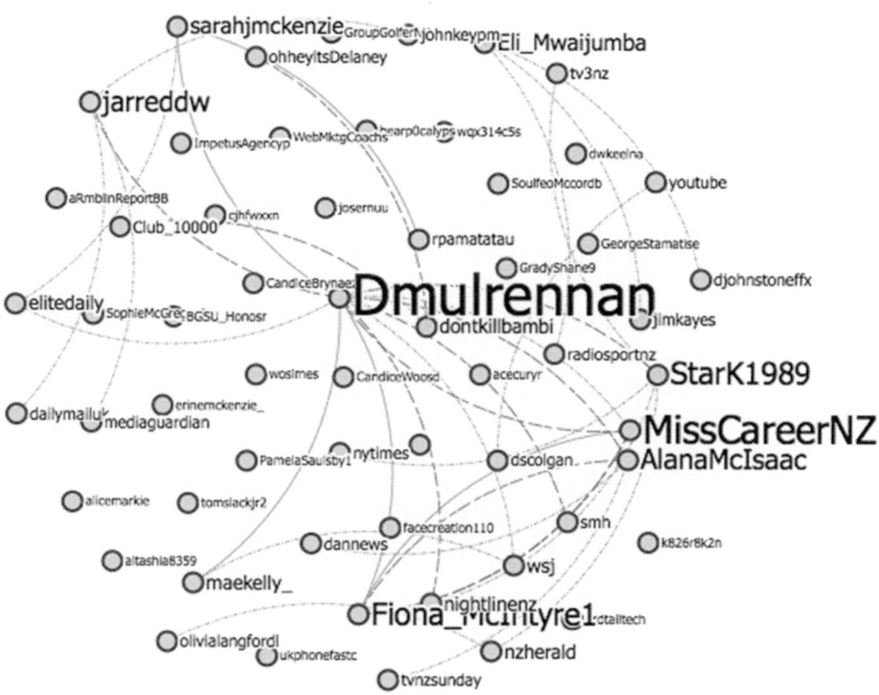

Fig. 3.2 TAGSExplorer for #autmojo2013

We explored the use of several mobile social media Apps such as Vine and Vyclone, giving the participants their first experience of becoming mobile content producers and publishers. Participants were asked to experiment with a variety of mobile social media. A Wordpress blog was created as a public face for each COP (for example http://ejeteam.wordpress.com), where participants can post project progress updates, examples of redesigned activities and assessments and examples of student work as part of the project outcomes. Google Docs was designated as a way of sharing and collaborating on our project outlines, goals and curriculum redesign plans. Other mobile social media curation tools used included a tagboard for the project hashtag to curate a social media stream from Twitter, Vine, Google plus and WhatsApp (for example https://tagboard.com/mojomlaw).

3.2.4 Social Scholarship

At the core of our msm framework is a new collaborative approach to curriculum design and reflective research informing ongoing iterative development of the participants teaching practice. Brokering these experiences and the resultant implementation strategies forms the basis of our social scholarship outputs. We briefly explore the implications of Greenhow and Gleason's (2014) approach to redefining Boyer (1990) in light of social media.

3.2.5 Scholarship of Design

"For scholars and graduate students who embrace these practices, the benefits may be a better contribution to the knowledge base, a more participatory research process, enhanced reputation, expanded definition of "expert" and democratized access to expertise" (Greenhow and Gleason 2014, p. 5). Utilizing Google Docs for collaborative research writing allows a new dimension to traditional publish or perish approach to academia. Our COPs include academic advisors as educational researchers as well as educational practitioners collaborating to improve practice via reflective and theoretically informed critique via peer-reviewed publication. We have COP participants establish their own research profiles on Researchgate.com for sharing and disseminating their reflective practice publications beyond the confines of traditional journal and book publishers. The use of a common hashtag for enabling curation of all associated social media created as part of each COP project provides a rich source of both quantitative and qualitative data (for example https://tagboard.com/mojomlaw).

3.2.6 Scholarship of Integration

"Epistemologically, Boyer's conception of SOI resonates with social constructivist values of knowledge as accessible and co-constructed by a broad base of users" (Greenhow and Gleason 2014, p. 6). Each COP is made up of an interdisciplinary team linked via mobile social media that provides the basis for a wide range of creativity to be brought into collaborative curriculum design. The use of mobile social media also allows input and peripheral participation from a global source of experts in a variety of contexts.

3.2.7 Scholarship of Teaching and Learning

"Scholars who practice a social SOTL may benefit from an increased ability to facilitate active, cocreated learning experiences through social analytic feedback. Students and instructors may also benefit from the diversity of perspectives that can be brought into the learning process via social media" (Greenhow and Gleason 2014, p. 7). We used both synchronous (Google Plus Hangouts) and asynchronous (Google Drive) mobile social media tools to collaborate upon curriculum redesign as a diverse interdisciplinary group.

3.2.8 Scholarship of Application

"Reconsidering SOA through the lens of social scholarship values and social media affordances, therefore, suggests expanded sites and methods for application scholarship that address community challenges" (Greenhow and Gleason 2014, p. 8). We used Google Spreadsheets to create an interactive curriculum design rubric based upon our mobile social media framework as a guide for all of the lecturers participating in the project.

3.2.9 Establishing a Technology Infrastructure

Because the msm framework is concerned with developing transferable strategies for BYOD mlearning, the choice of mobile social media tools was defined by availability across a range of mobile devices, including: smartphones and tablets. However, to keep the procurement logistics of the projects as simple as possible all of the participating lecturers were supplied with an iPhone 5S and an iPad Mini Retina. Other elements of a supporting infrastructure included the provision of a robust WiFi network throughout the institution, and the development and deployment of wireless screen mirroring facilities for mobile devices.

3.3 Discussion

Our developing mobile social media implementation framework involves three key elements: modelling a community of practice, redefining pedagogy and designing an appropriate technology support infrastructure.

Creating a sense of community was achieved utilizing the mobile social media tools that we wanted the participants to experience themselves and build on within their own teaching practice. The mobile-friendly Google Plus App and Hangouts App were particularly powerful for this purpose (Fig. 3.3).

In general, we find that the biggest conceptual shift for the COP participants is how to redesign curriculum activities and assessments to leverage the unique affordances of mobile social media without reverting to merely replicating current practice and pedagogies. In this section, we provide two brief examples of curriculum redesign of learning activities and assessments from two different professional development COPs.

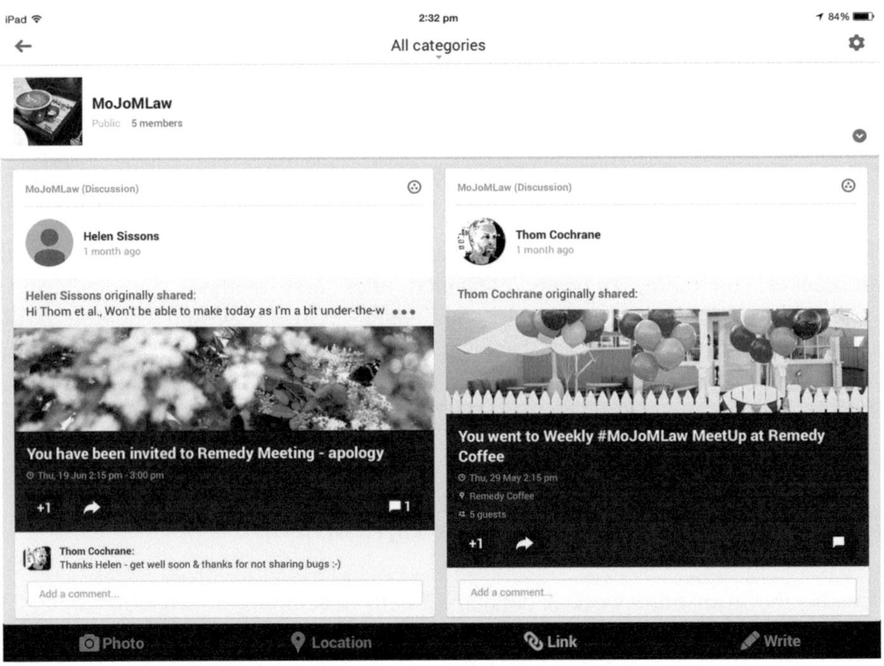

Fig. 3.3 Example MoJoMLaw google plus community

3.3.1 Journalism and Law

In 2011, we established the enhancing journalism education (EJE) mobile social media community of practice (COP) that continued throughout 2012 and 2013. The impact of the reified activities of the EJE COP was used to broker pedagogical change throughout the journalism department and has resulted in significant collaborative curriculum redesign, reconceptualizing learning from a predominantly teacher-directed pedagogy towards student-determined heutagogy. In 2014, we established a combined mobile journalism and mobile law lecturers COP (MoJoMLaw) to build up the momentum achieved within the journalism department throughout 2011–2013, and broker this pedagogical change into the context of the Law curriculum. The goal of the COP is to explore the potential of a cross-faculty collaboration to support pedagogical innovation, particularly in the shared overlap of the two curriculums. The MoJoMLaw COP met regularly face-to-face at a café that provided free WiFi and good coffee for nurturing the social fuel to sustain the COP. A Google Plus Community was established as a virtual community hub for the COP (http://bit.ly/mojomlaw), and a common hashtag was used to collate and curate the social media stream (for example using Tagboard to curate Twitter, Vine, Google Plus, Instagram at https://tagboard.com/mojomlaw/) around the project (#mojomlaw).

The MoJoMLaw COP also explored the relevance of blogging for creating student eportfolios, and invigorating lecture participation via facilitating backchannel conversations using Todaysmeet or Twitter. Another area of exploration included new modes of classroom presentation and collaboration via wireless screen mirroring from mobile devices. This was facilitated by the design and deployment of Mobile Airplay Screens (MOAs) (Cochrane et al. 2013).

The impact of the mojomlaw COP is illustrated by the resultant redesign of a cross-faculty course: Journalism Law and Ethics. The course began using Storify in 2013 to replace the traditional essay as well as bringing in a course blog on Wordpress. Both were popular among the students. We also began informal links with the law school that provided two lecturers to give the two lectures in the course, as well as two joint classes where a well-known lawyer and a leading judge gave guest talks. This has built into the beginnings a collaborative research project about the effect of social media on reporting the courts and the jury process. Table 3.3 outlines the changes to assessment practices as a result of implementing our framework for mobile social media within the course context. The integration of social media within the course enabled a more flexible link to authentic student projects beyond the classroom, and the establishment of shareable eportfolios of their work as outlined in Table 3.3.

Table 3.4 shows a comparison of the change in curriculum activities in the context of the redesigned Journalism law and ethics curriculum.

Table 3.3 Comparison of original and redesigned assessment activities

Previous assessment criteria	Redesigned assessment criteria
Assessment 1: Students each took part in a series of ethical scenarios through online proprietary software. They could compare their responses to other students but could not discuss these	Assessment 1: Students in groups of three select a case that has been considered by one of the media regulators. They then present that case in full to the class using Prezi or any online presentation tool of their choice and lead a class discussion. Following the class, the group members each write up the case including their response to it in a post on the Wordpress blog set up for the class. All other class members comment on one of the posts
Assessment 2: Essay of 1,500 words typed up and handed in as hard copy	Assessment 2: Essay of 1,500 words collated in and published on Storify.com and including material from at least three different social media platforms
Assessment 3: An in-class law test	Assessment 3: An in-class law test

Table 3.4 Mobile social media in the Journalism Law and Ethics curriculum

	Pedagogy	Andragogy	Heutagogy
Activity types	Assessment 1-3: 2013	Assessment 2: 2014	Assessment 1: 2014
	Assessment 3: 2014	Digital identity	Digital community building
	Digital assessment	Student-generated content	Student-generated contexts
	Teacher-delivered content	Student negotiated teams	Student negotiated projects
	Teacher-defined projects	Mobile device as content creation and curation tool	Mobile device as collaborative tool
	Simulated environment		

3.3.2 Communication Studies

This project was focused around the redesign of a course titled iCommunicate, and featured the use of wireless mobile device screen mirroring via our custom design MOAs. A COP was established with the two lecturers at the beginning of 2013 with an aim of helping them learn and utilize msm as a mechanism for enabling students to explore different communication genres and techniques. Prior to this project, the iCommunicate curriculum was taught by facilitating a series of media production lab sessions where the students were shown how to create podcasts as an outcome in the learning process. After three iterations over the last year of working with the lecturers and redesigning the course, the curriculum now features students negotiating a project in groups of two to three to explore an appropriate communication

Table 3.5 Example assessment redesign

Previous assessment criteria	Redesigned assessment criteria
Assessment 1	Assessment 1 (Sliding scale) 40 %
Production of a podcast for communication purposes	Creation of a 3–5 min preview of the project chosen by the group. Reflective documentation of the process by each member on the group blog
	Assessment 2 (60–100 %)
	Production of a movie that is entirely shot using a mobile device. Creation of an augmented layer using Wikitube and publication of the video on YouTube. Individual member reflections on the group blog that documents the process and learning

Table 3.6 Mobile social media in the curriculum

	Pedagogy	Andragogy	Heutagogy
Activity types	Assessment 1: 2013	Assessment 1: 2014	Assessment 2: 2014
	Digital assessment	Digital identity	Digital community building
	Teacher-delivered content	Student-generated content	Student-generated contexts
	Teacher-defined projects	Student negotiated teams	Student negotiated projects
		Mobile device as content creation and curation tool	Mobile device as collaborative tool

genre and how msm enables and enhances communication with an intended audience. As a result, the course now has embedded within its facilitation and assessment approach several msm tools including: Augmented Reality, a Google Plus Community that facilitates social and collaborative learning, a group WordPress blog and YouTube for students to publish their final project output.

Table 3.5 outlines the changes to the assessment in one of the course taught as part of the degree.

Table 3.6 provides an overview of the proposed changes to the curriculum in a course on the pedagogy-andragogy-heutagogy continuum.

3.3.3 Designing an Appropriate Technology Support Infrastructure

Key to supporting a BYOD strategy is the provision of a robust wireless infrastructure and interactive screen mirroring to allow inherently personal small-screen mobile devices to be used as collaborative group tools. The academic advisors negotiated with their institution's IT department to enable screen mirroring

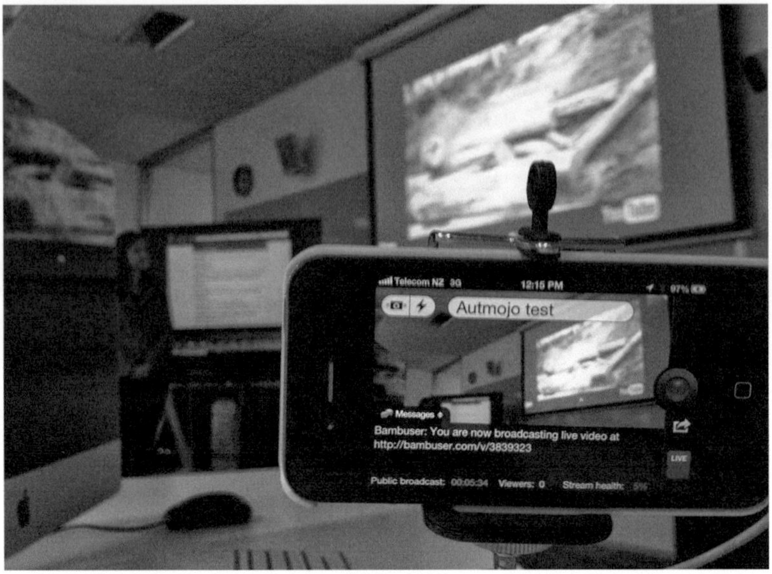

Fig. 3.4 MOA and mobile video streaming in action

protocols on the WiFi network (such as Apple Airplay, Miracast and Chromecast). In particular, we developed moveable large-screen wireless displays that we nick named MOAs or Mobile Airplay screens (Cochrane and Withell 2013) in reference to an extinct flightless bird native to New Zealand. These MOAs (in the background of Fig. 3.4) were then deployed as part of each COP project to enable new forms of msm interaction in classes, such as displaying a real-time discussion and back channel via Twitter or Todaysmeet, hosting remote participation via Hangouts or live streaming events via Apps such as Bambuser (Fig. 3.4).

3.3.4 Future Research

As the case studies develop and progress, they will provide rich data on the application of our mobile social media framework within a variety of educational contexts.

3.4 Conclusions

We have found that establishing communities of practice has been an effective methodology for supporting lecturers to explore the integration of mobile social media within the curriculum to enable new teaching and learning strategies, and to critique and refine these via collaborative scholarship on reflective teaching practice. We hope that our experiences will provide a useful example for others to follow. Our goal is to identify and articulate a developing theoretical framework for professional development implementing effective mlearning, resulting in a range of practical strategies for students and lecturers to utilize the unique attributes of mobile devices for transforming teaching and enhancing learning. Two case studies illustrate the three key elements of the framework: modelling a community of practice, redefining pedagogy and designing an appropriate technology support infrastructure. These are founded upon developing an explicit culture integrating the collaborative scholarship of teaching and learning. We argue that mobile social media provides a platform for developing both a culture of collaborative scholarship and student-determined learning.

References

Anderson, T., & McGreal, R. (2012). Disruptive pedagogies and technologies in universities. *Educational Technology & Society, 15*(4), 380–389.

Attewell, J., Savill-Smith, C., Douch, R., & Parker, G. (2010). Modernising education and training: Mobilising technology for learning. Retrieved from http://www.talpalink.co.uk/resources/Modernising+education+and+training$2C+mobilising+technology+for+learning.pdf.

Boyer, E. (1990). *Scholarship reconsidered: priorities of the professoriate*. Princeton: Carnegie Foundation for the Advancement of Teaching.

Chi, M., & Hausmann, R. (2003). Do radical discoveries require ontological shifts? In L. Shavinina & R. Sternberg (Eds.), *International handbook on innovation* (Vol. 3, pp. 430–444). New York: Elsevier Science Ltd.

Cochrane, T., & Bateman, R. (2013). A mobile web 2.0 framework: Reconceptualizing teaching and learning. In M. Repetto, & G. Trentin (Eds.), *Using network and mobile technology to bridge formal and informal learning* (10th ed.) (pp. 57–92). Oxford: Chandos Publishing.

Cochrane, T., & Narayan, V. (2013). Redesigning professional development: reconceptualising teaching using social learning technologies. *Research in Learning Technology, 21*, 1–19.

Cochrane, T., & Withell, A. (2013, December 1–4). *Do 21st Century students dream of electric sheep? A mobile social media framework for creative pedagogies*. Paper presented at the Electric Dreams: Proceedings 30th Ascilite Conference, Macquarie University, Sydney, Australia.

Cochrane, T., Munn, J., & Antonczak, L. (2013, November 20–22). *Design thinking for mlearning: Herding a flock of MOAs*. Paper presented at the 3rd Mobile Creativity and Innovation Symposium, AUT University, Auckland, New Zealand.

Garnett, F., & Ecclesfield, N. (2011). Towards a framework for co-creating Open Scholarship. In D. Hawkridge, K. Ng & S. Verjans (Eds.), *Proceedings of ALT-C 2011—Thriving in a Colder and more Challenging Climate: The 18th International Conference of the Association for Learning Technology* (pp. 199–216). University of Leeds, UK: ALT Association for Learning Technology.

Greenhow, C., & Gleason, B. (2014). Social scholarship: Reconsidering scholarly practices in the age of social media. *British Journal of Educational Technology, 45*(3), 392–402.

Gunn, C., & Steel, C. (2012). Linking theory to practice in learning technology research. *Research in Learning Technology, 20*(2012), 1–14.

Hawksey, M. (2011). *Twitter: How to archive event hashtags and create an interactive visualization of the conversation*. Retrieved from http://mashe.hawksey.info/2011/11/twitter-how-to-archive-event-hashtags-and-visualize-conversation/.

Herrington, J., Herrington, A., Mantei, J., Olney, I., & Ferry, B. (Eds.). (2009). *New technologies, new pedagogies: Mobile learning in higher education*. Wollongong: Faculty of Education, University of Wollongong.

Herrington, J., Reeves, T., & Oliver, R. (2005). Online learning as information delivery: Digital myopia. *Journal of Interactive Learning Research, 16*(4), 353–367.

International Telecommunication Union. (2014, April). *The World in 2014: ICT facts and figures*. Retrieved from http://www.itu.int/en/ITU-D/Statistics/Documents/facts/ICTFactsFigures2014-e.pdf.

Lave, J., & Wenger, E. (1991). *Situated learning: Legitimate peripheral participation*. Cambridge: Cambridge University Press.

Luckin, R., Clark, W., Garnett, F., Whitworth, A., Akass, J., Cook, J., et al. (2010). Learner-generated contexts: A framework to support the effective use of technology for learning. In M. Lee & C. McLoughlin (Eds.), *Web 2.0-Based E-Learning: Applying social informatics for tertiary teaching* (pp. 70–84). Hershey, PA: IGI Global.

Puentedura, R. (2006). *Transformation, Technology, and Education*. Retrieved February 18, 2013, from http://hippasus.com/resources/tte/puentedura_tte.pdf.

Sharples, M. (2001). Disruptive devices: mobile technology for conversational learning. *International Journal of Continuing Education and Lifelong Learning, 12*(5/6), 504–520.

Sternberg, R. J., Kaufman, J. C., & Pretz, J. E. (2002). *The creativity conundrum: A propulsion model of kinds of creative contributions*. Philadelphia: Psychology Press.

UNESCO. (2013). *UNESCO Policy guidelines for mobile learning (No. 978-92-3-001143-7)*. Paris, France: United Nations Educational Scientific and Cultural Organization.

Unterfrauner, E., & Marschalek, I. (2010). Appropriation of an online mobile community by marginalised young people: experiences from an Austrian case study. In M. Montebello, V. Camilleri & A. Dingli (Eds.), *Proceedings of MLearn 2010: The 9th International conference on mobile learning* (pp. 276–281). Valletta, Malta: University of Malta.

Wenger, E. (1998). *Communities of practice: Learning, meaning, and identity*. Cambridge: Cambridge University Press.

Wenger, E., McDermott, R., & Snyder, W. (2002). *Cultivating communities of practice: A guide to managing knowledge*. Boston: Harvard Business School Press.

Wenger, E., White, N., & Smith, J. (2009). *Digital Habitats: Stewarding technology for communities*. Portland, Oregon: CPsquare.

Wenger, E., White, N., Smith, J., & Rowe, K. (2005). Technology for Communities. In L. Langelier (Ed.), *Working, learning and collaborating in a network: Guide to the implementation and leadership of intentional communities of practice* (pp. 71–94). Quebec City: CEFIRO.

Author Biographies

Thomas Cochrane is an Academic Advisor at AUT University's Centre for Learning and Teaching (CfLAT). In 2011 he was awarded as an Ascilite Fellow. His research interests include mobile learning, web 2.0, and communities of practice. Thomas has implemented over 50 mobile learning projects. He has over 100 peer reviewed publications, receiving best paper awards at Ascilite 2009, ALT-C 2011, ALT-C 2012, and keynoted at several international conferences: 2012 m-Libraries conference (UK), the launch of UWS massive iPad project (2013), IBSA VET Practitioners Conference (Melbourne, 2014), and an invited speaker at EdMedia2014.

Vickel Narayan is a learning and teaching consultant at the Centre for Learning and Teaching (CfLAT) at the Auckland University of Technology. Previously, Vickel was an Academic Advisor (eLearning) at Unitec Institute of Technology from 2009 to 2011. He has a keen interest in Web 2.0 technologies and its potential to engage students and teachers in the teaching and learning process. Vickel is particularly interested in exploring mobile Web 2.0 tools for creating, nurturing and maintaining virtual communities, social connectedness, fostering social constructivism, student generated context and context. He is currently completing his Ph.D. in mobile learning.

Chapter 4
Interactivity and Mobile Technologies: An Activity Theory Perspective

Roy Rozario, Evan Ortlieb and Jennifer Rennie

Abstract Expert teachers are pragmatic in their curricular planning and instruction through embedding the use of mobile technologies towards providing their students with meaningful learning experiences. They use technology as a cornerstone within their instructional design. This study examined how pedagogy, professional learning and mobile technologies impact a teacher's ability to utilise a learner-centred interactive approach. Qualitative data were collected and analysed using the six-step activity theory in conjunction with a case study design were data was collected from four teacher participants through interviews, classroom observations and lesson plans. Data revealed that teaching and learning sequences involving mobile technologies were found to have varying degrees of learner–teacher interactivities, ranging from complete teacher control to total learner control. This range of interactivity can serve as a teacher guide to mobile learning design using appropriate pedagogy integrating apps in conjunction with other classroom resources to yield improved student outcomes.

4.1 Background

The concept of interactivity is so widespread that it embraces many facets of society including T.V., websites, online news, games, social networks, drama and mobile devices (m-devices) (Coursaris and Sung 2012; Downes and McMillan 2000; Larsson 2012). In recent years, the concept of interactivity has become firmly

R. Rozario (✉) · J. Rennie
Monash University, Melbourne, Australia
e-mail: roy.rozario@monash.edu

J. Rennie
e-mail: Jennifer.rennie@monash.edu

E. Ortlieb
St. John's University, New York, USA
e-mail: ortliebe@stjohns.edu

D. Churchill et al. (eds.), *Mobile Learning Design*, Lecture Notes in Educational Technology, DOI 10.1007/978-981-10-0027-0_4

entrenched within educational discourse with the increased utility of mobile learning devices, such as smart phones and tablet PCs in today's classrooms (Beauchamp and Kennewell 2010; Beauchamp and Parkinson 2005; Holzman 2006; Koolstra and Bos 2009; Larsson 2012; Tanner and Jones 2007; Ting 2013; Toteja and Kumar 2012). Yet, there remains an ongoing debate regarding the definition, theoretical nature, structure and significance of interactivity to student achievement (Coursaris and Sung 2012; Kiousis 2002; Kirsh 1997; Koolstra and Bos 2009; Larsson 2012; Smuts 2009). Although there has been a tremendous growth and potential in the use of wireless handheld mobile devices, the orchestration of mobile learning (m-learning) has still been scarcely explored (Motiwalla 2007).

While interactive whiteboards (henceforth IWB), an example of technology readily available in most classrooms, have been popular for over a decade, their use in conjunction with other mobile technologies is neither widespread nor thoroughly researched (Beauchamp and Kennewell 2010; Haydn 2010; Hennessy and London 2013; Maher 2012). Moss et al. (2007) call for more research to address the issue of how mobile technologies can widely contribute to best practices in multimodal learning experiences (Stokovski 2010). Further, there is a pressing need to examine these from an activity theory perspective using interactivity as a measure (Miller and Glover 2010). Its significance is paramount to preparing twenty-first century learners to thrive and to advancing pedagogies that return the focus to content acquisition 'rather than on searching for the next new technolog' (Elias 2011, p. 143).

Previous studies on mobile learning have predominantly focused on effectiveness rather than its design (Wu et al. 2012), thereby overlooking the need for orchestrating current classroom technologies with new m-devices. Understanding interactivities among technology, students and teachers are necessary to advance this agenda (Ting 2013). Research on interactivity has positioned it as a single point of interaction rather than continuum of teacher- and learner-centred interactivities (Banna 2011). This hybrid approach is invaluable in the design, application and review of teacher pedagogy in relation to classroom mobile learning design. This study attempts to identify the various points of interactivities by examining the teacher–student lived classroom practices with software apps on mobile devices towards achieving their learning goals using the activity theory framework.

4.2 Theoretical Framework

In the area of human–computer interaction (henceforth HCI), there is a dearth of theoretical research (Kaptelinin and Nardi 2012; Kuutti 1996). In recent years, research in 'HCI is afflicted by a Tower of Babel of incompatible, fragmentary terminologies and undisciplined, bricolage design strategies'. Activity theory as a methodological and theoretical tool has grown in importance especially in the field of information systems (Allen et al. 2013) and has much to contribute to situated learning pertaining to hybridising m-learning designs with other e-learning designs

(Peña-Ayala et al. 2014). That is to say, how IWB's can work in harmony with classroom mobile technologies to orchestrate better interactivity. Activity theory embraces interactivity while serving as an interactive teaching–learning approach as it encircles the idea of examining human interactions in achieving their end goal, placing emphasis on sociocultural elements of teaching and learning (Roschelle et al. 1998; Ryu and Parsons 2009; Spikol et al. 2008). As viewed from an activity theory perspective, activity is a valid indicator for measuring learner–teacher interactivity (Miller and Glover 2010). According to Engeström (2001), activity theory provides the ideal ground analysis for 'events in classroom discourse where the seemingly self-sufficient worlds and scripts of the teacher and the students occasionally meet and interact to form new meanings that go beyond the evident limits of both' (p. 135–136).

This concept is aptly summarised with Engeström's (1991, 1992) activity system model in relation to this study (see Fig. 4.1). It illustrates how the teacher and/or learner is the main focal point that drives all activities, actions and operations in the activity system (classroom) in order to achieve the end goal (lesson goal). Using activity as a unit of analysis, we can explore information behaviour within lived events (Allen et al. 2011). The teacher can pragmatically select apps along with other classroom technologies as a tool to help achieve their lesson goal. According to Engeström (1992), "an activity system does not exist in vacuum" (p. 19). Activities involve human actors, who are motivated towards an *object* (goal)

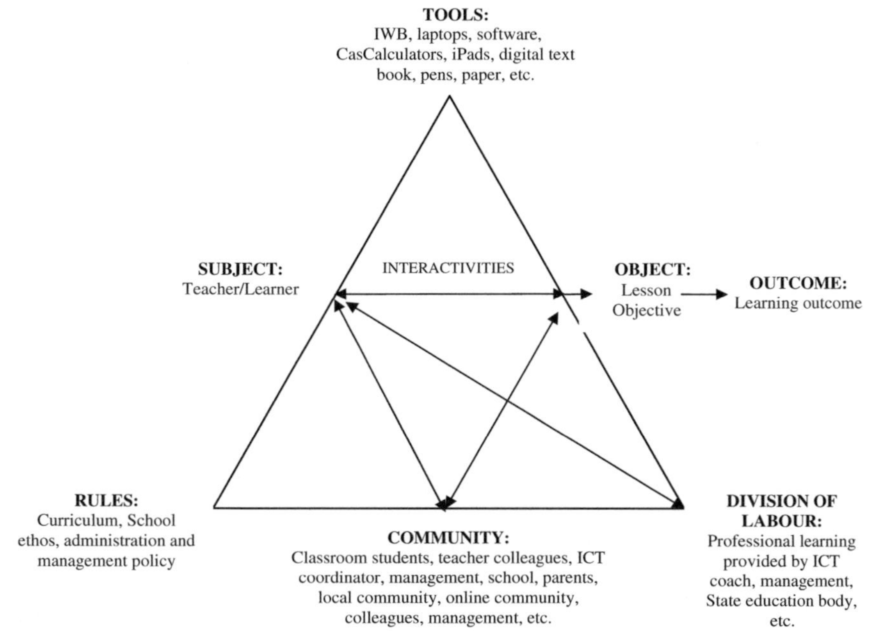

Fig. 4.1 Adapted model of Engeström (2001) Activity System in context of this research study

(Pietsch 2005) with their actions are impacted by *tools*, such as software in their *community* representing learners, parents and school management as shown in Fig. 4.1. The *object* of an activity system is the central characteristic of the system and the belief of goal-directed behaviour is vital to the concept of activity (Engeström 2001).

Therefore activity theory provided a common orientation, structuring method and a map that guided this research inquiry to capture interactivities during software use in complex classroom settings (Roschelle et al. 1998). Activity theory provides a lens for analysing activity and enables its use as a framework for determining the components of the activity system as represented in Fig. 4.1 (Jonassen and Rohrer-Murphy 1999; Karanasios et al. 2013). The various components *subjects*, *tools*, *objects*, *community*, *division of labour* and *rules* in the activity system help identify and explain the activities and interactivities taking place in the classroom. For instance, a classroom observation of this study examining a lesson conducted by a teacher brings to light the various components of the activity system and the diverse motivations and contradictions faced by the actors towards attaining it.

4.3 Review of Literature

Below we review current literature and significance of the concept of interactivity in relation to mobile apps in classroom technologies and how they affect pedagogy and professional learning. Literature gaps point to the need for mobile learning designs to embrace tenets of interactivity for better synthesis of new and current technologies in classroom settings.

4.3.1 Pedagogy

At the outset, it is important to make clear that although this study identifies three tensions, namely, pedagogy, professional learning and digital resources as factors that affect the shape of interactivity, the findings of this chapter focuses on the tensions that occur within pedagogy. Although the applications of mobile learning are widely accepted practice in primary, secondary and tertiary levels of educational settings, yet it still remains underdeveloped in terms of its pedagogical considerations (Park 2011). Central to any study on pedagogical interactivity is the notion that teachers possess the skills to make technology work. Haldane (2007) notes that digital technologies by themselves are not 'interactive' but "merely a medium through which interactivity may, to a greater or lesser extent, be afforded" (p. 258–259). While the software applications and Learning Objects enable interactivity, it is eventually the user of the app who chooses the extent to take full advantage of the software's interactive potential (Alyani and Shirzad 2011). Further, classroom technologies impact student-centred learning in ways previously not feasible

(Hennessy et al. 2010; Miller and Glover 2010) enabling learner interactions that capture experiences in physical and social realms (Ting 2013).

Studies show that some innovative and creative teachers attempt to seek digital resources in technologies to suit a learner-centred pedagogical style (see for example, Maher et al. 2012). While some teachers initially use software for teacher-directed activities, they generally move to using it collaboratively shifting towards learner-centred pedagogy (Moir 2014). However, improving access to technology alone does not translate to better pedagogy (Hennessy et al. 2010; Hennessy et al. 2007). Some teachers use the apps to promote 'controlled pre-communicative activities' (Gray et al. 2005, p. 43). That is, the learning funnel is controlled by the teacher, referred to as 'low-level funnelling questioning' (Tanner and Jones 2007, p. 38). What is needed is an 'interactive teaching' approach/pedagogy, which focuses on a dialogic rather than an authoritative interactivity to foster genuine learning (Beauchamp and Kennewell 2010; Hennessy and London 2013). Interactivity can be superficial or deep varying from 'intra-activity', 'surface interactivity', 'deeper interactivity', 'dialogical deep interactivity' and 'full interactivity' based on the teacher–pupil control of interaction (Tanner and Jones 2007, p. 38). Locally and globally, the current push is for a pedagogic change from a didactic to an interactive approach to learning and teaching and interactivity as a concept that drives this change (Miller and Glover 2010). Developing design principle of mobile learning that blend mobile and non-mobile technologies is important for the integration of technology into teacher pedagogy (Herrington et al. 2009). Hennessy and London (2013) aptly summarise "pedagogical change requires pedagogically oriented professional development" (p. 17).

4.3.2 Professional Learning

ICT has the potential to increase student performance and improve teacher pedagogy, however, teacher practitioners struggle to keep pace with the influx of tools within instructional technology in part due to inadequate professional development and training (Clarke and Fournillier 2012; Oigara and Wallace 2012). In many parts of the world, research has depicted that unidirectional investment on hardware such as IWBs without detailed professional development does not automatically translate to effective leaner-centred pedagogical practice (Becta 2003; Halford 2007; Lacina 2009; Moss and Jewitt 2010; Oigara and Wallace 2012; Somyurek et al. 2009). Piecemeal teacher training has long been a 'hit and miss' approach centred on the technical features of the equipment (Hennessy and London 2013; Miller and Glover 2007). What is needed is using software technologies that bolster curricular content in ways that are seamless and promote interactivity between and within teachers and students alike. Professional m-learning design models need to focus on pedagogical aspects of interactivity, rather than effective application of their use to specific technologies. This would enable teachers to be better equipped with the skills that focus on these pedagogical principles of interactivity making it easier to embrace newer technologies in synthesis with available m-devices in today's classroom.

According to Hooper and Rieber (1995), there are five stages of technology adaptation: familiarisation, utilisation, integration, reorientation and evolution. Generally, initial training is provided, but many teachers lack continued professional learning and support, which restricts the use and functionality of apps (Stein 2005a, b). Professional learning takes time; changes in pedagogy are subject to investment, nature and format of professional development (Hennessy and London 2013). Teachers should be allocated time for software application and integration with other classroom resources with emphasis on the principles of interactivity. The lack of professional learning time for teachers to learn how to use and integrate apps into their teaching can be a significant deterrent to interactive classrooms (Hedberg and Freebody 2007). ICT coordinators and ICT specialist teachers used their non-teaching time to assist teachers with curricular integration of apps (Hennessy and London 2013). Teachers need to be given opportunities to follow up these sessions with 'sandpit' time, that is, time to play, construct, trial, revise, collaborate, discuss and refine lesson plans infused with apps at their own pace (Halford 2007). Hennessy and London (2013) suggest formal training received outside the school was beneficial but in-house informal training and professional development conducted by colleagues was more effective and useful. Hennessy and London (2013) state that "technology by itself has no transformative power" (p. 24); however, when supplemented with appropriate professional learning, digital resources and learner-centred pedagogy, they become engines of change for connecting classrooms to outside lived worlds. Providing teachers ongoing professional learning skills focused on tenets of interactivity in the use of m-devices is the key to pedagogical change for successful integration of newer and current classroom technologies.

4.3.3 Digital Resources

To include a dialogic rather than authoritative interactivity approach along with the software features focussing on learner-centred interactivity is the key to pedagogical change (Beauchamp and Kennewell 2010, p. 759). Endeavours must be made to make available mobile digital resources such as games that enable active learner–technology interactivity so as to connect students' lived experiences in-class and out-of-class context (Masek et al. 2012; Wong and Looi 2011). Frameworks for evaluating and selecting applications for mobile technologies in learning settings are still at its developing stages (Sharples 2006) and should include and embrace elements of interactivity (Beauchamp and Kennewell 2010). Selecting appropriate applications for m-devices using rubrics and evaluative tools for teachers which focus on their pedagogical beliefs to integrate technology into classroom practices is absent in the literature (Green et al. 2014). Using iterative design models for mobile learning apps focusing on tenets of pedagogical interactivity could be a starting point of change towards this end (Marty et al. 2013).

Access to relevant educational websites and software applications is the first step towards positively impacting student learning in the digital age (Griffin and Woods 2006; Helfrich 2011). Selecting which apps to use must be done carefully and critically as apps having varying degrees of usefulness and age appropriateness (Sam 2012; Wang and Woo 2007). However, finding relevant and appropriate mobile apps are not easily accessible and decisions have to be made based on reviews (Hsu and Ching 2013). Teachers have even resorted to making their own apps using App Inventor due to unavailability of suitable apps (Hsu and Ching 2013). Sharing digital resources amongst the teacher community within and outside their school environment is a useful strategy to overcome difficulties posed by their learning curve (Ross 2011; Stein 2005a, b). Therefore, for successful implementation, efficacious linkages between pedagogical support systems are paramount.

Software applications allow teachers to attain their object; they determine how interactivities and information are transmitted and displayed (Maher 2012; Sam 2012). When choosing app, teachers should give preference to apps that enable active learner–technology interactivity so as to connect students' lived worlds outside of the school context (Masek et al. 2012). Apps provide great potential to facilitate simulative learning and less teacher talk and when used in conjunction with m-devices harness their pedagogical capacity. Interactivities have been found to be highest among learners in instances where simulation-based learning environments were used compared to expository learning environments (Beauchamp and Kennewell 2010). What follows is an investigation of how four classroom teachers selected and used apps within their pedagogical design and instruction, providing a window of the interactivities found from an activity theory perspective.

4.4 Methods

This research investigation utilised a case study design whereby data were collected through classroom observations, teacher interviews, samples teacher lesson plans using apps, journals and other resources used during class over a period of 4 months. A convenience sample was used for school selection, resulting in four teachers opting into the study based on a first-come first-serve criteria. Further, a purposeful sampling technique (Tongco 2007) was used to select lesson plans and software apps used by participatory teachers for data analysis in order to cater to a wide range of interactivities from learner-centred to teacher-centred approaches. Thematic analysis was used to identify emerging themes from field notes recorded from classroom observations and formal/informal interviews with the participant teachers.

An adapted model of the summary of six-step process by Jonassen and Rohrer-Murphy (1999) was used to collect data from an activity theory perspective.

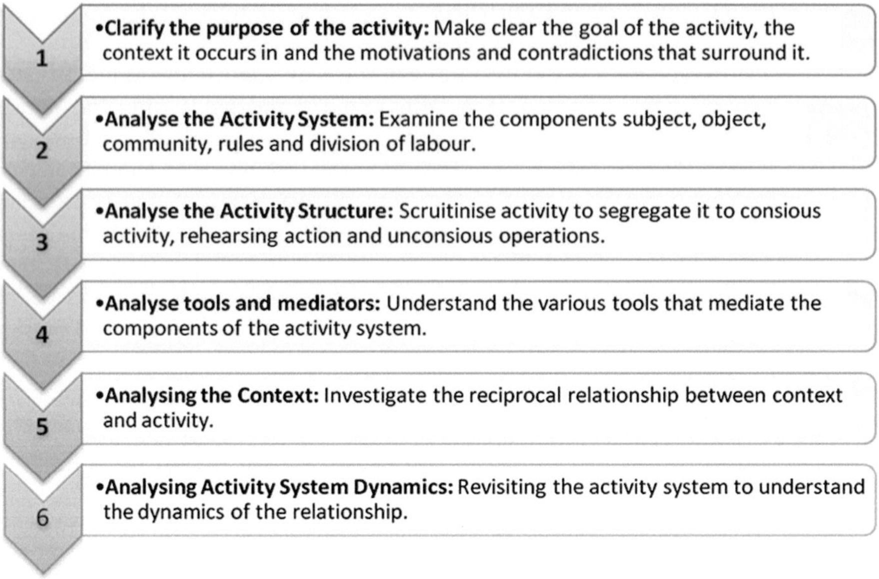

Fig. 4.2 Adapted six-step model of Jonassen and Rohrer-Murphy (1999)

Two research questions guided this inquiry: What types of learner–teacher interactivities does the teacher facilitate when using mobile learning apps? What teacher tensions affect interactivity within mobile learning app usage? (Fig. 4.2)

4.5 Findings/Discussion

Data analysis revealed that teachers tend to adopt various levels of learner–teacher interactivities in the use of software apps. These interactivities will fluctuate between teacher-centred to learner-centred approaches in classroom with the possibility of a contradictory situation of learner- and teacher-centred interactivity occurring simultaneously. This trend was noticed when observing various interactivities during classroom observations of all four teacher participants. All teachers also acknowledged during their interviews that with the use of apps by teachers and students reflects a spectrum of interactivities rather than one point. Therefore, it would be more appropriate to refer to interactivity as several points or a range of interactivities.

In relation to research question one, *What types of learner–teacher interactivities does the teacher facilitate in the use of mobile learning apps*, a range of points of interactivity were recognised, that is to say the amount of teacher and learner involvement with learning *tools* such as technologies and software applications slides constantly from active learner or teacher participation to passive learner or

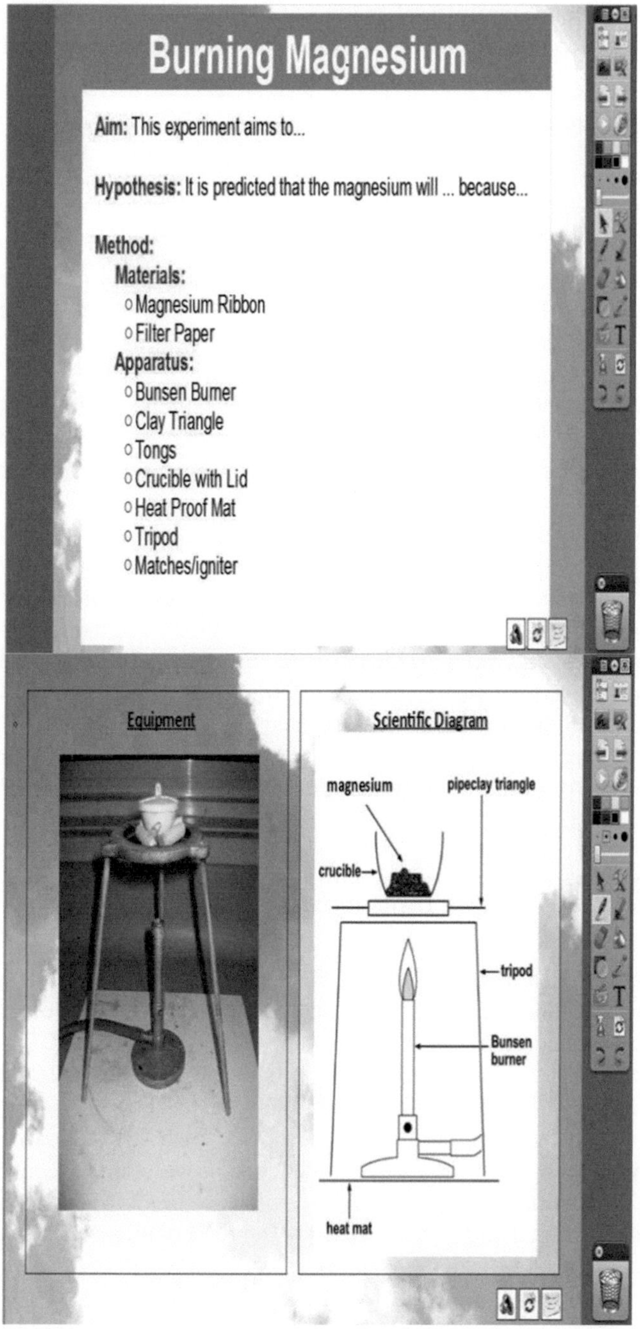

Fig. 4.3 Slides projected using the SlideShark app

teacher participation. During teacher interviews, this finding was regularly reiterated reflecting a continuum of pedagogical interactivities occurring in the classroom. As one teacher participant comments:

> Most lessons, particularly the initial part of the lesson will usually be teacher focused and hopefully moving towards student activity towards the second half of the lesson where to some extent there is a way of them applying the theory or the content. There are lessons that are much more student-focused and there are lessons that are teacher-focused but for me as a rule I would try and have a combination of both within each lesson.

Using the app, *SlideShark* by Brainshark, Inc., one participatory teacher displayed a slideshow from her iPad onto the IWB to demonstrate to students explicit directions for the science experiment, a visual model of what would look like, and then follow up pictures from each group's work (see Fig. 4.3).

Although the same slides were used at different junctures, the level of learner–teacher interactivity varied from complete teacher control to learner control and also resulting in a juxtaposition of both at one given time depending on the use of technology. For instance, during the first use of the slide the classroom teacher was in control of the teaching–learning process adopting a didactive teaching approach. However, when using the same slides at the second time, although in appearance the teacher was adopting a similar approach, the students were more learner-centred as they were in control of the technologies and tools they were using.

The spectrum of interactivity occurs with the use of SlideShark and the IWB ranging from teacher-centred, learner-centred, juxtaposition of learner and teacher-centred coexisting simultaneously and a blended approach either inclined towards active teacher control or learner control. Each of these categories and its features are discussed below using additional classroom-based examples.

4.5.1 Teacher-Centred Interactivity

Some teaching and learning processes are focussed on teachers and their authority of learning process (Tanner and Jones 2007), as they are actively involved and instrumental in classroom orchestration (Beauchamp and Kennewell 2013). This structure results in learners playing a more passive role in the activity system. Examples of this type of interactivity found from data analysis include: teachers controlling apps adopt an instructivist pedagogy; performing activities such as delivering theoretical knowledge with or without the use of embedded software in the technologies; providing overt instructions with or without the assistance of tools in the activity system, providing information using websites, expressing their opinions and inclinations and describing situations and scenarios. As the teacher exercised control over the use of the SlideShark app during the delivery of lesson object 'Ionic bonds', the teacher uses the iPad in conjunction with the app to explain this concept while learners are listening to instructions and demonstration of software applications by the teacher. The activity is controlled and exercised by the teacher. In a short moment, thereafter, teaching the same concept the teacher brings forward two learners to the IWB to drop, drag and attach 'ionic bonds' with other learners in the classroom providing feedback and suggestions. This example demonstrates how in a matter of few minutes complete teacher control can shift to learner control. Similar instances of this would be the teacher using the app in conjunction with the IWB as an expert to explain a mathematical concept while learners listen and thereafter apply that knowledge while using iPads or Cascalculators.

4.5.2 Learner-Centred Interactivity

The characteristics of this category of interactivity are opposite to teacher-centred interactivity and skew the interactivity towards the learner. Learners are more in control of their learning, while teachers are guided by learners to attain the learning goal. One of the teachers encouraged her students to use the app, *Geoboard* by The Math Learning Center, which can be used on almost any m-device (see Fig. 4.4). While the learners worked individually on making shapes and calculating area, the teacher's role involved facilitating, scaffolding, prompting with open-ended questions, observing, initiating discussions, enabling critical framing, learning from student observation, inquiring, and supplying feedback. Representatives use a selected iPad connected to the IWB and display the shapes, asking their peers questions for testing their knowledge. This illustrates how learners can lead the class in collaborative contributions to learning goals; learners conducting mathematical shape manipulation using necessary *tools*; and learners working independently on user-friendly and freely available software apps.

4.5.3 Blended Interactivity

In this category, both the learners and teacher are involved during the teaching–learning process and the level of interactivity can be slightly more inclined towards teacher- or learner-centred interactivity depending on the activity conducted. Multiple teachers utilised the app, *Twiddla by Expat Software,* to brainstorm and visually display graphics and print onto a blank canvas. This app essentially combines the functionality of an IWB with an iPad or mobile phone, allowing both the teacher and students' use to contribute to the overall classroom learning experience by overlaying mathematical formulas, graphs and lines (see Fig. 4.5). In this example, the teacher is, more or less, in control of IWB, while the students give verbal contribution and use Twiddla to relay input from their groups. That is, learners will be using the app while the teacher also facilitates its use and application by scaffolding the learning process. In this instance, the *tools, activities*, *actions* and *operations* in the activity system are juggled between learner and teacher orchestration, and in accordance to the lesson's objectives.

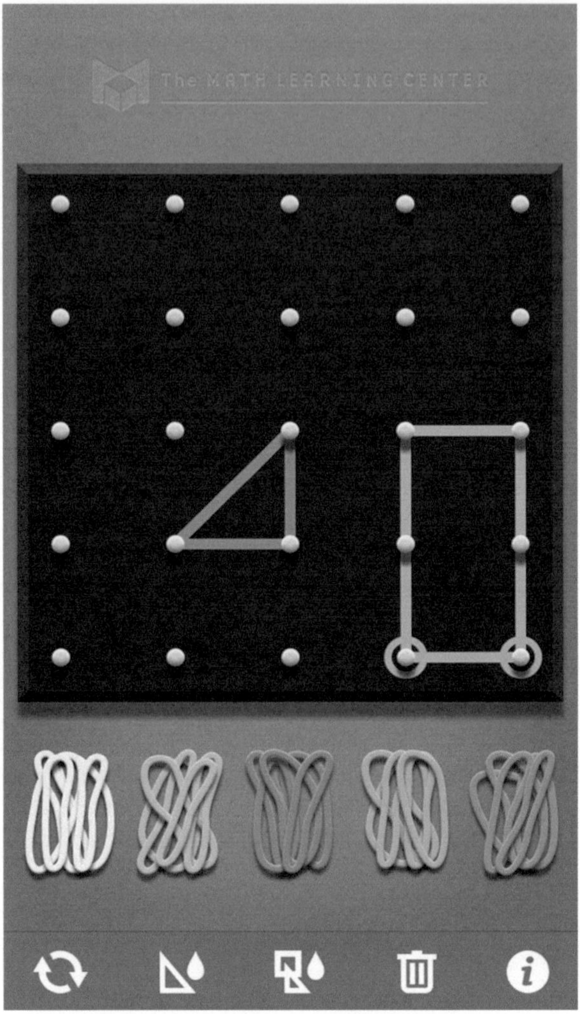

Fig. 4.4 Still image of Geoboard app showing an example of creating digital shapes and calculating area

4.5.4 Juxtaposition of Teacher- and Learner-Centred Interactivity

Sometimes teachers and learners have simultaneous access and control over mobile technologies. Here, learners will use *tools,* such as laptops, CasCalculators and software applications alongside teachers using IWBs. While teachers are in active control and use of their IWB, learners are in control of other classroom technologies. There is a coexistence of teacher-centred and learner-centred interactivity concurrently working in the classroom, resulting in a juxtaposition of pedagogies and interactivities. The teacher generally will be in control of the class with the use

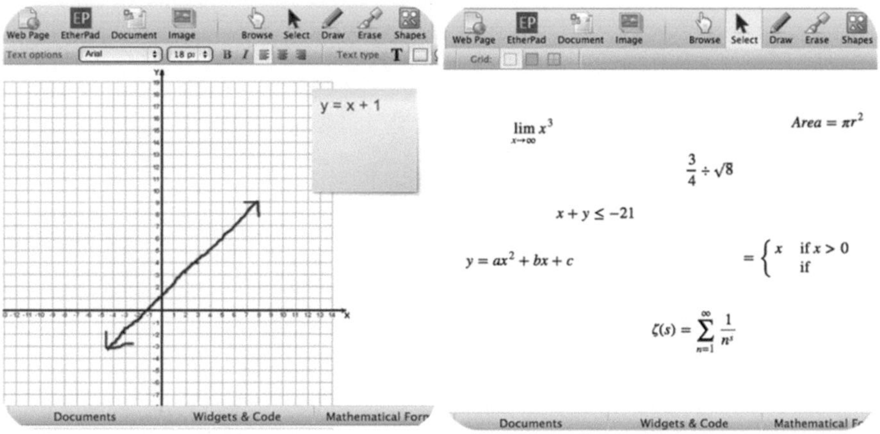

Fig. 4.5 Twiddla app showcasing teacher and student collective input

of the IWB and the related software applications using a teacher-centred approach; the learner will simultaneously be using other classroom *tools* and artefacts, such as laptops, iPads and software, applying his/her socioculturally and historically situated practices to attain the goal by using constructivist pedagogy.

Using the app, iPoe by iClassics Productions, S.L., one English teacher-embedded interactivity within her poetry lesson, an area that is often challenging for teachers to bring alive to reluctant readers. By combining visual imagery, motion and brilliant graphics, iPoe, encourages students to experience poetry rather than just read about it. This participatory teacher had her students use their iPads afterwards to write similarly themed poetry, of which some student samples would be displayed by the teacher onto the IWB for whole class sharing.

This juxtaposition of interactivity results in a position where the teacher is adopting instructivist pedagogy, while the learners are applying constructivist pedagogy in the learning process. This brings about a harmonising effect between the teacher and the learner in relation to interactivity rather than discord sometimes experienced at other stages of the interactivity spectrum. Harmony is experienced as a result of a hybrid of pedagogies being achieved. Another example of this type of interactivity was found when using the Hoodamath website, as learners were in full control of their iPads and the teacher using the IWB led discussions on the concept 'similar triangles'. In this instance, the learners explored the online game and learning by comparing similar triangles using their prior knowledge, while the teacher used the IWB at the same time explaining concepts related to this topic.

It is clear that the level of interactivity between learners, teacher and technology in classroom settings is constantly changing to reshape and develop to meet the learning objectives. In fact, the precept of Engeström's theoretical framework is that activity system is constantly developing (Engeström 2010; Kaptelinin and Nardi 2012). This dynamic series of activities comes together to form the many points of

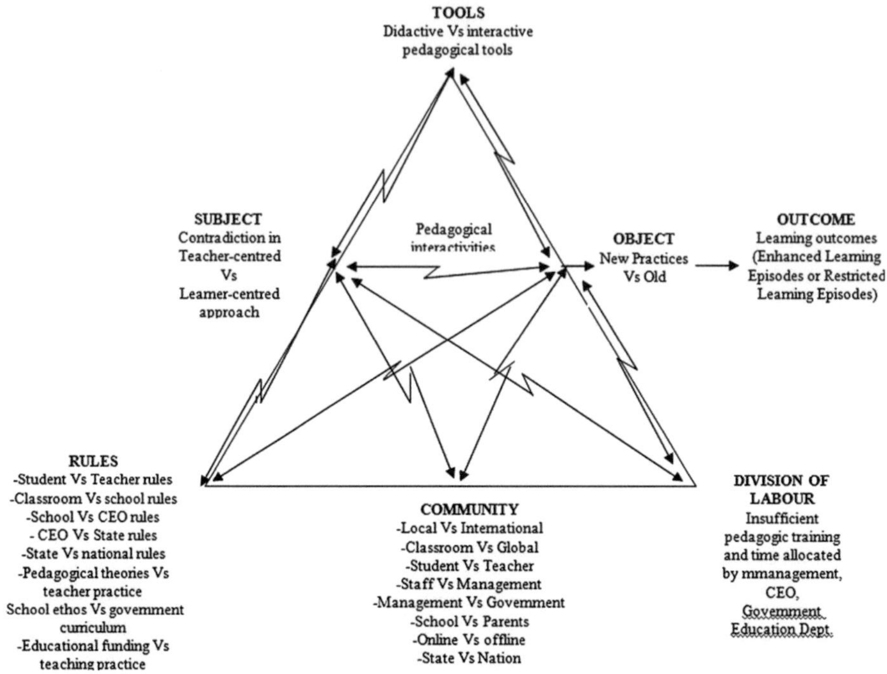

Fig. 4.6 Pedagogy as an activity system tension

interactivity possible within teaching and learning contexts which is represented in the Engeström's activity theory as shown in Fig. 4.6. The pedagogical tensions faced by teachers in the activity system. Here the teacher is constantly faced with the tension of making pedagogical choices of having a leaner or teacher-centred approach to teaching. The components such as the *rules, community, division of labour* and *tools* either facilitate or hinder the teachers' pedagogical approach and thereby influence the shape of interactivity. For instance the teacher has the tension of constantly having contradictions between the rules of the classroom, school and Catholic Education Office (CEO, the Australian governing body for Catholic schools) which affects the pedagogical choices. Many activity systems similar to the below figure helped the researcher identify and analyse in step two of the theoretical framework.

4.6 Conclusion

Using activity theory as a theoretical framework for this inquiry into the interactivities found when teachers and students use digital technologies and apps, this study sought to scrutinise, clarify and describe the many relationships between the learner, teacher and technology. The key findings of this study may be summarised as: first, interactivity in the classroom is not a single point of interaction that represents teacher-centred (didactive) or learner-centred (interactive) interactivity but rather a spectrum or continuum of points that represents many shades of learner–teacher interactions. The learner–teacher interactivity is ever-fluctuating when apps are used in conjunction with other m-devices. Second, when software apps are used in conjunction with other classroom technologies, there can be a coexistence of a learner-centred and a teacher-centred approach, resulting in a hybrid teaching and learning style. This juxtaposition of interactivities is more often made possible with the use of m-devices resulting in a tool kit of pedagogical practices. Third, the coexistence of learner and teacher-centred teaching–learning styles harmonises the inbuilt contradictions that could exist within their interactivity. Fourth, the spectrum points of interactivity are dynamic and ever-changing over time and can fluctuate so dramatically that the likelihood of two learning episodes, keeping all factors constant, are minimal, thereby making the contours of the interactivity unique and unpredictable. Teachers have to constantly work on short and long-term changes to keep up with this changing nature of the context and shape of interactivity. And finally, a variety of teacher tensions affects teachers' app usage along with other classroom technologies and the level of interactivity in lessons.

Given the inconsistency, voids in existing literature, and novelty of the concept of measuring interactivity in relation to teacher, learner and classroom technologies in the field of ICT, there is ample opportunity for research amongst scholars to clarify, append, amend, improve and make relevant this concept in our twenty-first century learning environments. There are many factors directly and/or indirectly influencing the shape of interactivity in relation to teacher pedagogy. For instance, the study observed there are tensions encountered by teachers in finding and preparing digital resources that are content relevant and interactive; teachers constantly resorted to teacher-centred approaches in order to cover vast amounts of curricula and in turn, impacted interactivity; teacher attitude and its relationship to the shape of interactivity; availability and access to m-devices and supported apps; developing an instrument to measure various levels of interactivity for reliable and valid data collection and analysis. The juxtaposition of learner- and teacher-centred interactivity appears to constantly reoccur when apps are used in conjunction with m-devices. Therefore, the need for studies to determine these occurrences and their influence on the orchestration of interactivities between apps and m-devices and their impact on teacher pedagogy m-learning design has invaluable potential for future scope of research in this area.

One of the most important considerations for selecting apps to use in the classroom is the element of interactivity. Betcher and Lee (2009) point out that "if you don't learn to tap into the interactive aspect of technology, you may as well not use [it]" (p. 68). The proliferation of mobile technologies (Toteja and Kumar 2012) provides more potential to tap into the interactivity and integration of the apps within pedagogical planning and delivery (Churchill et al. 2014). However, in order to achieve this feat, it is important that teachers can understand interactivity and determine the factors that influence them. Using activity theory as a lens provides a prominent starting point towards unearthing new discoveries related to context and collaborative learning and its interactivity with m-devices and software applications (Owen 2009). This study suggests that it is a difficult path; one that is not without challenges. It requires researchers, teachers, learners, policy makers, governments and others alike to play a significant role in supporting teachers towards overcoming these tensions related to bolstering interactivity in their classrooms.

References

Allen, D., Karanasios, S., & Slavova, M. (2011). Working with activity theory: Context, technology, and information behavior. *Journal of the American Society for Information Science and Technology, 62*, 776–788. doi:10.1002/asi.21441.

Allen, D. K., Brown, A., Karanasios, S., & Norman, A. (2013). How should technology-mediated organizational change be explained? A comparison of the contributions of critical realism and activity theory. *MIS Quarterly, 37*(3), 835–854.

Alyani, N., & Shirzad, S. (2011, September). Learning to innovate in distributed mobile application development: Learning episodes from Tehran and London. In *Federated Conference on Computer Science and Information Systems (FedCSIS), 2011* (pp. 497–504). Middlesex, NJ: IEEE.

Banna, S. (2011). *The evolving design of online health websites: An interpretive study of different users' activities.* Doctor of Philosophy thesis, University of Wollongong.

Beauchamp, G., & Kennewell, S. (2010). Interactivity in the classroom and its impact on learning. *Computers & Education, 54*(3), 759–766.

Beauchamp, G., & Kennewell, S. (2013). Transition in pedagogical orchestration using the interactive whiteboard. *Education and Information Technologies, 18*(2), 179–191.

Beauchamp, G., & Parkinson, J. (2005). Beyond the 'wow' factor: developing interactivity with the interactive whiteboard. *School Science Review, 86*(316), 97–103.

Betcher, C., & Lee, M. (2009). *The interactive whiteboard revolution: Teaching with IWBs.* Australian Council for Education Research.

BECTA. (2003). *What the research says about interactive whiteboards.* Retrieved from http://dera.ioe.ac.uk/5318/1/wtrs_whiteboards.pdf.

Churchill, D., Lu, J., & Chiu, T. K. (2014). Integrating mobile technologies, social media and learning design. *Educational Media International, 51*(3), 163–165.

Clarke, P. A. J., & Fournillier, J. B. (2012). Action research, pedagogy, and activity theory: tools facilitating two instructors' interpretations of the professional development of four preservice teachers. *Teaching and Teacher Education, 28*(5), 649–660.

Coursaris, C. K., & Sung, J. (2012). Antecedents and consequents of a mobile website's interactivity. *New Media & Society, 14*(7), 1128–1146.

Downes, E. J., & McMillan, S. J. (2000). Defining interactivity a qualitative identification of key dimensions. *New Media & Society, 2*(2), 157–179.

Driver, R., Asoko, H., Leach, J., Scott, P., & Mortimer, E. (1994). Constructing scientific knowledge in the classroom. *Educational Researcher, 23*(7), 5–12.

Elias, T. (2011). Universal instructional design principles for mobile learning. *The International Review of Research in Open and Distributed Learning, 12*(2), 143–156.

Engeström, Y. (1991). Non scolae sed vitae discimus: Toward overcoming the encapsulation of school learning. *Learning and Instruction, 1*(3), 243–259. doi:10.1016/0959-4752(91)90006-T.

Engestrom, Y. (1992). Interactive expertise: Studies in distributed working intelligence. *In Research Bulletin, 83*. Department of Education, University of Helsinki, Bulevardi 18, SF-00120. Helsinki, Finland.

Engeström, Y., & Sannino, A. (2010). Studies of expansive learning: Foundations, findings and future challenges. *Educational Research Review, 5*(1), 1–24. doi:10.1016/j.edurev.2009.12.002.
Engeström, Y. (2001). Expansive learning at work-toward an activity theoretical reconceptualization. *Journal of Education and Work, 14*(1). doi: 10.1080/13639080020028747.
Gray, C., Hagger-Vaughan, L., Pilkington, R., & Tomkins, S. (2005). The pros and cons of interactive whiteboards in relation to the key stage 3 strategy and framework. *Language Learning Journal, 32*, 38–44.
Green, L. S., Hechter, R. P., Tysinger, P. D., & Chassereau, K. D. (2014). Mobile app selection for 5th through 12th grade science: The development of the MASS rubric. *Computers & Education, 75*, 65–71.
Griffin, P., & Woods, K. (2006). *Interactive whiteboards in Victorian schools: Installation and processes of use*. Parkville: Assessment Research Centre, The University of Melbourne.
Haydn, T. (2010). History teaching and ICT. In D. Ian (Ed.), *Debates in history teaching* (pp. 236–248). New York: Routledge.
Hennessy, S., Deaney, R., & Tooley, C. (2010). Using the interactive whiteboard to stimulate active learning in school science. In M. Thomas, & E. Schmid (Eds.), *Interactive whiteboards for education: Theory, research and practice* (pp. 102–117). Hershey, PA: IGI Global. doi:10.4018/978-1-61520-715-2.ch007.
Hennessy, S., & London, L. (2013). Learning from international experiences with interactive whiteboards: The role of professional development in integrating the technology. *OECD Education Working Papers*, No. 89, OECD Publishing. http://dx.doi.org/10.1787/5k49chbsnmls-en.
Hennessy, S., Deaney, R., Ruthven, K., & Winterbottom, M. (2007). Pedagogical strategies for using the interactive whiteboard to foster learner participation in school science. *Learning, Media and Technology, 32*(3), 283–301.
Hedberg, J., & Freebody, K. (2007). *Towards a disruptive pedagogy: Exploring classroom practices with interactive whiteboards and TLF digital content.* Retrieved from http://www.ndlrn.edu.au/verve/_resources/towards_a_disruptive_pedagogy_2007.pdf.
Haldane, M. (2007). Interactivity and the digital whiteboard: weaving the fabric of learning. *Learning, Media and Technology, 32*(3), 257.
Halford, B. (2007). Interactive whiteboards: The future is already here. *Teacher, 183*, 32–35.
Helfrich, J. (2011). *The influence of learning object interactivity on student achievement*. Idaho State University.
Hooper, S., & Rieber, L. P. (1995). Teaching with technology. In A. C. Ornstein (Ed.), *Teaching: Theory into practice* (pp. 154–170). Needham Heights, MA: Allyn and Bacon.
Holzman, L. (2006). What kind of theory is activity theory? *Introduction. Theory & Psychology, 16*(1), 5–11. doi:10.1177/0959354306060105.
Hsu, Y. C., & Ching, Y. H. (2013). Mobile app design for teaching and learning: Educators' experiences in an online graduate course. *The International Review of Research in Open and Distributed Learning, 14*(4), 117–139.
Jonassen, D. H., & Rohrer-Murphy, L. (1999). Activity theory as a framework for designing constructivist learning environments. *Educational Technology Research and Development, 47*, 61–79.
Karanasios, S., Thakker, D., Lau, L., Allen, D., Dimitrova, V., & Norman, A. (2013). Making sense of digital traces: An activity theory driven ontological approach. *Journal of the American Society for Information Science and Technology, 64*(12), 2452–2467.
Kaptelinin, V., & Nardi, B. (2012). Activity theory in HCI: Fundamentals and Reflections. *Synthesis Lectures Human-Centered Informatics, 5*(1), 1–105.
Kiousis, S. (2002). Interactivity: a concept explication. *New Media & Society, 4*(3), 355–383.
Kirsh, D. (1997). Interactivity and Multimedia Interfaces. *Instructional Sciences, 25*, 79–96.
Koolstra, C. M., & Bos, M. J. (2009). The development of an instrument to determine different levels of interactivity. *International Communication Gazette, 71*(5), 373–391.

Kuutti, K. (1996). Activity theory as a potential framework for human-computer interaction research. In B. Nardi (Ed.), *Context and consciousness: activity theory and human-computer interaction* (pp. 17–44). Cambridge, MA: MIT Press.

Lacina, J. (2009). Interactive whiteboards: Creating higher-level, technological Thinkers? *Childhood Education, 85*(4), 270–272.

Larsson, A. O. (2012). Interactivity on Swedish newspaper websites: What kind, how much and why? *Convergence: The International Journal of Research into New Media Technologies, 18* (2), 195–213. doi:10.1177/1354856511430184.

Masek, M., Murcia, K., & Morrison, J. (2012). Getting serious with iPads: The intersection of game design and teaching principals. *Australian Educational Computing, 27*(2), 34–38.

Maher, D. (2012). Learning in the primary school classroom using the interactive whiteboard. In J. Jiyou (Ed.), *Educational stages and interactive learning: from kindergarten to workplace training: from kindergarten to workplace training* (pp. 150–162). Hershey, PA: Information Science Reference.

Maher, D., Phelps, R., Urane, N., & Lee, M. (2012). Primary school teachers' use of digital resources with interactive whiteboards: The Australian context. *Australasian Journal of Educational Technology, 28*(1), 138–158.

Marty, P. F., Mendenhall, A., Douglas, I., Southerland, S. A., Sampson, V., Kazmer, M., et al. (2013). The iterative design of a mobile learning application to support scientific inquiry. *Journal of Learning Design, 6*(2), 41–66.

Moir, T. (2014). Getting in touch with technology without losing touch with early childhood pedagogy. *Educating Young Children: Learning and Teaching in the Early Childhood Years, 20*(1), 34–37.

Moss, G., & Jewitt, C. (2010). Policy, pedagogy and interactive whiteboards: What lessons can be learnt from early adoption in England? In M. Thomas & E. Schmid (Eds.), *Interactive whiteboards for education: Theory, research and practice* (pp. 20–36). Hershey, PA: IGI Global.

Moss, G., Jewitt, C., Levaãiç, R., Armstrong, V., Cardini A., & Castle, F. (2007). *The interactive whiteboards, pedagogy and pupil performance evaluation: An evaluation of the schools whiteboard expansion* (SWE). (Report No. 816). Project: London Challenge DfES, London.

Motiwalla, L. F. (2007). Mobile learning: A framework and evaluation. *Computers & Education, 49*(3), 581–596.

Miller, D., & Glover, D. (2007). Into the unknown: The professional development induction experience of secondary mathematics teachers using interactive whiteboard technology. *Learning, Media and Technology, 32*(3), 319–331. doi:10.1080/17439880701511156.

Miller, D., & Glover, D. (2010). Interactive whiteboards: A literature survey. In M. Thomas & E. Schmid (Eds.), *Interactive whiteboards for education: Theory, research and practice* (pp. 1–19). Hershey, PA: IGI Global. doi:10.4018/978-1-61520-715-2.ch001.

Oigara, J. N., & Wallace, N. (2012). Modelling, training, and mentoring teacher candidates to use SMART board technology. *Issues in Information Science and Information Technology, 9*, 297–315.

Owen, M. (2009). From individual learning to collaborative learning—Location, fun, and games: Place, context, and identity in mobile learning. In H. Ryu & D. Parsons (Eds.), *Innovative mobile learning: Techniques and technologies* (pp. 102-122). Hershey, PA: IGI Global.

Park, Y. (2011). A pedagogical framework for mobile learning: Categorizing educational applications of mobile technologies into four types. *The International Review of Research in Open and Distributed Learning, 12*(2), 78–102.

Peña-Ayala, A., Sossa, H., & Méndez, I. (2014). Activity theory as a framework for building adaptive e-learning systems: A case to provide empirical evidence. *Computers in Human Behavior, 30*, 131–145.

Pietsch, J. R. (2005). *Collaborative learning in mathematics.* Doctor of Philosophy, University of Sydney, Sydney. Retrieved from http://hdl.handle.net/2123/1088

Roschelle, J., Kaput, J., Stroup, W., & Kahn, T. M. (1998). Scaleable integration of educational software: Exploring the promise of component architectures. *Journal of Interactive Media in Education*, *2*, Art-6.
Roth, W. M., & Lee, Y. J. (2007). Vygotsky's neglected legacy: Cultural-historical activity theory. *Review of Educational Research, 77*(2), 186–232.
Ross, P. E. (2011). *Teachers and interactive whiteboards: Accessing, creating, sharing and storing resources within a school community (Masters by Coursework & Shorter thesis).* Melbourne: The University of Melbourne.
Ryu, H., & Parsons, D. (2009). *Designing learning activities with mobile technologies.* Hershey, PA: IGI Global.
Sam, C. (2012). Activity theory and qualitative research in digital domains. *Theory into Practice, 51*(2), 83–90.
Sharples, M. (2006). Big issues in mobile learning. report of a workshop by the kaleidoscope. network of excellence mobile learning initiative. < hal-00190254 >. Retrieved from https://telearn.archives-ouvertes.fr/hal-00190254/document.
Smuts, A. (2009). What is interactivity? *The Journal of Aesthetic Education, 43*(4), 53–73.
Somyurek, S., Atasoy, B., & Ozdemir, S. (2009). Board's IQ: What makes a board smart? *Computers & Education, 53*(2), 368–374.
Spikol, D., Kurti, A., & Milrad, M. (Eds.) (2008). Collaboration in context as a framework for designing innovative mobile learning activities. In H. Ryu, & D. Parsons (Eds.), *Innovative mobile learning: Techniques and technologies* (pp. 170–194). Hershey NJ: Information Science Reference.
Stein, G. (2005a). *Pedagogy, practice and ICT: Snapshots of practice.* Canterbury: Canterbury University.
Stein, G. (2005b). *Pedagogy, practice & ICT.* Canterbury: Canterbury Christ Church University.
Stojkovski, T. (2010). *Computer-mediated learning in a social constructivist environment.* Doctor of Education: University of Wollongong, New South Wales.
Tanner, H., & Jones, S. (2007). How interactive is your whiteboard? *Mathematics Teaching, 200*, 37–41. doi:1299085551.
Ting, Y. L. (2013). Using mobile technologies to create interwoven learning interactions: An intuitive design and its evaluation. *Computers & Education, 60*(1), 1–13.
Toteja, R., & Kumar, S. (2012). Usefulness of m-devices in education: A survey. *Procedia-Social and Behavioral Sciences, 67*, 538–544.
Tongco, M. D. C. (2007). Purposive sampling as a tool for informant selection. *Ethnobotany Research & Applications, 5*, 147–148.
Wu, W. H., Wu, Y. C. J., Chen, C. Y., Kao, H. Y., Lin, C. H., & Huang, S. H. (2012). Review of trends from mobile learning studies: A meta-analysis. *Computers & Education, 59*(2), 817–827.
Wang, Q., & Woo, H. L. (2007). Comparing asynchronous online discussions and face-to-face discussions in a classroom setting. *British Journal of Educational Technology, 38*(2), 272–286.
Wong, L. H., & Looi, C. K. (2011). What seams do we remove in mobile-assisted seamless learning? A critical review of the literature. *Computers & Education, 57*(4), 2364–2381.

Author Biographies

Roy Rozario is a doctoral student and sessional staff member at Monash University with over 20 years of teaching experience at University and school level. He has a Masters and M.Phil. degree in Economics, M.Ed (ICT), Grad. Dip. Ed. and Grad. Cert. Mathematics. His current doctoral work is on mobile learning and interactivity

Evan Ortlieb is a Professor and Coordinator of the Literacy Program in the Department of Education Specialties at St. John's University in New York City. He is an internationally recognized leader in the field of literacy education withprevious work experience in Singapore and Australia and whose expertise centers on struggling readers, literacy teacher preparation, language diversity, and differentiated literacy instruction. He has published over 100 manuscripts that substantiate some of his contributions to the field including a book series entitled, Literacy, Research, Practice and Evaluation, as well as new instruments and refined instructional practices now being used worldwide.

Jennifer Rennie is a Senior Lecturer in literacy education at Monash University. Prior to working in higher education she worked for several years as a primary and high school teacher in the Northern Territory. Her current research interests are related to Indigenous literacies and reading pedagogy. She has published widely in this area and has been the Managing Editor of the Australian Journal of Language and Literacy since 2009. Her current research draws from both cognitive and sociocultural theories and involves developing and implementing new ways for thinking about how we work with struggling and disengaged readers.

Chapter 5
Educational Apps Ontology

Michele P. Notari, Michael Hielscher and Mark King

Abstract This chapter provides an overview of educational mobile apps as well as different models of classification. It also sheds light on the complex mobile app ecosystem from four perspectives: the learner, the teacher, the developer and the distributor. It highlights the contradictions that arise from the different goals and expectations from each perspective. The chapter concludes with a brief discussion of the motivational aspects of games and their current role for educational mobile apps as well as the future development of apps.

5.1 Introduction

Educational apps are transforming the education sector and to date have had a significant impact on both formal and non-formal learning ecosystems. Mobile devices permeate our daily lives, providing unparalleled access to communication and information. Looking over the horizon into the next decade and beyond, it is clear that mobile learning will be embedded in an ecosystem that is increasingly accessible, affordable and connected, and its impact on humankind will be profound. Abundant literature has provided evidence of the uses, advantages, consequences and concerns about mobile apps and their effectiveness in the education domain.

M.P. Notari (✉)
HKU, The University of Hong Kong, Hong Kong SAR China, PHBern, University of Teacher Education, Bern, Switzerland
e-mail: michele.notari@phbern.ch

M. Hielscher · M. King
The Schwyz University of Teacher Education, Schwyz, Switzerland
e-mail: michael.hielscher@phsz.ch

M. King
e-mail: mark.king@unsw.edu.au

The University of New South Wales, Hong Kong, SAR China

D. Churchill et al. (eds.), *Mobile Learning Design*, Lecture Notes in Educational Technology, DOI 10.1007/978-981-10-0027-0_5

The mobile app market is growing rapidly in the various 'app-ecosystems' (like the Apple App Store™ with 1.5 million apps, the Google Play Store™ with 1.6 million apps, Windows phone store™ with 310'000 apps, Amazon appstore™ with 250'000 apps and the BlackBerry world with 135'000 apps; total amount of apps counted in July 2015, Statista (2015)). In the Apple App Store, 21.8 % of all apps are games, 10.3 % categorized as business, and 9.8 % categorized as 'education' with about half of the educational apps being free (Statista 2015). McKinsey and Company and GSMA (2012) revealed that 270 million apps linked to education were downloaded in 2011—more than a tenfold increase from 2009. These statistics include the educational apps derived from educational web sites like Khan Academy as well as 'standalone' apps like 'Wheels on the bus' (Wheels on the bus 2015). The categorization of the various educational and learning apps in the major app stores is aligned to consumers and their shopping habits.

From a pedagogical perspective, 'Apps for learning' can be classified in various ways. One classification might focus on 'instructional design' criteria or address the 'learning goals' of a given app (e.g. information transmission, communication or collaboration, assessment centred, 'drill and practice', situated, knowledge building). Another classification system might focus on motivational domains via 'gamification', 'reward systems' or the amount of 'infotainment'.

This work aims to create an educational profile that can be used by learners, educators, researchers, parents, and app developers in order to empower them to analyse Apps on the basis of clear and convincing principles (Wartella 2015). It is built upon the philosophical definition of 'ontology' (i.e., the nature of being, becoming, existence, or reality, as well as the basic categories of being and their relations) and not the definition used by the computer science community (i.e. formal naming and definition of the types, properties, and interrelationships of the entities that really or fundamentally exist).

5.2 What Is an App/What Is an Educational App?

A mobile app can be defined as a mobile technology used by an end-user for a particular purpose (Nickerson et al. 2007). We use the term 'mobile App' or 'App' as a self-contained program or piece of software designed to fulfil a particular purpose; an application, especially as downloaded by a user to a mobile device (the Oxford dictionary 2015) or as self-contained program downloadable from one of the common app stores. Various approaches to categorize or classify apps by different criteria are in the literature, for example Nickerson et al. (2007) describe the following seven dimensions for general app taxonomy:

- Temporal dimension, defining when the user interact (synchronous/asynchronous);
- Communication dimension, where the information flow can be uni- or bidirectional (unidirectional: the information flows from the application to the user);

- Transaction dimension, where the app provides the capability to purchase goods or services
- Public dimension, where the app is public or limited to a specific group of users
- Multiplicity or participation dimension, where the user interacts just with the app or the user interacts with other users using the app.
- Location dimension, where the app provides customized information or functionality based on the user's location.
- Identity dimension, where the information is adjusted based on an awareness of who the user is.

From an educational perspective those dimensions have limited use to the classification of apps for specific learning situations. One issue is that multiple definitions of educational apps exist like. 'Apps that provide an interactive learning experience on a specific skill or subject' (Mobile Roadie 2015). More specifically for this work, an educational mobile app has a designed purpose relevant for learning either to build up factional knowledge, to share and collaborate with others, or even just helping with ordinary administration and organization tasks in daily school life.

5.3 Educational App Typologies

Several studies have attempted to classify educational apps and build taxonomies from different pedagogical perspectives. Cherner et al. (2014) used three dimensions: skill-based app (mainly factional knowledge building), content-based apps (provide information like dictionaries or maps) and function-based apps (mainly tools for presentations, sketches. communication and collaboration). Goodwin and Highfield (2012) classified apps based on their instructional design: instructive, manipulable and constructive.

There are many other classifications in specific subject areas like math or languages focusing more on typical tasks solved with a given app (e.g. Handal et al. 2013). Some organizations and institutions offer mobile app rating databases like Children Technology Review (About Rating 2015). There are also various websites like Langwitches.org (Langwitches Blog 2011) where teachers attempted to categorize educational apps to help other teachers in the same area find appropriate apps. Most of those classifications and reviews are a bit fuzzy and often imply a specific use-case of an app, which might be used quite differently as well.

From a pedagogical point of view, the aim or purpose of an app can be categorized into the following six groups:

- *Knowledge & skill building apps*—The largest group of educational apps uses formalized content that can be easily checked by a computer. Most of those apps have a well-defined setting with a specific instructive design. It became common practice to use game mechanics-like levels, points and high scores to increase motivation for solving repetitive tasks (drill & practice). There is a broad range from simple calculation training apps to more complex learning games with a story and different tasks to master. They most likely address the first two levels from the bloom taxonomy: Remember (identify, recall) and Understand (compare, match, classify).

- *Collaboration apps*—Several apps can help when students have to produce a text or other media in a group. Also brainstorming or other resource collecting tasks can benefit from collaboration apps like the Google Docs App or Dropbox, even so they are not categorized as educational apps in major app stores.
- *Learning and teaching support apps*—Timetable and homework schedule apps to organize student's task are quite common in major app stores. Especially for language learning there exist various apps using the flashcard method. All those apps allow the student to provide his/her own content to be trained. Support and assisting apps do not provide learning content but help to apply learning strategies. Teachers can find a lot of administration and organization apps from digital class books to learning material management. There are also several classroom activity apps for teachers like clicker tools for polls that can be used to engage students in teacher-centred lessons and lectures.
- *Communication apps*—Students communicate a lot through apps like WhatsApp, Skype or Facebook. Many classes for example have an own WhatsApp channel/group nowadays. Timetable changes or information about the next school-trip travels fast through such social networks. Such apps are not listed as education apps in major app stores. Even so, they can play an important role in digital learning scenarios. For teachers communication apps like Twitter can also help to stay informed or to share teaching material.
- *Other tools and reference apps*—Calculators, periodic tables, lexica, maps

Goodwin and Highfield (2012) propose a classification for preschool children following the pedagogical design of the app based on the learner's locus of control over the activities presented in the app and their level of cognitive investment. The three broad classifications are as follows:

- *Instructive apps*—have elements of 'drill-and-practice' design, whereby the app delivered a predetermined 'task' that elicit a homogenous response from the user. These apps require minimal cognitive investment on behalf of the learner. 'Math Bingo' by BCya.com is an exemplar of instructive design.
- *Manipulable apps*—allow for guided discovery and experimentation, but within a predetermined context or framework. These apps require more cognitive involvement than instructive apps, but less than constructive apps. An example of a manipulable tool is 'Toontastic' by Launchpad Toys.
- *Constructive apps*—are characterized by a more open-ended design that allow users to create their own content or digital artefact using the app. Musical apps and drawing apps are emblematic of constructive apps. 'Drawing Pad' by Darren Murtha is an example of this pedagogical design. Goodwin and Highfield (2012).

After all, synthesized from psychology, linguistics, computer science, animal behaviour, machine learning, brain imaging, neurobiology and other areas, this newly minted field asks not merely what we should teach children—that is, what content—but also how children best learn the strategies they will need to cope flexibly and creatively in the twenty-first century world (e.g. Benassi et al. 2014; Golinkoff and Hirsh-Pasek 2015; Pellegrino 2012; Pellegrino and Hilton 2013; Sawyer 2006)' Cited from Hirsh-Pasek et al. (2015).

5.4 Educational Apps from Different Perspectives

In this section we would like to shed light on the educational mobile app market without delving too deeply into an app categorization schema or model. Therefore, our analysis provides perspectives from four different user groups: learners, teachers, developers and distributors (store companies). In the following sections, we will provide analysis from the above-mentioned perspectives to issues such as 'affordance', 'needs', 'expectations', and 'constraints' of educational apps. In order to situate the varying perspectives, we will first provide a short outline of the analytical process (what we call 'learning') used when interacting with the educational app.

Nowadays, learning is described, defined, and interpreted in very different ways. Therefore, we first provide an overview of the many concepts of learning. We have chosen a 'concept-mix' (Notari and Doebeli 2010) that is relevant for the evolving characteristics of educational apps. Learning is a social and active process (e.g. Vygotsky 1978; Hirsch-Pasek et al. 2015; Dillenbourg 1999; Crawford 1996) where social interactions play an essential role for development of cognition (Kearsley 1994) and the cognitive theory for intervention strategies proposed by Bandura (1977). Motivational learning strategy and self-determination described by Deci and Ryan (2002), where the learner needs competence, autonomy and relatedness is crucial for the acquisition of skills in a world of increasing complexity and the fast pace of change, is a lifelong learning process. For this reason, a distinction is drawn between formal learning (scheduled and taught in a curriculum) and informal learning (happening intentionally or inadvertently) (Cross 2006, p. 16). Paavola et al. (2004) and Scardamalia (2003a) pointed out the importance of collaboration and knowledge building stating that knowledge building is a social act performed by a certain number of people sharing and exchanging information on/through their created artefacts. Collaboration is seen as mechanisms of interaction among the people involved, whether directly or through the created artefacts (using technology) (Scardamalia 2003a; Scardamalia and Bereiter 2003b)

5.4.1 The Learner's View

The learner is the person who interacts with an educational app and learns something by doing so. The 'learner' might have quite different needs and goals depending on their age and topic being learned. Hirsch-Pasek et al. (2015) suggest that children of the age between 0 and 8 years learn best when they are cognitively active and engaged, when learning experiences are meaningful and socially interactive, and when learning is guided by a specific goal. Visual presentation, entertainment, easy of use and narrative guidance e.g. by an appealing character are typical characteristics of thousands of apps available in the app stores that are aimed at the early childhood market.

At the age between 5 and 7 parents often try to foster their children to acquire basic reading, writing and calculating skills as preparation for school. Such children start learning using apps with written symbols or easy calculation apps. While the entertaining part of such apps may be reduced, parents will seek well-tested apps reviewed by professional educators.

For elementary school kids (7–12 years) many textbook publishers currently develop mobile apps to accompany their books (Bird 2011). Apps for this age often focus on factual knowledge building for math and language (especial first foreign language). They are often designed to accompany traditional textbooks and to help learners to build skills. The learner will no longer only deal with an app for fun. Aspects like efficiency, completeness and a general fit to current classroom activities will become increasingly important.

For high school students apps for the organization of the homework like Myhomework (My Homework, Student planner 2015), timetables and note taking apps like Evernote start to dominate the app stores. Also for the age of vocational college and university education, self-organization and education management are the dominating topics on the app market.

In the area of adult educational apps focus on self-controlled just-in-time learning for improving language skills or prepare for tests like driving licence or various business certificates.

For retired and older people the paradigm switches again from a just-in-time skill learning to rather interest-driven learning.

Noessel (2003) describes the learner's needs as representing the gap between what the learner wants to get out of the learning experience and his or her current state of knowledge, skill, and enthusiasm. He identifies potential learning needs in four different domains: cognitive, social, affective and psychomotor (Table 5.1).

Learners needs within a learning setting also depend on their motivation toward the specific topic to learn their age and their mood. Unfortunately, there is no universal method to increase children's motivation to learn (Which factors affect motivation 2015). Elementary school programs are focused on knowledge acquisition and learning process in general. By the end of elementary school, learning interest may be decreasing due to a range of psychological factors, one of them being inability to find practical appliance of theoretical knowledge (Eccles et al. 1998).

Table 5.1 Learner's needs (Noessel 2003)

Cognitive	Social	Affective	Psychomotor
Recognize good questions	Communicate with peers	Attain goals	Be in a comfortable setting
Ask good questions	Give and receive support	Nurture positive attitudes	Have transportation
Get help from experts	Experience external motivation	Be open to feedback from others	Have child care
Practice problem solving	Make a difference	Have time for reflection and self-assessment	Get enough sleep
Think independently	Interact while problem solving	Possess well-founded self-confidence	Have good diet/adequate energy level
Create work products	Explore and challenge conventions	Define and respond to locus of control	Exercise
Process new information	Grow with friends	Have a sense of belonging	Have access to equipment and tools
Use learning resources	Manage time and tasks	Understand motivations of others	Engage in appropriate and timely demonstrations

In general, younger learners prefer gamified, not too repetitive content constructed in a narrative way. For adult learners the possibility to learn in small chunks, so called microlearning and nanolearning (Masie 2006) might be more relevant for the choice of a specific app. At any age the financial factor might be relevant for the choice of specific app; the cost factor will not be as relevant from the point of view of the learner as it is from the point of view of the teacher or parent.

5.4.2 *The Teacher's View*

A teacher will have a specific learning scenario or task in mind and wants to use a suitable educational mobile app to support it. The teacher is often responsible to choose adequate learning apps for his learners. In early childhood also parents might be in the teacher's role, at least for the selection process. Most of the parents are no experts in media pedagogy and have to rely on what friends say, what they get to know from their usual media channels or what they find in the specific app store. Low costs, a good visual presentation, user ratings and other non-pedagogical attributes will most likely influence a parent's choice.

Schoolteachers will look for apps with high learning value and good fit to current curriculum content. Depending on the type of activity and learning goal a wide range of mobile apps will be suitable or not. The pedagogical typologies and classification models mentioned in the beginning of this chapter can help teachers to find and evaluate an appropriate learning app. Most published research papers on app classification models are teacher focused. From a very practical view, teachers might look for cheap, easy to use and distribute, platform-independent, and learn-time efficient apps.

Textbook publishers already offer digital content in form of mobile apps in combination with classic textbooks. Based on the best-selling educational apps in Apples App store, an investment study recommends publishers to invest in collections of small short-time activities that fit in the typical class time schedule and allow more freedom for teachers to use them in personal and adaptive learning environments. This might also be an indicator of the preferred teacher's use of educational apps today.

The teacher's profession also involves various administration, planning and organization tasks that can be supported by mobile apps. There exists plenty of class administration, homework schedule and binder apps that address teachers rather than learners. Choosing appropriate tools is not easy and often legally restrained, existing school infrastructure (like learning management systems) and missing synchronization or export mechanisms will hold back teachers from using them. Even so such apps are filed in the education category of major app stores they are no real learning apps and might need an own classification schema to help teachers and developers to choose and categorize them better.

5.4.3 The Developer's View

Production of mobile apps is a very new challenge for traditional educational publishers. Technology is changing rapidly over the last few years. The first Apple iPad was introduced 2010 and only a few years later we see entire schools buying tablet devices for the classroom. A wide range between schools with 1:1 tablet infrastructure (each child has an own device), notebook pools, computer poolrooms or just 2–3 computers in each classroom makes production of mobile apps almost unpredictable for publishers. Schools with 1:1 equipment want content for their devices.

Almost every school kid (>12 years) has a smart phone in Switzerland (JAMES Studie 2014) but still, many schools prohibit the use of such devices. As if this was not enough, the technical systems are incompatible (iOS vs. Android) and force publishers either to choose one platform or produce multiple versions at greater expense. From a technical perspective some incompatibility problems can be solved by switching to HTML5 and so called hybrid apps (running on all platforms) in the future. There still are some limitations on HTML5 and mobile web apps (i.e. making use of hardware like the microphone) but situation continues to get better with each major update of mobile device operating system and development frameworks. Today as a traditional textbook publisher you can either wait until the market and infrastructure situation in schools becomes a bit more settled or take the risk and invest in an unsteady market and so being the first and bet on the long run. Right now it is almost impossible to make a positive return on investment calculation while the target group in schools is still small and initial expenses for building up frameworks and tools for content production are high. Today many publishers try to combine traditional paper-based books with additional digital content to reduce the risk and distribute the costs.

Besides the traditional textbook publishers, there are many new companies producing educational apps very profit driven. Specialized on programming, designing and marketing, such publishers often miss a pedagogical foundation and evaluation. Hirsch-Pasek et al. (2015) surmised that only a handful of apps are really designed with an eye toward how children actually learn. Fast-produced and fast-selling content for a broad target audience is the key for high profits in the low priced app store market. As a result many apps offer the same content with just a slightly different audio visual presentation (Princess Lilifee's Numbers vs. Captain Sharky's Numbers, Animal touch sounds vs. Vehicle touch sounds and so on) sold for a few or even less than a dollar. Competition on the market is hard and new companies try to get into the market from all over the world every month. A study from Shuler, Levine and Ree indicated in 2012 that within 2 years about 80 % of all publishers with educational apps in the top 100 were replaced by new competitors. The success or fail of a mobile app is not just a result of development quality but also on how many users will notice an app in the huge collection of hundred thousands of apps in stores. The most successful, best-selling learning app might not be the most useful from a pedagogical perspective.

5.4.4 The Distributor's View

Almost the entire distribution of mobile apps is handled via app stores preinstalled with a given operating platform like Apple (iOS, App Store), Google (Android, Play Store), Microsoft (Windows Phone Store) or Amazon (Amazon Appstore). The stores act as gate keepers to the mobile app world and decide which content is allowed or not and, almost with the same effect, which content is promoted and shown to the store users and which is hard to find. In all major app stores a share of about 30 % of all incomes will directly go to the shop operator company. This also includes any in-app-purchases after already installing an app for a fee or for free. Some apps offer a subscription model that will require monthly or yearly payments through in-app-purchases. The shop owner often prohibits selling subscriptions or any additional content without taking his 30 % share which is one often stated reason why textbook (and newspaper) publishers held back publishing in the stores. The shops are organized in classic categories and a few top 100 lists e.g. of most viewed, most bought or highest rated apps. The categories offered by stores like Apple are still of little use for educators to choose appropriate apps and many of the more useful application exist outside the education category (Murray and Olcese 2012). The iLearn II study (Shuler et al. 2012) reveals that 86 % of the most popular apps in the education category in Apples App Store are not intended by their developers to be used in school. The marketing and target audience are mainly parents looking for apps to foster their young children especially in basic math, reading and writing skills at home. The study also showed that the major part (80 %) of all educational categorized apps is made for children.

Almost 60 % of them are targeting very young children before and up to pre-school age. While the average price for educational apps in Apples App Store has increased from 1.13$ in 2009 to 2.14$ in 2011 it is still a market where mass marketing (high sales, low prices) is the only key to success. This model is almost carved in stone by the store operators to keep their devices attractive (high device price but low content prices). For example Apple only offers fixed prices to developers to choose from and everything above a given price will require a special approval. As a shop operator a major goal is to maximize the 30 % share profit by advertising high selling apps. As a result apps for specific learning topics will be less prominent advertised and harder to find than those with a broad topic addressing potentially many more users. As a result profit-driven developers will try to concentrate on apps for broad topics as well, to increase their chance to get their apps on the most valuable top 100 lists created by the shop operators (often based on statistical data). For traditional textbook publishers it is very hard to be successful in this marketing system, while not only providing mainstream content.

5.5 Why so Many Educational Apps Are Games?

From the view of developers and distributors we see the major part of the educational app market addresses young children in doodle and preschool age. The use of such apps is most likely intrinsic motivation by the child who either likes to deal with it or not. Therefore, in this target age almost all educational apps use some sort of gaming mechanisms to engage children. We need to distinguish between different forms of educational games. The term 'digital game-based learning' describes the idea to build fully functional video games (with game rules and an own game world) and cover one or multiple learning topics by deeply integrating them in the game. The motivational parts of playing a game, like exploration, challenges and competition, are used to encourage students to learn something or to apply their knowledge; even without knowing that they actually learn while playing the game. In contrast 'Serious Games' transform real-world problems into a playable game world (typical a simulation) without hiding or mask the learning topic behind a fictional game world. A good example for a successful serious game is *Democracy 3* (Democracy 3 2015) with over 200,000 copies sold. The game allows the player to be president of a simulated democratic country who has to make all kind of decisions. Keeping all parties of a society happy is almost impossible and a real challenge. Also some video games cover school relevant topics like economy cycles in Anno 1604 (Anno 1604 2015) or traffic simulation in Transport Tycoon (Transport Tycoon Deluxe (TTD)—Online 2015) or SimCity (2015). Under the term 'Serious Play' there are various publications covering the use of such non-educational games in formal learning settings. A different approach is 'Gamification': the use of typical game mechanics in non-game environments.

Instead of building a complete game, only mechanics-like high scores, points, levels, badges and achievements are placed on top of more or less traditional learning materials. The learning content is not hidden behind a complex game world making the development much easier and cheaper.

While using games for educational software is an old concept since many decades, mobile games and the way people play them on smartphones and tablets opens new possibilities for adult education and lifelong learning scenarios. Prensky briefly summarized the ideas, concepts and success factors in his book Digital Game-Based Learning in 2004 (Prensky 2004).

Games and educational apps seem to be inevitable connected and market development indicates a continuing growth of educational apps with various game mechanics. As an educational publisher completely new skills and employees like game designers, animators and programmers are required when building content for the mobile app market.

5.6 Conclusion

In this chapter we pointed out different types of apps sustaining specific needs and requirements of the target audience. We also tried to show different points of view and different needs of learners, teachers, developers and distributors. In fact it turns out to be difficult to find the appropriate educational app for the specific need in today's app store market. Educator web pages, reviewers and guides try to build up rankings in order to help to find the best and most valuable apps.

The goal of educational apps for preschool children proposed by Chau (2014) might be also adaptable for all age categories, distinguishing apps for building competencies, fostering confidence, encouraging caring, promoting connection, fostering character, and for encouraging contribution. Gamification will have a notable impact on development of future educational apps specially affecting motivational factors (Huotari and Hamari 2012). In future, gamification will not be the only approach enhancing intrinsic motivation for the use of the apps. Due to the increasing importance of non-formal education factors rising intrinsic motivation will become increasingly important.

Another interesting future development of apps and educational apps may be the trend called micro-moments, where people use their smart phone during small amount of time performing relevant situated actions and take pertinent decisions in that 'Micro-Moments' (Consumers in the Micro–Moments 2015). Such 'Micro-Moment', where education might take place situated, in relevant moments with appropriate tailored and individualized interaction.

To conclude, we contend that educational technologies are also subject to constant theoretical transformations and are prone to inherit instability (Handal et al. 2013). It is therefore highly predictable that future advances in human–machine relations, as Facer and Sandford (2010) have proposed, will undoubtedly keep rearranging conceptualizations on a continual basis.

References

About The Ratings (2015). (n.d.). Retrieved from http://childrenstech.com/about/ratings.

Anno 1604. (n.d.). Retrieved from http://en.softonic.com/s/anno-1604:windows-mac-iphone-android-windows-phone-7-web-apps/windows.

Bandura, A. (1997). *Self-efficacy: The exercise of control*. New York: W.H. Freeman.

Benassi, V., Overson, C. E., & Hakala, C. (Eds.). (2014). Applying science of learning in education: Infusing psychological science into the curriculum. Retrieved from http://teachpsych.org/ebooks/asle2014/index.php.

Bird, E. (2011). Planet App: Kids' book apps are everywhere. But are they any good? *School Library Journal, 57*(1), 26–31.

Chau, C. L. (2014). Positive technological development for young children in the context of children's mobile Apps. *Dissertation submitted at the Tufts University in May 2014*. Retrieved from http://ase.tufts.edu/DevTech/resources/Theses/CChau_2014.pdf.

Cherner, T., Dix, J., & Lee, C. (2014). Cleaning up that mess: A framework for classifying educational apps. *Contemporary Issues in Technology and Teacher Education, 14*(2), 1–61. Retrieved from: http://www.citejournal.org/

Consumers in the Micro-Moment (2015) Google/Ipsos, US, March 2015, n = 5,398, based on internet users. Retrieved from https://www.thinkwithgoogle.com/micromoments/.

Crawford, K. (1996). Vygotskian approaches to human development in the information era. *Educational Studies in Mathematics, 31*, 43–62. doi:10.1007/BF00143926.

Cross, J. (2006). *Informal learning: Rediscovering the natural pathways that inspire innovation and performance*. San Francisco: Pfeiffer.

Deci, E., & Ryan, R. (Eds.). (2002). *Handbook of self-determination research*. Rochester, NY: University of Rochester Press.

Democracy 3. (2015). (n.d.). Retrieved from http://www.positech.co.uk/democracy3/.

Dillenbourg, P. (1999). What do you mean by collaborative learning? In P. Dillenbourg (Ed.), *Collaborative-learning: Cognitive and computational approaches* (pp. 1–19). Oxford, UK: Elsevier.

Eccles, J. S., Wigfield, A., & Schiefele, U. (1998). Motivation. *See Eisenberg 1998*, pp. 1017–95.

Facer, K., & Sandford, R. (2010). The next 25 years? Future scenarios and future directions for education and training. *Journal of Computer Assisted learning, 26*(1), 74–93.

Goodwin, K., & Highfield, K. (2012). *iTouch and ilearn: an examination of 'educational' Apps*. Paper presented at the Early Education and Technology for Children conference, March 14–16, Salt Lake City, Utah.

Golinkoff, R. M., & Hirsh-Pasek, K. (2015). *The learning illusion*. Washington, DC: APA Press.

Handal, B., El-Khoury, J., Campbell, C., & Cavanagh, M. (2013). A framework for categorising mobile applications in mathematics education. In *Australian Conference on Science and Mathematics Education*. Retrieved from http://goo.gl/zBkqMT.

Hirsh-Pasek, K., Zosh, J. M., Golinkoff, R. M., Gray, J. H., Robb, M. B., & Kaufman, J. (2015). Putting education in "Educational" apps: lessons from the science of learning. *Psychological Science in the Public Interest, 16*(1), 3–34. doi:10.1177/1529100615569721. Retrieved from http://www.psychologicalscience.org/index.php/publications/educational-apps.html.

Huotari, K., & Hamari, J. (2012). Defining gamification: A service marketing perspective. In *Proceedings of the 16th International Academic MindTrek Conference,* Tampere, Finland, October 3–5, 2012, New York: ACM Press, New York, NY, 2012.

James Studie (2014). Nationales Programm zur Foerderung von Medienkompetenzen. *Eidgenoessisches Departement des Inneren. Bundesamt fuer Sozialwissenschaften BSV*. Retrieved from http://www.jugendundmedien.ch/de/home.html.

Kearsley, G. (1994). Social learning theory. Retrieved May 1, 2009, from http://www.gwu.edu/~tip/bandura.html.

Langwithces blog: Bloom's Taxonomy and iPad Apps. (2011, August 21). Retrieved http://langwitches.org/blog/2011/08/21/blooms-taxonomy-and-ipad-apps/.

Masie, E. (2006). Nano Learning. A miniaturisation of Design. ©2006 Media Tec Publishing. Retrieved from http://www.cedma-europe.org/newsletter%20articles/Clomedia/Nano-Learning%20-%20Miniaturization%20of%20Design%20(Jan%2006).pdf.

McKinsey and Company & GSMA. (2012). http://www.gsma.com/connectedliving/gsma-and-mckinsey-transforming-learning-through-meducation/

Mobile roadie: Mobile Marketing In One Powerful Solution. (2015). Retrieved from http://mobileroadie.com/.

Murray, O., & Olcese, N. (2012). Teaching and Learning with iPads, Ready or Not? *Tech trends, 55*(60). 42–50.

MyHomework Student Planner. (n.d.). Retrieved from https://myhomeworkapp.com/.

Nickerson, R. C., Varshney, U., Muntermann, J., & Isaak, H. (2007). Towards a taxonomy of mobile applications. In *Proceedings of the Thirteenth Americas Conference on Information Systems*, Keystone, Colorado, USA, August 09th–12th. Retrieved from http://courses.cs.washington.edu/courses/csep590b/11wi/readings/taxonomy.pdf.

Noessel, C. (2003). Free Range learning Support. Thesis at the Interaction Design Institute Ivera. Retrieved from http://www.chrisnoessel.com/freerangelearning/images/2003_thesis_fresh_cnoessel.pdf.

Notari, M., & Döbeli, B. (2010). Learning in an active collaborative space. In M. Ebner & M. Schiefner (Eds.), *looking toward the future of technology-enhanced education: Ubiquitous learning and the digital native*. IGI global, Hershey USA, 2009.

Oxford dictionary (2015). Definition of app in English: (n.d.). Retrieved September 21, 2015 from http://www.oxforddictionaries.com/definition/english/app?searchDictCode=all.

Paavola, S., Lipponen, L., & Hakkarainen, K. (2004). Models of innovative knowledge communities and three metaphors of learning. *Review of Educational Research, 74*(4), 557–576.

Pellegrino, J. W. (2012). From cognitive principles to instructional practices: The devil is often in the details. *Journal of Applied Research in Memory and Cognition, 1*, 260–262.

Pellegrino, J. W., & Hilton, M. L. (Eds.). (2013). *Education for life and work: Developing transferable knowledge and skills in the 21st century*. Washington, DC: National Academies Press.

Prensky, M. (2004). *Game-based learning*. McGraw-Hill Pub. Co. ISBN: 007145004. Retrieved from http://dl.acm.org/citation.cfm?id=1209307.

Sawyer, R. K. (Ed.). (2006). *The Cambridge handbook of the learning sciences*. New York, NY: Cambridge University Press.

Statista (2015). Retrieved from http://statista.com.

Scardamalia, M. (2003a). Knowledge building environments: Extending the limits of the possible in education and knowledge work. In A. DiStefano, K. E. Rudestam, & R. Silverman (Eds.), *Encyclopedia of distributed learning* (pp. 269–272). Thousand Oaks, CA: Sage Publications.

Scardamalia, M., & Bereiter, C. (2003b). Knowledge Building. In *Encyclopedia of Education*. (2nd ed., pp. 1370–1373). New York: Macmillan Reference, USA, Retrieved from http://ikit.org/fulltext/2003_knowledge_building.pdf.

Shuler, C., Levine, Z., & Ree, J. (2012, January). iLearn II: An analysis of the education category of Apple's app store. In New York: The Joan Ganz Cooney Center at Sesame Workshop.

SimCity. (2015). (n.d.). Retrieved from http://www.simcity.com/.

Transport Tycoon Deluxe (TTD)—Online. (2015). Retrieved from http://epicport.com/en/ttd.

Vygotsky, L. S. (1978). *Mind in society: The development of higher psychological processes*. Cambridge, MA: Harvard University Press.

Wartella, E. (2015). Educational apps: What we do and do not know. *Psychological Science in the Public Interest, 16*(1), 1–2. doi:10.1177/1529100615578662. Retrieved from http://psi.sagepub.com/content/16/1/1.full?ijkey=.iIo2GcGXzknM&keytype=ref&siteid=sppsi.

Wheels on the Bus—Android Apps on Google Play. (n.d.). Retrieved September 21, 2015, from https://play.google.com/store/apps/details?id=com.duckduckmoosedesign.bus.
Which factors affect motivation? | CoCreat. (2015). Retrieved from http://cocreat.purot.net/32_which_factors_affect_motivation.

Author Biographies

Michele P. Notari is professor in educational technologies at the University of Teacher Education in Bern, Switzerland and actually (2015–2016) spends a research year at the Faculty of Education, the University of Hong Kong. He has published articles in key journals in the area of technology-enhanced learning, a book and several book chapters related to collaborative learning using participative technologies. He is guest editor of a special edition of the Journal of Educational Research and Evaluation, board member of Wikisym, the international symposium on open Collaboration, the international Conference on Mobile Learning, and International Mobile Learning Festival. He holds a Ph.D. in education, master degree in biology and computer sciences from the University of Berne, and a master degree in educational technologies from the University of Geneva. His actual research focus is on using mobile and wearable technologies for conceptual learning.

Michael Hielscher is a developer and researcher in the area of computer based learning environments with a special focus on computer science education since 2005. He holds a Ph.D. in Computer Science from the University of Mainz and currently works at the Schwyz University of Teacher Education in Switzerland. He developed various educational software and mobile applications widely used around the globe.

Mark King is a psychologist and cognitive ethnographer working in the areas of human interactivity and learning. He is currently developing methodologies for studying human learning within distributed cognitive systems, including an eye-tracking methodology for investigating the co-synchronization of language behaviour and visual perception between persons in learning events. He currently serves as the Head of Educational Design and Development at UNSW Australia.

Chapter 6
Augmented Learning with Augmented Reality

Susan Herrington Kidd and Helen Crompton

Abstract Perhaps no other digital technology has the potential for revolutionizing the educational experience as augmented reality (AR). In this chapter the philosophical, pedagogical, and conceptual underpinnings are unpacked regarding learning with AR. Specifically, AR is defined and the evolution detailed. Next, some of the common usages of the technology are described, recommendations given, and finally the future educational implications are presented.

6.1 Introduction

Perhaps no other digital technology has the potential for revolutionizing the educational experience as augmented reality (AR). AR is an interactive technology which applies computer-generated information to incorporate detailed information about locations or activities from the real world (Yuen et al. 2011). In this chapter the philosophical, pedagogical, and conceptual underpinnings regarding learning with AR are unpacked. Specifically, AR is defined and the evolution detailed. Next, some of the common usages of the technology are described, recommendations given, and finally the future educational implications are presented.

For some, AR is considered to be the realm between reality and virtual reality where countless educational opportunities exist (Pasaréti et al. 2011). For example, in New Zealand at the Arts Center of Christchurch, visitors are exposed to an AR experience as they walk down into the basement room. As they enter the room, they hear a voice of a man telling them to "Come closer into the darkness." As the visitor moves forward, a life-sized 3D image of an old man appears to be floating in front of where they stand. This man explains what it was like working in that dark place a

S.H. Kidd (✉) · H. Crompton
Old Dominion University, Norfolk, VA, USA
e-mail: Susanherringtonkidd@gmail.com

H. Crompton
e-mail: Crompton@odu.edu

D. Churchill et al. (eds.), *Mobile Learning Design*, Lecture Notes in Educational Technology, DOI 10.1007/978-981-10-0027-0_6

100 years ago. This old man is a virtual image of Ernest Rutherford, New Zealand's Nobel Prize winning physicist who performed his initial research as an undergraduate at the University of Canterbury. Through the use of AR, the empty room is turned into a unique learning experience (Billinghurst 2002). In the past 5 years, AR applications have become increasingly more portable and available through mobile devices (Yuen et al. 2011). The availability and portability of AR can provide students with on-the-spot access to multi-sourced, location-specific information which will foster continuous and universal instruction (Yuen et al. 2011).

6.2 Defining Augmented Reality

AR is a 3D technology that fuses the physical and digital world in real time (Pasaréti et al. 2011). In other words, digital information, such as text, images, and video are layered and blended into our perception of the real world (Yuen et al. 2011). AR differs from virtual reality in that AR permits the user to view the real world while simultaneously viewing the virtual layered imagery (Billinghurst 2002); virtual reality provides a digitalized representation of the real world. AR is typically utilized and viewed through either a handheld or head-mounted display unit. These handheld and head-mounted units can be used outside the classroom, thereby eliminating the need for instruction to be limited to a specific environmental context.

AR is unique in comparison to other computer interfaces as it can be used to embellish real-world experiences, as opposed to simply separating the user from the real world and thrusting them into a virtual reality (Billinghurst 2002). These augmentations enhance an individual's perception and comprehension of what is occurring around them (Yuen et al. 2011). The additional over-laid information flows smoothly together as one visual, not appearing incongruous to the user (Yuen et al. 2011).

6.3 Evolution of Educational Augmented Realty

Educational technology is a rapidly growing and evolving field. Schools today must prepare their students for a society that does not currently exist. As the world becomes increasingly complex, this becomes more and more difficult (Ohidi 2006). Ivan Sutherland created one of the first head-mounted 3D displays in 1968, projecting a rudimentary framed graphical image into a room (Caudell and Mizell 1992). Tom Caudell, an engineer working for Boeing in 1992, designed a method that could display cables and other parts of the aircraft, virtually, without having to remove the shield of the machine (Caudell and Mizell 1992). Applications were later developed that housed entire interactive translucent screens providing airmen with basic flight information (Pasaréti et al. 2011). AR first appeared for the average consumer during a live sporting broadcast on television as the country's flag could be seen over the video of a sporting victory (Pasaréti et al. 2011).

In the last 5 years, various new headsets have been made available to the public. Google Glass is a type of Augmented Reality device worn on a spectacle type frame. As the user wears the Google Glass, they can see in their top right field of vision a small screen that provides text feedback to verbal commands or tapping and swiping the frame. Microsoft is advertising the Hololens that claims to enable you to operate various computer programs, such as email and calendar, while viewing them in real-world environments. For example, your calendar could appear over your fireplace so you could check your schedule for the day.

The majority of headsets work with mobile phones that are slotted into the headset. Cases are purchased for the headset to match the particular brand and version of the phone. The user can download various VR applications to view with the headset; some of these are specific to the headset and others can be accessed via multiple devices. At the time that this chapter was written, there are four main headsets available to the public; Occulus Rift, VR One, Poppy3D, and Google Cardboard. The *Occulus Rift* can be used for playing immersive games and VR movies. At the end of 2014, the *VR One* was available with opportunities to take 3D sightseeing tours to famous destinations and tour VR museums. The Occulus Rift and VR One offer 360° × 360° vision. Therefore, you can turn to your left or right in a complete circle to look all around you, and you can also look up at the sky and down at the floor.

AR devices can be expensive for whole class one-to-one purchases. For this reason, other solutions have been made available. The *Poppy3D* provides 360° × 360° experiences and the ability to record video in one direction. Nonetheless, the cheapest option at this time is to purchase *Google Cardboard*. As the name suggests, this is a cardboard headset with a Velcro panel access to place your phone. Using Google Cardboard some 360° × 360° experiences can be gained; however, the quality does decrease to match the decrease in the cost of the headset.

6.4 Augmented Reality in Education

Technology has changed the way people work and socialize. A new generation of students has emerged, who fully engage in the technological affordances available. These technologies are seeping into educational practice as Sharples (2005) describes:

> Every era of technology has, to some extent, formed education in its own image. That is not to argue for the technological determinism of education, but rather that there is a mutually productive convergence between main technological influences on a culture and the contemporary educational theories and practices. (p. 147)

Pedagogies are also changing due to pressure from educators and governments advocating for educational reforms to utilize these technologies for educational purposes (Common Core State Standards Initiative 2010; Greenhow and Robelia 2009; Jonassen et al. 2008).

AR technology has matured to the point where it can be applied to a much wider range of application domains and education is an area where this technology could be especially valuable (Billinghurst 2002). An educator's vision of omnipresent learning could become a reality with AR, as students are able to access a vast array of location-specific information that has been assembled and supplied by an assortment of resources (Yuen et al. 2011). However, AR experiences must be aligned with the student's interests if they are to be educationally effective (Bujak et al. 2013).

6.4.1 Educational Affordances

Using AR in K-12 can provide many pedagogical affordances (Billinghurst 2002; Klopfer 2008) as AR can engage, inspire, and induce students to explore educational concepts from different perspectives (Kerawalla et al. 2006). As students use the AR they are connected to a very stimulating multi-sensory experience. Learning is hightened as the students engage in hands-on manipulation of materials and handle the acquired knowledge in a new and interactive fashion (Wu et al. 2013).

Additionally, researchers have worked to implement AR within those educational genres which have historically proven difficult for students to acquire real-world, first-hand experience, such as physics and astronomy (Lee 2012). AR technologies enable students to take greater control of the speed and direction of their education, while also generating a genuine educational atmosphere conducive to individuals with diverse learning styles (Hamilton and Olenewa 2010). In today's schools, educators are requiring students to be critical consumers of knowledge and not passive learners focused on memorization. Klopfer and Yoon (2004) found that AR can be used to develop active learners as it can be used to develop and enhance critical twenty-first century IT skills.

It is suggested that the role-playing intrinsic to the AR experience may result in an increased sense of confidence and efficacy relative to the field of study (Wasko 2013). While partaking in virtual activities and by learning to assume virtual personas, students can learn to separate themselves from negative self-concepts which could potentially hamper their education (Steinkuehler and Williams 2006). Dede (2008) posits that students show an increased ability to apply what they learn into a variety of scenarios including both AR activities and real-life situations (Yuen et al. 2011).

Motivation is a reported benefit as students' attention is maintained throughout the class and students reach higher levels of participation in educational activities with less cognitive efforts (Di Serio et al. 2013). Students reported that using AR programs increased their motivation for learning, made them feel more like active investigators, increased their interest in the physical settings and content of the experience, and helped them view issues from multiple vantage points (Wasko 2013).

The instructional practice AR provides is unique for three main reasons; it offers seamless interaction between real and virtual environments, it facilitates a tangible interface metaphor for the handling of items, and it offers a smooth shift between reality and virtually (Billinghurst 2002).

6.4.2 Learning with Others

Billinghurst (2002) conducted studies on collaboration and AR. These are findings come from that work. Within the classroom, students' collaborative work efforts improve when they are utilizing a shared workspace. However, this proves difficult to accomplish when working on computer-based instruction. Inkpen (1997) postulates that students' achievement improves when they are crowded around a solitary computer, as opposed to working on individual machines. Group communication patterns also change when students are seated in front of one computer compared to when students have an open communication space between them such as at a worktable. Students alter their gaze, mannerisms, and other nonverbal actions when they are no longer facing each other around a workspace, but instead sitting side-by-side in front of a joint computer workstation. The use of AR allows students to be seated in a round table fashion, while simultaneously situated around the virtual image. Billinghurst found that AR marries the desired group communication patterns of students seated around a joint workstation, with the instinctively physical collaborative nature of students in front of a solitary computer. The end result is an authentic conversation through technology.

6.4.3 Tangible Interface Metaphor

To express understanding in an instructional environment, tangible items are often used. In a collaborative setting, these tangible items are utilized to launch a shared understanding (Gay and Lentini 1995). Billinghurst (2002) found that the relationship between virtual and tangible items is very personal in AR and these tangible items can be embellished in ways that are impossible in traditional settings. The advantage is that students with little to no technology experience will still enjoy a fulfilling interactive experience.

6.4.4 Transition

The extent to which the user's world is digitally created can be used as the defining range on a visual spectrum of computer interactions (Milgram and Kishino 1994). Virtual imagery increases and the level of interaction with reality decreases, moving from the left to the right on the spectrum (Billinghurst 2002). This spectrum illustrates how AR technology can be transitioned from being introduced to gradually increasing virtual depth.

6.5 Augmented Reality Programs for Education

AR programs could potentially enhance a myriad field of study, from geometry lessons, to 3D representation of cells in biology, to displaying molecular structure in chemistry, and simulating a sport in physical education (Pasaréti et al. 2011). AR allows any subject to become more vibrant, appealing, and interactive (Pasaréti et al. 2011). Educational AR games and AR books are two such platforms.

6.5.1 AR Educational Gaming

Games are routinely used in the classroom to facilitate instruction in concepts that are confusing or complicated (Yuen et al. 2011). AR technology has the potential to assist educators by presenting difficult information, associations, and relationships in alternative ways (Yuen et al. 2011). Educational AR games insert a layer of information which augments 'users' experience of reality (Klopfer and Squire 2008, p. 205). This is often accomplished through handheld devices. Klopfer and Squire refer to these AR applications as "augmented reality educational gaming" (2008, p. 203). This layer of information connects the learner to a particular place, location, or time. Educators are then able to manipulate this technology for instruction in specific geographic locations, events in history, or science and mathematical concepts (Klopfer and Squire 2008). Some of the most common examples of AR gaming involve smart-phone applications which incorporate GPS information, effectively connecting real-world information and virtual images (Yuen et al. 2011).

6.5.2 Augmented Realty Books

AR books may very well be the books that close the gap between the digital and physical world and AR books can offer a vital conduit for students because through the use of AR gear, users are able to experience three different levels of reality while utilizing an interactive AR story book (Yuen et al. 2011). In the first level, a simple book could be utilized collaboratively by several users while actually holding and using the book itself. In the second level, several users could view 3D or animated-added AR content in an AR pop-up book. In the third level, AR gear is used to allow users "to 'fly' or 'teleport' into the 3D environment produced by the book, and then participate in the story as it unfolds, interacting with virtual objects, characters, or even other 'readers'" (p. 132). This is the moment when users no longer exist in the real world, but have transitioned to acting within a virtually augmented, real-world setting and then become completely engrossed in an interactive and wholly virtual setting (Yuen et al. 2011).

Digital native students are attracted to the 3D appearance and interactive activities that AR brings to the educational activities (Yuen et al. 2011). When these 3D appearances and interactive activities are melded with literacy, as in the AR book, *"The Future is Wild: The Living Book,"* developed by Meatio in Germany and launched at the Frankfurt Book Fair in 2011, it demonstrates the potential for readers to develop deeper connections in the book (Yuen et al. 2011).

6.5.3 Book Applications

There are various AR books available to the public. The *MagicBook* is an AR book that Lee (2012) described as an enhanced version of a traditional pop-up book. This AR interface system permits AR content to be produced for a traditional book and then brings the story to life with animated and interactive models drawn from the text or illustrations already in the book. Young children can imagine themselves as active participants in a story as the MagicBook brings this dream to fruition using a traditional book as the main interface object. These books remain traditional in every way; turning pages, observing the illustrations, and reading the text without the requirement of any additional technological device. However, with the use of a handheld AR display, readers will now see 3D virtual models projecting from the pages.

ZooBurst is another system which allows students to design their own 3D pop-up books. Storytellers choose one of the books on the Website and then simply hold the ZooBurst marker in front of their Webcam. The on-screen book is entirely interactive and customizable; from arranging characters and props to uploading personalized artwork, to changing the page, clicking on characters to see the dialogue, or tipping the pages in different directions to see it from different angles, audio files can even be recorded. Another AR pop-up book is the *Digilog Books* and when students wear the appropriate eye wear, 3D characters launch from the pages. AR books can be used at the elementary school level to supplement instruction on subjects such as geology; demonstrating the earth's layers, their relationships, differences, and roles (Yuen et al. 2011).

AR books will change the way stories are experienced; commanding increased awareness from the storyteller on an array of concerns, such as the book's structure, value, and immersiveness and "The potential of AR books to appeal to many types of learners, through many paths, is undeniable and exciting for educators" (Yuen et al. 2011, p. 128). These AR interfaced books can change the way students and teachers view and use traditional textbooks. Textbooks no longer need to be stagnant wells of information (Billinghurst 2002). AR can transform the printed page into an avenue with the capacity to transfer students into animated, interactive virtual environments.

6.6 Teaching with Augmented Reality

AR for educational purposes has a foundation in constructivism and situated learning theories (Wasko 2013). Collaborative assignments can be improved through the use of AR (Kesim and Ozarslan 2012). AR can be used to position students outside of the classroom in authentic situations and the capacity to offer learners opportunities to test potential solutions to situations or problems and to explore the outcomes for acceptability (Wasko 2013). The ability to experience these types of issues is not available in our current environment, but through AR we will be able to understand our current environment better (Wasko 2013).

6.6.1 Student-Centered Learning

Today's learners are encouraged to be active, reflective thinkers in the learning process. This cultural and societal pedagogical shift from passive learners to active participants has been driven by reactions to behaviorism, minority rights movements, wider access to education, linguistic pragmatism, and increased internationalism (Gremmo and Riley 1995). With the diachronic pedagogical advancements towards student-centered learning, a concomitant progression can be found in technological affordances manipulated to support those learning philosophies (Crompton 2013). From the discovery learning of the 1970s, to the socio-constructivist learning approach of the 1990s, educators have been working towards designing curriculum and utilizing technologies to enable students to connect with the material and understand the concepts being taught.

With the introduction of AR, educators have been provided with a way to push the boundaries of traditional pedagogies. Components of constructivism are clear within the design of AR learning environments as students are given the opportunity to connect educational concepts to form their own understandings (Wasko 2013). Students can investigate the world around them, learn through their successes and mistakes, and find multiple outcomes. The problem-based style of many of the educational settings, along with the goal of reducing the margins between learning and doing, are evidence of the sway the situated learning theory has within AR learning environments (Wasko 2013). Another key element of this instructional approach is that students experience the augmented version of the reality in the real world, not as an avatar in a virtual world in their computer (Wasko 2013).

Barab and Duffy (2000) posit that once students enter the AR world or "practice fields," teachers' objectives transition from conceptual learning to an authentic educational experience that is very likely to require the use of learned skills or ideas. Students have the opportunity to experience and train for situations or problems without the added risk of injury or mishap, such as natural disasters, hazardous material concerns, or any other task which may be logistically or physically too dangerous to perform in the real world (Wasko 2013).

6.6.2 *Designing for Learning*

Kirkley and Kirkley (2005) stated that "with advances in computer technologies and networked learning, we have exciting opportunities to design learning environments that are realistic, authentic, engaging and extremely fun" (p. 20). Additionally, with the improvements in required hardware and software, an increased number of students and instructors have the capacity to develop and utilize AR enhanced instructional environments (Wasko 2013). Teachers can now create interactive environments to enhance their lesson plans on everything from specific historical landmarks or locations, specific time periods, or environmental or weather situations like volcanoes or hurricanes which would be exceptionally dangerous to explore first-hand.

Cuendet et al. (2013) report of three things that must be remembered when designing AR learning activities. The first is that the AR system must be flexible enough so that teachers can make necessary adaptations to meet the needs of their students. Second, AR lessons should be the same content size and length as traditional lessons and from the same curriculum. Third, the system must take into consideration the limits of the context.

6.7 The Future of Augmented Reality in Education

While AR technology is not new, its use in education is still in its infancy. In order to determine how to best utilize this technology in the school environment, educators must continue to work with researchers within the field (Billinghurst 2002). Current research suggests that AR technology has potential as a practical extension to textbooks and exercise workbooks, allowing for hands-on experiences to facilitate the lessons (Pasaréti et al. 2011). AR is becoming more commonplace in today's society. The accessibility and affordability of mobile devices and other hardware with the capability to process and display information at rapid speeds has made the potential use of AR possible (Yuen et al. 2011). However, as the tools facilitating AR continue to evolve, so must the research and development of educational AR applications.

Experts reported in the 2015 Horizon Report K-12 Edition that AR as a visualization tool is an important technological development and a way of teaching complex thinking (Johnson et al. 2015). However, additional scaffolding and support would be required to assist educators in developing a suitable instructional framework, identifying possible answers to their issues, and decoding the clues provided by the technological devices and embedded in the real-world environment (Wu et al. 2013). It will become necessary to find instructional designers who can create the learning activities for AR systems in the future (Kesim and Ozarslan 2012).

6.8 Recommendations

As educators consider implementing AR in their educational activities, the following recommendations can help get the most out of the experience.

- Provide multiple opportunities for students to collaborate and share their AR experiences.
- Provide opportunities such as field trips to optimize AR's inherent mobile capabilities.
- Utilize AR as an additional learning platform in conjunction with other visual, auditory, and tactile opportunities.
- Connect AR experiences to educational standards.
- Think outside the box, rather than try to fit AR to a traditional approach.

6.9 Conclusion

AR offers extensive opportunities for educators to create authentic, engaging, and customizable learning experiences for their students (Yuen et al. 2011). The unique affordances of AR have pushed the boundaries of traditional pedagogies to enable educators to provide rich, student-centered, learning experiences. AR technology has the potential to be a powerful remediation and special education tool because of its inherent mobile capabilities. Future researchers may consider how AR programs could be used to help students who struggle with dyslexia or other reading disabilities to read clearly. Perhaps AR programs could be individually designed as a corrective layer for each student's particular reading disability. As AR technologies evolve future researchers could discover new affordances to learning using AR.

This chapter offers researchers, educators, and policy makers' insight into the educational opportunities for AR in K-12 learning. In addition, a set of recommendations are provided to those who seek to employ these methods to enhance learning opportunities. From one of the early AR devices developed by Ian Sutherland in 1968, AR developers have made great leaps in technological advancements. However, AR is only still emerging with many researchers, practitioners, and policymakers eagerly waiting to see what AR learning opportunities will be provided in future years.

References

Barab, S. A., & Duffy, T. M. (2000). From practice fields to communities of practice. In D. H. Jonassen & S. M. Land (Eds.), *Theoretical foundations of learning environments* (pp. 25–55). Mahwah, NJ: Lawrence Erlbaum Associates.

Billinghurst, M. (2002). Augmented reality in education. *New Horizons for Learning, 12.*

Bujak, K. R., Radu, I., Catrambone, R., MacIntyre, B., Zheng, R., & Golubski, G. (2013). A psychological perspective on augmented reality in the mathematics classroom. *Computers & Education, 68*, 536–544.

Caudell, T. P. & Mizell, D.W. (1992). Augmented reality: An application of heads-up display technology to manual manufacturing processes. In *Proceedings of IEEE Hawaii International Conference on Systems Sciences* (pp. 659–669).

Common Core State Standards Initiative. (June, 2010). Reaching higher: The common core state standards validation committee. In *A report from the National Governor's Association Center for Best Practices and the Council of Chief State School Officers.* Retrieved June, 2010, from http://www.corestandards.org/assets/CommonCoreReport_6.10.pdf.

Crompton, H. (2013). A historical overview of mobile learning: Toward learner-centered education. In Z. L. Berge & L. Y. Muilenburg (Eds.), *Handbook of mobile learning* (pp. 3–14). Florence, KY: Routledge.

Cuendet, S., Bonnard, Q., Do-Lenh, S., & Dillenbourg, P. (2013). Designing augmented reality for the classroom. *Computers & Education, 68*, 557–569.

Dede, C. (Speaker) (2008). Immersive interfaces for learning: Opportunities and perils [motion picture]. Retrieved from http://cyber.law.harvard.edu/interactive/events/luncheon/2008/12/dede.

Di Serio, A., Ibanez, M., & Kloos, C. (2013). Impact of an augmented reality system on students' motivation for a visual art course. *Computers & Education, 68*, 586–596.

Gay, G. & Lentini, M. (1995). Use of communication resources in a networked collaborative design environment, Journal of Computer-Mediated Communication, 1(1). doi: 10.1111/j.1083-6101.1995.tb00320.x.

Greenhow, C. & Robelia, B. (2009). Web 2.0 and classroom research: What path should we take now? *Educational Researchers, 38*(4), 246–259.

Gremmo, M. J., & Riley, P. (1995). Autonomy, self -direction and self-access in language teaching and learning: The history of an idea. *System, 23*(2), 151–164.

Hamilton, K. & Olenewa, J. (May, 2010). Augmented reality in education [PowerPoint slides]. Retrieved May, 2010, from http://www.authorstream.com/Presentation/k3hamilton-478B23-augmented-reality-in-education/.

Inkpen, K. (1997). *Adapting the human computer interface to support collaborative learning environments for children.* Ph.D. Dissertation, Department of Computer Science, University of British Columbia.

Johnson, L., Adams-Becker, S., Estrada, V., & Freeman, A. (2015). *NMC Horizon Report: 2015 K-12 Edition.* Austin, Texas: The New Media Consortium.

Jonassen, D., Howland, J., Marra, R., & Crismond, D. (2008). *Meaningful learning with technology* (3rd ed.). Upper Saddle River, NJ: Pearson.

Kerawalla, L., Luckin, R., Seljeflot, S., & Woolard, A. (2006). "Making it real": Exploring the potential of augmented reality for teaching primary school science. *Virtual Reality, 10*, 163–174.

Kesim, M., & Ozarslan, Y. (2012). Augmented reality in education: Current technologies an the potenial for education. *Procedia—Social and Behavioral Sciences, 47*, 297–302.

Kirkley, S., & Kirkley, J. (2005). Creating next generation blended learning environments using mixed reality, video games and simulations. *TechTrends, 49*(3), 42–54.

Klopfer, E. (2008). *Augmented learning: Research and design of mobile educational games.* Cambridge, MA: MIT Press.

Klopfer, E., & Squire, K. (2008). Environmental Detectives—the development of an augmented reality platform for environmental simulations. *Educational Technology Research and Development, 56*(2), 203–228.

Klopfer, E., & Yoon, S. (2004). Developing games and simulations for today and tomorrow's tech savvy youth. *TechTrends, 49*(3), 33–41.

Lee, K. (2012). Augmented reality in education and training. *Techtrends, 56*(2), 13–21. doi:10.1007/s11528-012-0559-3.

Milgram, P. & Kishino, F. A. (1994). Taxonomy of Mixed Reality Visual Displays. Institute of Electronics, Information, and Communication Engineers Trans. *Information and Systems (IECE special issue on networked reality), E77-D* (12), 1321–1329.

Óhidi, A. (2006). Az élethosszig tartó tanulás és az iskola, Új Pedagógiai Szemle.

Pasaréti, O., Hajdú, H., Matuszka, T., Jámbori, A., Molnár, I., & Turcsányi-Szabó, M. (2011). Augmented reality in education. *INFODIDACT Informatika Szakmódszertani Konferencia.*

Sharples, M. (2005). Learning as conversation: Transforming education in the mobile age. In *Proceedings of Conference on Seeing, Understanding, Learning in the Mobile Age*, Budapest, Hungary (pp. 147–152).

Steinkuehler, C., & Williams, D. (2006). Where everybody knows your (screen) name: Online games as "third places". *Journal of Computer-Mediated Communication, 11*(4), 885–909.

Wasko, C. (2013). What teachers need to know about augmented reality enhanced learning environments. *Techtrends, 57*(4), 17–21. doi:10.1007/s11528-013-0672-y.

Wu, H., Less, S., Chang, H., & Liang, J. (2013). Current status, opportunities and challenges of augmented reality in education. *Computers & Education, 62*, 41–49.

Yuen, S., Yaoyuneyong, G., & Johnson, L. (2011). Augmented reality: An overview and five directions for AR in education. *Journal of Educational Technology Development and Exchange, 4*(1), 119–140.

Author Biographies

Susan Herrington Kidd is an Elementary Education graduate student at Old Dominion University, Virginia. She gained her Bachelor of Arts in Anthropology from Arizona State University. Kidd's study and research interests include the utilization of technology to assist students on the autism spectrum and the effective utilization of technology throughout K-12 education. Kidd exploits the affordances of technologies as she works as a substitute teacher for Chesapeake Public Schools Virginia. This teaching experience has provided Kidd with opportunities to work with diverse groups of students to better understand how technology can be used to support all learners.

Helen Crompton (Ph.D.) is an Assistant Professor of Instructional Technology at Old Dominion University, Virginia. Professor Crompton is a keen researcher and educator in the field of instructional technology. She gained her Ph.D. in educational technology and mathematics education from the University of North Carolina at Chapel Hill. Her research is focused on mobile learning and the effective integration of technology into K-12 education. Dr. Crompton works as a consultant for two United Nations Agencies (United Nations, Educational, Scientific, and Cultural Organization: UNESCO and International Telecommunication Union: ITU) to research, author and edit publications summarizing research on mobile learning. Dr. Crompton is also a faculty member for the International Society for Technology in Education (ISTE), teaching the ISTE Standards academy, consulting, and recently designing ISTE's self –paced Mobile Learning Academy and Verizon's Mobile Learning Academy.

Part II
Mobile Learning Adoption and Student Perception

Chapter 7
Mobile Technology and Interactive Lectures: The Key Adoption Factors

Vimala Balakrishnan and Chin Lay Gan

Abstract Lecture classes are fundamental and essential for teaching and learning in higher education. The objective of this study is to investigate adoption factors for promoting interactive lectures in higher education from reviews of technology acceptance models, motivational factors, and cultural dimension theory. The study aims to elicit key factors influencing mobile technology adoption in the classrooms as an interaction tool, focusing on the notion of communication barriers caused by classes with large number of students. Survey involving higher education students enrolled in academic courses in Malaysia was conducted with a sample size of 396. Factor analysis produced three key factors: User system perception (USP), system and information quality (SIQ) and user uncertainty avoidance (UUA). Results of regression analysis revealed UUA as the strongest significant predictor of adoption (beta = −0.225, $p < 0.001$), and a high proportion of UUA was strongly explained by USP ($r = -0.513$) and SIQ ($r = -0.537$). This study underscores the need for researchers to further explore blended learning pedagogies using mobile technology.

7.1 Introduction

Despite huge advancement in mobile technology sophistication and its role in the fields of e-learning and mobile learning, lecture classes are still fundamentally important in higher education institutions. Face-to-face lecture classes where students congregate at scheduled venues to listen and participate in learning activities provide a myriad of learning opportunities for the students. The ability to engage in

V. Balakrishnan (✉)
University of Malaya, Kuala Lumpur, Malaysia
e-mail: vimala.balakrishnan@um.edu.my

C.L. Gan
Multimedia University, Cyberjaya, Malaysia
e-mail: gchinlay@gmail.com

D. Churchill et al. (eds.), *Mobile Learning Design*, Lecture Notes in Educational Technology, DOI 10.1007/978-981-10-0027-0_7

real-life discourses with their peers and lecturers are invaluable. Effective pedagogy principles emphasize the importance of prompt feedback to students' enquiries, active participation and conducting collaborative activities in the classroom (Reeves 2006; Chickering and Gamson 1987). However, lecture classes with large number of students conducted in huge theatre halls are problematic for a number of reasons. Lack of opportunities for the students to ask questions or for lecturers to encourage feedback and to engage in discussions with their students due to time constraint is chief barriers (Dobson-Mitchell 2011; Tesch et al. 2011). In addition, students' personality traits such as shyness or introversion, and low language proficiency compound the problem further (Gan and Balakrishnan 2014; Stowell et al. 2010).

Incorporating effective use of the right technology in the classrooms can be the solution to reduce some of the barriers preventing interactions in large lecture classes. Students and lecturers alike are already using mobile technology for numerous academic activities, for instance downloading learning resources on the Internet, accessing their institutions' learning management system to download learning materials, to read the latest announcements and to open materials downloaded using tablets or laptops during lectures are common occurrences observed among higher education students (Balakrishnan and Gan 2013). Technology-enabled lecture halls that promote interactions and real-time feedback in problem-solving scenarios suggest that benefits gained outweighed possible technology distractions (Donovan and Loch 2013). Venema and Lodge (2013) study on the use of digital ink technology to promote interactions in large lecture classes produced promising results. Similarly, using instructional tools in the classroom with the aim of promoting active learning resulted in increased students' satisfaction in aiding their participation during lectures, although such tools do not increase their motivation level to study (Oigara and Keengwe 2013).

Similar findings were reported by Chen and Lan (2013)'s study on the use of a personal response system in large lecture classes where the perceived benefits of incorporating technology in the classrooms to improve students' learning experiences were inconclusive. Other drawbacks observed were technology-induced disruptions during lecture classes, the temptation among students to engage in personal conversations using their mobile messaging applications, or discretely playing online computer games (Scornavacca et al. 2009).

Therefore, using mobile technology in the classrooms bring has its benefits as well as disadvantages. Concerns of possible disruptions are serious and warrant in-depth investigation towards drafting an implementation guideline for responsible use of mobile technology in the classroom. Maturity among the students is important to ensure students' readiness for responsible use of such technology in the classroom. Alzaza and Yaakub (2011), and Mahat et al. (2012) investigated Malaysian higher education students' readiness to use mobile technology, and results suggested that students possessed sufficient knowledge and maturity to use such technology responsibly. The findings point to a growing awareness towards use of mobile technology inside the classroom to facilitate students' and lecturers' interactions in order to overcome communication barriers of large lecture classes which are oftentimes unavoidable. Consequently, the present study aims to develop and evaluate a conceptual framework for acceptance of mobile technology for promoting interactive lectures by subjecting key determinant factors of adoption intention elicited for statistical analysis.

7.2 Background Study and Hypotheses Development

This section presents the main theories from literature of past research studies on technology acceptances, motivational theories, and cultural dimension theory. The focus of the discussion is mainly on computing technology and information system adoption studies across a wide range of domains.

7.2.1 Technology Acceptance Model (TAM, TAM2, TAM3)

The well-known technology acceptance model (TAM) hypothesized that perceived usefulness and perceived ease of use impact user attitude and behavioural intention toward subsequent acceptance of information system (Davis 1989). Perceived usefulness is the degree in which one believes that using an information system will improve productivity, whereas perceived ease of use is the degree in which one believes using an information system will require minimal effort (Davis 1989). A field study (two existing systems with a sample size of 112 employees from IBM) and a laboratory study (two new IBM PC-based graphics system with a participant size of 40) to test the reliability and validity of perceived ease of use and perceived usefulness were conducted by Davis (1989). The study validated perceived ease of use and perceived usefulness as strong predictors of system acceptance. TAM was subsequently adopted and replicated in areas of email, voice mail, word perfect, Lotus 123 and Harvard graphics system (Adams et al. 1992), database and spreadsheet applications (Hendrickson et al. 1993), and voice mail and dial-up system (Subramanian 1994).

TAM was later extended to TAM2, and included subjective norm, image, job relevance, output quality, result demonstrability, experience and voluntariness (Venkatesh and Davis 2000). TAM2 was found to be appropriate for predicting acceptance in both voluntary and mandatory technology usage environment (Venkatesh and Davis 2000). Venkatesh and Bala (2008) introduced TAM3 that integrated TAM2 (Venkatesh and Davis 2000) and the extended TAM2 version of Venkatesh (2000). The authors postulated that TAM3 does not have the cross-over effects by suggesting that determinants of perceived ease of use and perceived usefulness will have no influence on one another. TAM3 retained the determinants of perceived ease of use and incorporated subjective norm, image, job relevance, output quality and result demonstrability as determinants of perceived usefulness.

Recent studies proved that ease of use and usefulness are pivotal predictors of technology acceptances. Calisir et al. (2014) study on web-based learning system acceptance among college students revealed usefulness to be the strongest determinant of behavioural intention, which was strongly influenced by information and system quality, and ease of use. Similar study on e-learning technology validated the importance of ease of use and usefulness as important determinants (Tarhini et al. 2014; Lin and Wang 2012; Šumak et al. 2011). Ease of use and usefulness are also vital as mediating factors, such as in the study of blended learning approaches where usefulness strongly predicted attitude among the male respondents (Padilla-Meléndez et al. 2013). In another study by Huang et al. (2014), ease of use and usefulness strongly influenced attitude towards behavioural intention to use a disaster prevention education system.

Therefore, it is hypothesized that user acceptance of mobile wireless technology (MWT) for promoting interactive lectures will be determined by ease of use and usefulness. The conceptual framework proposed is presented in Fig. 7.1.

7.2.2 Intrinsic Motivator Theory

Social cognitive theory (SCT) is based on cognitive mechanisms that mediate changes in events. According to the theory, people are neither driven by inner forces nor driven by external motivations. They are, however, explained by a model

Fig. 7.1 Proposed conceptual framework

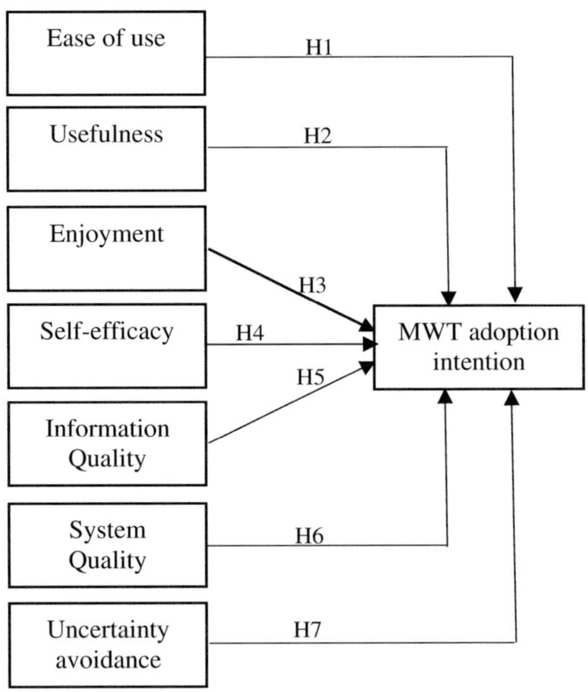

where a person's cognitive, behaviour, personal factors and environmental factors act as determinants of each other. SCT was proposed as an attempt to explain human behaviour by placing importance on the intrinsic factor of self-efficacy as a direct determinant of a person's behaviour (Bandura 1977, 2001). In the context of information system, a strong relationship between self-efficacy and system use was found in a survey among Canadian managers and professionals (Compeau and Higgins 1995). Self-efficacy was further tested in another study by Compeau et al. (1999), and their results further validated self-efficacy as a key determinant of computer acceptance. Self-efficacy was also deemed an important factor in a study among older people or people with disabilities when using everyday technologies (Laver et al. 2012). Self-efficacy also emerged as significant predictor of usefulness and behavioural intention in studying user behavioural intention to use Youtube (Lee and Lehto 2013).

Another form of intrinsic motivator is enjoyment in performing a set of task, without any need for external positive rewards or reinforcements (Scott et al. 1988). Enjoyment and usefulness were revealed as strong determinant variables of intention to use mobile social network games (Park et al. 2014). A recent study on e-commerce identifies enjoyment and self-efficacy as significant mediator variables influencing perceived value towards consumer purchase intention for online content services (Wang et al. 2013). The present study incorporates the constructs of self-efficacy and enjoyment into the proposed conceptual framework to examine its influence on mobile technology adoption for interactive lectures.

7.2.3 Delone and McLean Information System Success Model

Various studies have acknowledged the importance of system and information quality factors, and one of the model advocating both system and information quality as predictors of system acceptances is Delone and McLean (D&M) Information System (IS) success model. In the D&M IS success model, the determinants that lead to intention to use and user satisfaction are information quality, system quality, service quality, user satisfaction, and intention to use as distinct factors but related dimensions of information system success (DeLone and McLean 2003). Study by Lin and Wang (2012) that integrated D&M IS success model with TAM reported information quality and usefulness as strong determinants of e-learning. Both service and the content quality were found to be significant contributors in a study of an e-government website (Tan et al. 2013).

Information quality was also proved to be vital towards influencing use of an online community municipal portal (Detlor et al. 2013). Information and system quality were found to directly affect perceived user benefits and satisfaction, which in turn determined user continuance intention to consume and to provide information in an information-exchange virtual community (Zheng et al. 2013). It is therefore hypothesized that user acceptance of mobile technology for promoting interactive lectures will be predicted by the quality of information and quality of the mobile application functionalities.

7.2.4 Uncertainty Avoidance

One of the cultural national dimensions which is of particular interest in the study of technology acceptance model is uncertainty avoidance. Uncertainty avoidance is characterized as the absence of predictability, composition, and information. Hofstede et al. (2010) defined uncertainty avoidance as the degree in which people feel uncomfortable with the presence of uncertainties or doubts. The effect of cultural influences such as uncertainty avoidance is gaining traction as a key determinant of information system adoption in recent years. According to Ayoun and Moreo (2008), people with high level of uncertainties have a higher stress level in dealing with the future than those with lower level of uncertainties. Lee et al. (2007) researched how people from various countries differ in uncertainty avoidance and product uncertainty. Their findings revealed a significant relationship between product uncertainty and cultural uncertainty avoidance.

Lin (2014) study revealed cultural differences influenced physicians' perception towards knowledge management system acceptance in healthcare organizations. Differences in cultural background between Korean students and U.S. students revealed Korean students to be more apprehensive towards new Web 2.0 technologies compared to their counterparts in the U.S. despite similar personal characteristics (Yoo and Huang 2011). Similar results were also found in a study of e-commerce adoption where respondents from different nationalities revealed cultural influences to be significant predictors (Ashraf et al. 2014). Cultural factors were also significant in the area of mobile health applications (Mohamed et al. 2011). Lastly, negative correlations were revealed between uncertainty avoidance and cell phone and Internet subscription study by Matusitz and Musambira (2013). Therefore, it is expected that the presence of uncertainty avoidance will influence adoption intention in the present study.

7.3 Research Methods

The survey comprised of two main sections—the demographics section, followed by five item statements for each of the constructs identified (usefulness, ease of use, self-efficacy, enjoyment, uncertainty avoidance, system quality, information quality, adoption intention). Respondents can rate their level of agreement for each item statement using a 5-point Likert scale (1 = "Strongly disagree", 3 = "Neutral", and 5 = "Strongly agree"). The study survey instrument was subjected to pilot testing involving ten selecting students. All students completed the survey within 10 min,

Table 7.1 Respondents' gender and education background

		Frequency (%)
Gender	Male	198 (50.0 %)
	Female	198 (50.0 %)
Education	Foundation	44 (11.1 %)
	Diploma	70 (17.7 %)
	Bachelor/Undergraduate	258 (65.2 %)
	Master/Ph.D./Postgraduate	24 (6.1 %)

and feedback was gathered to redefine ambiguous statements. Respondents were recruited via email invitations sent to students of higher learning institutions in Malaysia, and data were collected from April till November 2014. Students were invited to fill up an online survey hosted by Google drive. Data from the online survey was transferred into IBM SPSS Statistics 21 software and organized for statistical analysis.

A total of 396 Malaysian students of higher education participated in the online survey. Average age of respondents was 21 years old. Majority of the respondents were undergraduates ($N = 258$, 65.2 %). None of the respondents are using mobile technology to interact with their lecturers during lectures. Majority of the respondents are using their mobile devices for learning purposes ($N = 335$, 84.6 %). Table 7.1 tabulates the gender and education background of the respondents.

7.4 Results

7.4.1 Principal Component Analysis

Constructs identified from reviews of acceptance models, motivational and cultural theories for examining adoption of mobile wireless technology (MWT) for interactive lectures, i.e. usefulness (U), ease of use (EU), self-efficacy (SE), enjoyment (E), uncertainty avoidance (UA), system quality (SQ) and information quality (IQ) with five research items each were subjected to principal component analysis (PCA). Examination of the correlation matrix revealed the presence of many coefficients of 0.3 and above (with the exception of research item SE2 and research item SQ2). Kaiser-Meyer-Olkin value was 0.964, indicating sample size adequacy and exceeding the recommended value of 0.6. Bartlett's test of sphericity reached statistical significance, supporting the factorability of the correlation matrix ($x^2 = 12979.32$, $df = 561$, $p < 0.001$).

Principal components analysis (PCA) revealed the presence of four factors with eigenvalues exceeding 1, explaining 54.2, 7.2, 4.1 and 2.9 % of the variance, respectively. Examination of the scree plot revealed an uncertain break after the third factor. Parallel analysis was then conducted and results revealed two factors with eigenvalues clearly exceeding and the third factor just slightly exceeding a randomly generated data matrix (35 variables and 396 respondents). Therefore, three factors were then retained for further analysis and results explained a total of 67.07 % of the variance, with factor 1 contributing 55.52 %, factor 2 contributing 7.38 %, and factor 3 contributing 4.17 %.

Oblimin rotation was performed to aid the interpretation and solution revealed the presence of simple structure with all research items loading substantially on only one factor. The factors are named user system perception (USP), system and information quality (SIQ) and user uncertainty avoidance (UUA). Solution revealed strong loadings for USP from usefulness (U), ease of use (EU), self-efficacy (SE) and enjoyment (E), suggesting that user perception towards system usefulness, ease of use and their intrinsic motivations (enjoyment and self-efficacy) are tightly interrelated. Only one research item from self-efficacy (SE) was removed for further empirical testing (loading < 0.4). Pattern and structure coefficients for the three factors are presented in Table 7.2.

According to Pavot et al. (1991), the survey instrument scale has good internal consistency, and results are presented in Table 7.3.

The proposed conceptual framework was then updated. Figure 7.2 illustrate the updated framework. The resulting hypotheses to determine adoption intention of mobile technology for promoting interactive lectures are:

H1: User system perception positively influences MWT adoption intention.
H2: System and information quality positively influences MWT adoption intention.
H3: User uncertainty avoidance negatively influences MWT adoption intention

7.4.2 Pearson Product-Moment Correlation

The relationship between user system perception (as measured by USP) and adoption intention was investigated using Pearson product-moment correlation coefficient. There was a small, positive correlation between the two factors, $r = 0.151$, $n = 396$, $p < 0.005$. The relationship between system and information quality (as measured by SIQ) and MWT adoption intention echoed similar results, with $r = 0.171$, $n = 396$, $p < 0.005$, suggesting small positive correlation. User uncertainty avoidance (as measured by UUA) revealed moderate negative correlation with MWT adoption intention ($r = -0.254$, $n = 396$, $p < 0.005$), with high levels of uncertainty avoidance being associated with lower levels of MWT adoption intention. Correlations of the factors are shown in Table 7.4.

Table 7.2 Pattern and Structure Matrix for PCA with Oblimin Rotation of three-factor solution of research items

Item	Pattern coefficients			Structure coefficients			Communalities
	USP	SIQ	UUA	USP	SIQ	UUA	
U1	**0.866**	−0.007	−0.090	**0.815**	0.465	0.350	0.671
U2	**0.864**	−0.001	−0.018	**0.854**	0.509	0.425	0.730
U3	**0.672**	−0.149	0.211	**0.690**	0.368	0.476	0.509
U4	**0.792**	0.080	−0.025	**0.827**	0.543	0.424	0.688
U5	**0.866**	−0.106	0.056	**0.831**	0.445	0.443	0.698
EOU1	**0.881**	0.009	−0.046	**0.863**	0.514	0.411	0.747
EOU2	**0.862**	−0.052	0.014	**0.837**	0.473	0.428	0.703
EOU3	**0.927**	0.014	−0.102	**0.882**	0.516	0.380	0.786
EOU4	**0.697**	0.070	0.005	**0.742**	0.492	0.401	0.554
EOU5	**0.858**	0.111	−0.052	**0.898**	0.599	0.448	0.814
SE1	**0.748**	0.174	−0.062	**0.821**	0.591	0.416	0.692
SE3	**0.779**	0.150	−0.015	**0.861**	0.610	0.464	0.755
SE4	**0.583**	0.086	0.156	**0.714**	0.520	0.500	0.538
SE5	**0.703**	0.232	−0.004	**0.840**	0.652	0.481	0.740
E1	**0.770**	0.025	0.064	**0.818**	0.522	0.472	0.673
E2	**0.685**	0.061	0.164	**0.806**	0.561	0.548	0.676
E3	**0.484**	−0.033	0.328	**0.632**	0.434	0.558	0.474
E4	**0.715**	−0.006	0.190	**0.809**	0.526	0.554	0.681
E5	**0.787**	0.059	0.078	**0.862**	0.574	0.513	0.752
UA1	0.287	0.176	**0.426**	0.612	0.578	**0.668**	0.562
UA2	0.011	0.090	**0.711**	0.430	0.478	**0.765**	0.591
UA3	−0.084	0.056	**0.903**	0.413	0.491	**0.890**	0.797
UA4	0.088	0.063	**0.805**	0.539	0.548	**0.884**	0.793
UA5	0.173	0.124	**0.689**	0.601	0.599	**0.845**	0.760
SQ1	−0.076	**0.714**	0.044	0.376	**0.692**	0.389	0.483
SQ2	−0.210	**0.808**	0.100	0.327	**0.736**	0.427	0.569
SQ3	0.027	**0.713**	0.137	0.526	**0.803**	0.534	0.660
SQ4	0.287	**0.540**	0.076	0.651	**0.754**	0.513	0.633
SQ5	0.347	**0.510**	0.019	0.663	**0.728**	0.470	0.610
IQ1	0.118	**0.681**	0.062	0.560	**0.785**	0.488	0.631
IQ2	0.245	**0.669**	0.007	0.651	**0.821**	0.492	0.712
IQ3	0.083	**0.755**	0.023	0.549	**0.818**	0.472	0.674
IQ4	0.199	**0.739**	0.013	0.650	**0.866**	0.512	0.776
IQ5	0.337	**0.601**	−0.052	0.671	**0.775**	0.443	0.669

Table 7.3 Reliability analysis of each factor

Factors		Number of items	Cronbach's Alpha	Mean	Standard Dev.
1.	User system perception	19	0.97	4.05	0.93
2.	User uncertainty avoidance	5	0.89	3.56	0.99
3.	System and information quality	10	0.93	3.73	0.88

Fig. 7.2 Updated conceptual framework

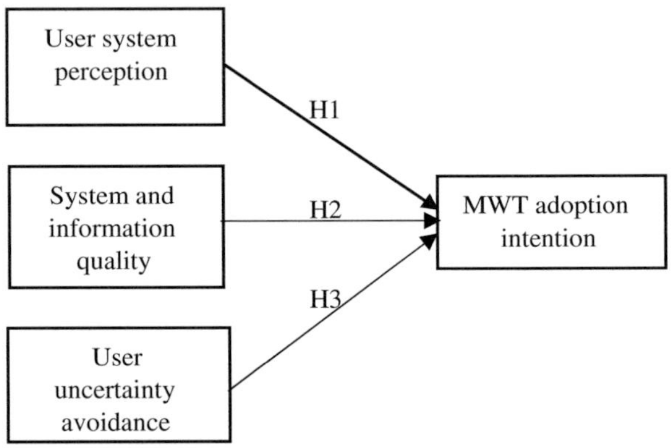

Table 7.4 Correlations of the variables (N = 396)

Variable	2	3	4
1. MWT adoption intention	0.151	0.171	−0.254
2. User system perception	–	0.601	−0.513
3. System and information quality	–	–	−0.537
4. User uncertainty avoidance	–	–	–

Note All correlations were statistically significant ($p < 0.005$)

7.4.3 Standard Multiple Regression

Standard multiple regression was used to assess the ability of USP, SIQ and UUA to predict levels of MWT adoption intention. The prediction model was statistically significant, $F(3, 392) = 9.240$, $p < 0.001$. However, the variables accounted for approximately only 7 % of the variance of MWT adoption intention ($R^2 = 0.066$, Adjusted $R^2 = 0.059$). MWT adoption intention was primarily predicted by user uncertainty avoidance recording the highest beta value (beta = −0.225, $p < 0.001$). User uncertainty avoidance accounts uniquely for about 3 % of the variance of MWT adoption intention. Standardized regressions coefficients of the predictors and their squared semi-partial correlations are shown in Table 7.5.

Table 7.5 Standard regression results

Model	Beta	sr^2
User system perception	0.009	0.00005
System and information quality	0.045	0.00116
User uncertainty avoidance*	−0.225	0.03312

Note Dependent variable was MWT adoption intention, sr^2 squared semi-partial correlation, $*p < 0.005$

7.5 Discussions

The objective of this study is to determine crucial determinants of mobile technology adoption during lecture classes to promote interactivity between students and lecturers of higher learning institutions. Review of literatures proved the importance of the usefulness and ease of use factors (widely replicated and validated across technology acceptances studies) from Technology Acceptance Model; intrinsic motivators conceptualized as enjoyment from motivational model and self-efficacy from SCT; system quality and information quality from Delone and McLean Information System success model; and uncertainty avoidance from Hofstede's national culture dimensions. An online survey was distributed to Malaysian higher education students to elicit perceptions of mobile technology adoption during lecture classes to promote interactivity.

Results from principal component analysis produced three main factors (independent variables). Good system quality features, for instance fast response speed and clean, pleasing interface design are of utmost importance. Distinction between system quality and information quality are presented in the proposed acceptance framework, as information quality represents the quality of the data output produced by the system. Studies have included system quality and information quality constructs (Wixom and Todd 2005). Predictably, both system quality and information quality survey items loaded together, and the factor was named System and Information Quality (SIQ). This can be attributed to the fact that the focus of the study is on harnessing the use of an appropriate mobile messaging application to increase interactions between lecturers and students in the classroom. With mass and widespread downloads and use of mobile applications, distinction between system quality and information quality may not be as pivotal as it may be for more complex information system, such as decision support systems or knowledge management systems used by organizations. Therefore, there may exist an expectation among the younger generation for system quality and information quality to go hand in hand. Uncertainty avoidance items loaded together and its name retained as a factor.

Interestingly, survey items from usefulness, ease of use, enjoyment and self-efficacy loaded together. This is in contrast to many past and current technology acceptances study in the area of information system adoptions, and is suggestive of the lessening distinction of ease of use, usefulness, and intrinsic motivators of self-efficacy and enjoyment as determinants of mobile technology such as a messaging application. The vast majority of the study's respondents (84.6 %) are already using mobile devices in their learning activities affirms the convenience and practicality of mobile devices among the students. Park et al. (2012) study among Korean universities' students revealed attitude, which encompasses their beliefs or perceptions towards mobile learning, as the most important factor for successful implementation. Similar findings were reported by Shroff et al. (2011) in their study examining students' behavioural intention to use an e-portfolio system. Perceptions towards new technology may be representative of the existing confidence towards a system's expected usefulness, ease of use and enjoyment. Therefore, the resulting factor is named as user system perception (USP).

The study hypothesizes that USP and SIQ positively influences MWT adoption intention. Results from correlation and regression analysis do not support these hypotheses. This disputes previous findings by Chong et al. (2011) on mobile learning adoption in Malaysia where quality of system functionality were revealed to be significant, and findings by Pay and Huang (2011) and Calisiri et al. (2014) where system service and content, usefulness, and ease of use positively influences adoption intention. However, Wang and Wang (2009) study on user acceptance of web-based learning challenges the significance of ease of use, and findings concedes weak role of ease of use as a predictor. Evidence was also found of the weak significance of ease of use towards predicting re-purchase intention (Jang and Noh 2011). Lee and Lehto' (2013) study to determine the intention to use YouTube for learning among students affirms the weak role of the ease of use as a predictor.

However, user uncertainty avoidance is revealed to be conclusive, supporting the hypotheses that UUA negatively influences MWT adoption intention, thereby providing a vital insight into the role of cultural and social influence towards the study of technology acceptance, as evidence by the findings by previous studies (Ashraf et al. 2014; Matusitz and Musambira 2013; Lin 2014; Yoo and Huang 2011). Strong correlations were revealed between USP and SIQ towards UUA, suggesting that UUA is intricately influenced by USP and SIQ even though USP and SIQ do not directly influence MWT adoption.

7.6 Conclusion and Future Works

Lecture classes with large number of students suffer from low level of participation and interaction among students and lecturers. Contributing factors are the layout design of lecture halls, i.e. vast hall size and close seating arrangements (Geske 1992), students individual traits and background, i.e. shyness (Stowell et al. 2010), language barriers (Krause 2005), cultural background (Beekes 2006; Van Dijk et al. 2001), and time constraints which limit lecturers opportunity to encourage interactivity in order to complete the required syllabus (Allen and Tanner 2005). In order to overcome these barriers preventing interactivity in large lecture classes, it is vital to exploit the numerous functionalities and advantages that MWT have to offer.

Widespread use of mobile applications on increasingly sophisticated mobile devices is fundamental across all walks of life. The effect across the higher education landscape is enormous. The purpose of proposing a technology acceptance framework for interactive lectures via MWT is to explore, understand and predict factors that influence students' and lecturers' acceptance and use of MWT during large lecture classes in order to increase the level of interactivity and participation of the students with their lecturers.

Existing validated factors such as ease of use, usefulness, enjoyment and self-efficacy may no longer be distinct entities, particularly in the area of mobile technology and mobile learning and thus warrants further investigations. Factors of cultural influence, students' background, personalities or attitudes are growing in prominence as antecedents of technology acceptance. Though the regression analysis revealed user system perception, and system and information quality as insignificant and non-predictive of MWT adoption for interactive lectures, strong correlation among user system perception, and system and information quality with user uncertainty avoidance suggest that user perceptions and quality of the system functionalities and data output are pivotal in the area of mobile technology acceptance.

Therefore, further analyses are justified, and reviews of new acceptances framework or theories pertaining to mobile technology adoption are necessary. Existing technology models such as TAM may no longer be sufficient in areas of mobile technology acceptances among the younger generation. Future works will focus on using advance statistical methods, namely confirmatory path analysis and structural equation modelling to validate and strengthen the framework model factors and proposed hypotheses. Common-source bias analysis should also be conducted in order to address concerns over same-source biases when data are collected via self-reported online survey. As the respondents were sourced from large established higher learning institutions located in urban areas with strong technology infrastructure support, results cannot be generalized as representative of Malaysia's higher education. Sampling of respondents from smaller institutions or institutions located in rural areas should be included in the future. The findings from this study hopes to serve as a catalyst for future researches of mobile technology acceptances in higher education.

Acknowledgments The authors extend their gratitude to University of Malaya (RP028A-14AET) and Multimedia University for supporting the study.

References

Adams, D., Nelson, R., & Todd, P. (1992). Perceived usefulness, ease of use and usage of information technology: A replication. *MIS Quarterly, 16*(2), 227–247.

Allen, D., & Tanner, K. (2005). Infusing active learning into the large-enrollment biology class: Seven strategies, from the simple to complex. *Cell Biology Education, 4*(4), 262–268.

Alzaza, N. S., & Yaakub, A. R. (2011). Students' awareness and requirements of mobile learning services in the higher education environment. *American Journal of Economics and Business Administration, 3*(1), 95–100.

Ashraf, A. R., Thongpapanl, N., & Auh, S. (2014). The application of the technology acceptance model under different cultural contexts: The case of online shopping adoption. *Journal of International Marketing, 22*(3), 68–93.

Ayoun, Baker M., & Moreo, Patrick J. (2008). The influence of the cultural dimension of uncertainty avoidance on business strategy development: A cross-national study of hotel managers. *International Journal of Hospitality Management, 27*(1), 65–75.

Balakrishnan, V., & Gan, C. L. (2013). Mobile wireless technology and its use in lecture room environment: An observation in Malaysian Institutes of higher learning. *International Journal of Information and Education Technology, 3*(6), 635–637.

Bandura, Albert. (1977). Self-efficacy: Toward a unifying theory of behavioral change. *Psychological Review, 84*(2), 191–215.

Bandura, Albert. (2001). Social cognitive theory: An agentic perspective. *Annual Review of Psychology, 52*, 1–26.

Beekes, W. (2006). The 'Millionaire' method for encouraging participation. *Active Learning in Higher Education, 7*(1), 25–36.

Calisir, F., Altin Gumussoy, C., Bayraktaroglu, A. E., & Karaali, D. (2014). Predicting the intention to use a web-based learning system: Perceived content quality, anxiety, perceived system quality, image, and the technology acceptance model. *Human Factors and Ergonomics in Manufacturing and Service Industries, 24*(5), 515–531.

Chen, T. L., & Lan, Y. L. (2013). Using a personal response system as an in-class assessment tool in the teaching of basic college chemistry. *Australasian Journal of Educational Technology, 29* (1).

Chickering, Arthur W., & Gamson, Zelda F. (1987). Seven principles for good practice in undergraduate education. *New Direction for Teaching and Learning, 47*, 1–7.

Chong, J. L., Chong, A. Y. L., Ooi, K. B., & Lin, B. (2011). An empirical analysis of the adoption of m-learning in Malaysia. *International Journal of Mobile Communications, 9*(1), 1–18.

Compeau, D. R., & Higgins, C. A. (1995). Computer self-efficacy: Development of measure and initial test. *MIS Quarterly, 19*(2), 189–211.

Compeau, D. R., Higgins, C. A., & Huff, S. (1999). Social cognitive theory and individual reactions to computing technology: A longitudinal study. *MIS Quarterly, 23*(2), 145–158.

Davis, F. D. (1989). Perceived usefulness, perceived ease of use, and user acceptance of information technology. *MIS Quarterly, 13*(3), 319–340.

DeLone, W. H., & McLean, E. (2003). The DeLone and McLean model of information systems success: A ten-year update. *Journal of Management Information Systems, 19*(4), 9–30.

Detlor, B., Hupfer, M. E., Ruhi, U., & Zhao, L. (2013). Information quality and community municipal portal use. *Government Information Quarterly, 30*(1), 23–32.

Dobson-Mitchell, S. (2011). Are big classes really a problem? Retrieved November 10, 2014, from http://oncampus.macleans.ca/education/2011/12/16/are-big-classes-really-a-problem/.

Donovan, D., & Loch, B. (2013). Closing the feedback loop: Engaging students in large first-year mathematics test revision sessions using pen-enabled screens. *International Journal of Mathematical Education in Science and Technology, 44*(1), 1–13.

Gan, C. L., & Balakrishnan, V. (2014). Determinants of mobile wireless technology for promoting interactivity in lecture sessions: An empirical analysis. *Journal of Computing in Higher Education, 26*(2), 159–181.

Geske, J. (1992). Overcoming the drawbacks of the large lecture class. *College Teaching, 40*(4), 151–154.

Hendrickson, A. R., Massey, P. D., & Cronan, T. P. (1993). On the test-retest reliability of perceived usefulness and perceived ease of use scale. *MIS Quarterly, 17*(2), 227–230.

Hofstede, G., Hofstede, G. J., & Minkov, M. (2010). *Cultures and organizations: Software of the mind* (3rd ed.): McGraw-Hill.

Huang, Y. M., Liu, C. H., Huang, Y. M., & Yeh, Y. H. (2014). Adopt technology acceptance model to analyze factors influencing students' intention on using a disaster prevention education system. In *Advanced technologies, embedded and multimedia for human-centric computing* (pp. 197–202). Springer Netherlands.

Jang, H. Y., & Noh, M. J. (2011). Customer acceptance of IPTV service quality. *International Journal of Information Management, 31*(6), 582–592.

Krause, K. (2005). Understanding and promoting student engagement in university learning communities. *Paper presented at the Sharing Scholarship in Learning and Teaching: Engaging Students*, James Cook University, Townsville/Cairns, Queensland. http://learningspaces.edu.au/herg/assets/resources/StudengKrause.pdf.

Laver, K., George, S., Ratcliffe, J., & Crotty, M. (2012). Measuring technology self efficacy: Reliability and construct validity of a modified computer self efficacy scale in a clinical rehabilitation setting. *Disability and Rehabilitation, 34*(3), 220–227.

Lee, D. Y., & Lehto, M. R. (2013). User acceptance of YouTube for procedural learning: An extension of the technology acceptance model. *Computers and Education, 61*, 193–208.

Lee, J. A., Garbarino, E., & Lerman, D. (2007). How cultural differences in uncertainty avoidance affect product perceptions. *International Marketing Review, 24*(3), 330–349.

Lin, H. C. (2014). An investigation of the effects of cultural differences on physicians' perceptions of information technology acceptance as they relate to knowledge management systems. *Computers in Human Behavior, 38*, 368–380.

Lin, W. S., & Wang, C. H. (2012). Antecedences to continued intentions of adopting e-learning system in blended learning instruction: A contingency framework based on models of information system success and task-technology fit. *Computers and Education, 58*(1), 88–99.

Mahat, J., Ayub, A. F. M., & Luan, S. (2012). An assessment of students' mobile self-efficacy, readiness and personal innovativeness towards mobile learning in higher education in Malaysia. *Procedia-Social and Behavioral Sciences, 64*, 284–290.

Matusitz, J., & Musambira, G. (2013). Power distance, uncertainty avoidance, and technology: Analyzing Hofstede's dimensions and human development indicators. *Journal of Technology in Human Services, 31*(1), 42–60.

Mohamed, A. H. H., Tawfik, H., Al-Jumeily, D., & Norton, L. (2011, December). MoHTAM: A technology acceptance model for mobile health applications. In *Developments in E-systems Engineering (DeSE), 2011* (pp. 13–18). IEEE.

Oigara, J., & Keengwe, J. (2013). Students' perceptions of clickers as an instructional tool to promote active learning. *Education and Information Technologies, 18*(1), 15–28.

Padilla-Meléndez, A., del Aguila-Obra, A. R., & Garrido-Moreno, A. (2013). Perceived playfulness, gender differences and technology acceptance model in a blended learning scenario. *Computers and Education, 63*, 306–317.

Pai, F. Y., & Huang, K. I. (2011). Applying the technology acceptance model to the introduction of healthcare information systems. *Technological Forecasting and Social Change, 78*(4), 650–660.

Park, E., Baek, S., Ohm, J., & Chang, H. J. (2014). Determinants of player acceptance of mobile social network games: An application of extended technology acceptance model. *Telematics and Informatics, 31*(1), 3–15.

Park, S. Y., Nam, M. W., & Cha, S. B. (2012). University students' behavioral intention to use mobile learning: Evaluating the technology acceptance model. *British Journal of Educational Technology, 43*(4), 592–605.

Pavot, W., Diener, E., Colvin, C. R., & Sandvik, E. (1991). Further validation of the satisfaction with life Scale: Evidence for the cross-method convergence of well-being measures. *Journal of Personality Assessment, 57*(1), 149–161.

Reeves, T. C. (2006). How do you know they are learning? The importance of alignment in higher education. *International Journal of Learning Technology, 2*(4), 294–309.

Scornavacca, E., Huff, S., & Marshall, S. (2009). Mobile phones in the classroom: If you can't beat them, join them. *Communications of the ACM, 52*(4), 142–146.

Scott, W. E., Farh, J., & Podaskoff, P. M. (1988). The effects of "intrinsic" and "extrinsic" reinforcement contingencies on task behavior. *Organizational Behavior and Human Decision Processes, 41*(3), 405–425.

Shroff, R. H., Deneen, C. D., & Ng, E. M. (2011). Analysis of the technology acceptance model in examining students' behavioural intention to use an e-portfolio system. *Australasian Journal of Educational Technology, 27*(4), 600–618.

Stowell, J. R., Oldham, T., & Bennett, D. (2010). Using student response systems ("Clickers") to combat conformity and shyness. *Teaching of Psychology, 37*(2), 135–140.

Subramanian, G. H. (1994). A replication of perceived usefulness and perceived ease of use measurement. *Decision Sciences, 25*(5/6), 863–874.

Šumak, B., HeričKo, M., & Pušnik, M. (2011). A meta-analysis of e-learning technology acceptance: The role of user types and e-learning technology types. *Computers in Human Behavior, 27*(6), 2067–2077.

Tan, C. W., Benbasat, I., & Cenfetelli, R. T. (2013). IT-mediated customer service content and delivery in electronic governments: An empirical investigation of the antecedents of service quality. *MIS Quarterly, 37*(1), 77–109.

Tarhini, A., Hone, K., & Liu, X. (2014). Measuring the moderating effect of gender and age on e-learning acceptance in England: A structural equation modeling approach for an extended technology acceptance model. *Journal of Educational Computing Research, 51*(2), 163–184.

Tesch, F., Coelho, D., & Drozdenko, R. (2011). The relative potency of classroom distracters on student concentration: We have met the enemy and he is us*. *Paper presented at the ASBBS Annual Conference*, Las Vegas.

Van Dijk, L. A., Van Der Berg, G. C., & Keulan, V. (2001). Interactivies lectures in engineering education. *European Journal of Engineering Education, 26*(1), 15–18.

Venema, S., & Lodge, J. M. (2013). Capturing dynamic presentation: Using technology to enhance the chalk and the talk. *Australasian Journal of Educational Technology, 29*(1).

Venkatesh, V. (2000). Determinants of perceived ease of use: Integrating control, intrinsic motivation, and emotion into the technology acceptance model. *Information Systems Research, 11*(4), 342–365.

Venkatesh, V., & Bala, H. (2008). Technology acceptance model 3 and a research agenda on interventions. *Decision Sciences, 39*(2), 273–315.

Venkatesh, V., & Davis, F. D. (2000). A theoretical extension of the technology acceptance model: Four longitudinal field studies. *Management Science, 46*(2), 186–204.

Wang, W. T., & Wang, C. C. (2009). An empirical study of instructor adoption of web-based learning systems. *Computers and Education, 53*(3), 761–774.

Wang, Y. S., Yeh, C. H., & Liao, Y. W. (2013). What drives purchase intention in the context of online content services? The moderating role of ethical self-efficacy for online piracy. *International Journal of Information Management, 33*(1), 199–208.

Wixom, B. H., & Todd, P. A. (2005). A theoretical integration of user satisfaction and technology acceptance. *Information Systems Research, 16*(1), 85–102.

Yoo, S. J., & Huang, W. H. D. (2011). Comparison of Web 2.0 technology acceptance level based on cultural differences. *Educational Technology and Society, 14*(4), 241–252.

Zheng, Y., Zhao, K., & Stylianou, A. (2013). The impacts of information quality and system quality on users' continuance intention in information-exchange virtual communities: An empirical investigation. *Decision Support Systems, 56*, 513–524.

Chapter 8
University Student Conceptions of M-learning in Bangladesh

Md. Shahadat Hossain Khan, Benadjih Oiriddine Abdou and Che Kum Clement

Abstract This article presents emerging results from a phenomenographic study that examines Bangladeshi university students' experience of using mobile devices in their learning. Three students from one renowned university participated in the semi-structured interviews to explore their experiences of m-learning. The findings revealed that university students viewed mobile learning (m-learning) in four qualitatively different ways that were: (i) storing learning materials; (ii) accessing information and knowledge; (iii) effective learning tool; and (iv) effective tool for collaboration. This study is constructed on previous studies of university students' conceptions of learning. However, the focus taken in this research was on the experience of m-learning, as an emerging research area, which revealed new facets of university learning. The findings of this study play a significant role in the faculty development program and have an impact on the teaching and learning practices in university education.

8.1 Introduction

Over the last few decades, most of the developing countries have been trying to introduce Information Communication Technology (ICT) in their education sector (Kafyulilo 2014). As such, ICT, in recent years, has gone on to become one of the most crucial components that determine the basic competence of student learning (Noor-Ul-Amin 2013; Potyrala 2001). This has been possible because of a number of reasons. For example, Hammond (2014) claimed that some of the major reasons

Md. S.H. Khan (✉) · B.O. Abdou · C.K. Clement
Islamic University of Technology, Dhaka, Bangladesh
e-mail: skha8285@iut-dhaka.edu; skha8285@uni.sydney.edu.au

B.O. Abdou
e-mail: obenadjih@gmail.com

C.K. Clement
e-mail: ieres2001@yahoo.com

D. Churchill et al. (eds.), *Mobile Learning Design*, Lecture Notes in Educational Technology, DOI 10.1007/978-981-10-0027-0_8

behind the introduction of ICT in education being promoted in England was because of "the belief that ICT can have an impact on the standards, and provide more vocational relevance in the curriculum and can be a catalyst for curriculum reform" (p. 192). On the other hand, Hammond, Reynolds and Ingram (2011) reported that it offers a number of benefits such as "supporting personalized pathways; monitoring progress; providing for 'anytime, anywhere' learning; enabling independent and collaborative learning and developing new modes of learning" (p. 191). All these reasons have led to the successful introduction of ICT at different levels of education in almost all the countries in the world.

The trend goes well beyond this with the rapid advancement of mobile devices. This new trend has led to the formulation of mobile learning or simply, m-learning (Serin 2012). Considering its importance, the current Government of Bangladesh has already introduced several initiatives to integrate different forms of ICT (*mobile devices for teaching and learning is one of the emerging areas of ICT*) in both higher education and secondary education (Karim 2010). Moreover, the present Government realized that ICT is the key element to eradicate poverty from the society, so is taking steps by integrating ICT in education. Due to this, the Government has introduced a charter for change in the form of a long-term development strategy called "Digital Bangladesh". In order to make this vision a success, universities have been deemed to be one of the pivotal areas in which different forms of ICT, for instance mobile learning, could be put to full effect. This emerging area is predicted to contribute enormously, particularly for producing technologically rich manpower that could meet industrial requirement. This mass of future graduates can also work in a technology-integrated environment for the development of their country. It is important to point out that these benefits will show themselves only when the university students will start using mobile devices effectively in their everyday learning process. To support this claim, research has discovered that technology alone cannot lead students to learn (Koehler and Mishra 2005). Therefore, research needs to be conducted on how the students could use mobile devices effectively in their learning. Considering this emerging demand, the present study is proposed to investigate the experiences of university students on applying mobile devices in their learning.

8.2 M-Learning and Related Literature to the Research Problem

As m-learning is a recent concept in student learning, at first we attempt to clarify the term along with its possible benefits from previous literature. According to Park, Nam, and Cha (2012) m-learning can be claimed to be "any educational provision where the sole or dominant technologies are handheld or palmtop devices" (p. 592). These devices can facilitate learning at anytime and anywhere (Ozdamli and Uzunboylu 2014; Serin 2012). In this study, m-learning is considered as a learning platform by using portable (mobile) devices such as cell phones, smartphones, palmtops, tablets, and portable multimedia players. The use of these devices in learning has eased the geographical barriers that existed among students as well as provided a learning environment that is collaborative among different groups of students (Ozdamli and Cavus 2011).

The advantages in using mobile devices in student learning are quite enormous. The biggest advantages of using mobile devices are to provide student-oriented teaching and learning contexts where the learning of the students generally depends on their active involvement, and where teachers are generally seen as a facilitator. For instance, Sha, Looi, Chen, Seow and Wong (2012) claimed that the collaborative environment afforded to them by m-learning enables them to learn at their own pace. It is considered as playing a vital role in simulating critical and logical thinking from the students. Thus, the use of mobile devices provides myriad ways of offering student learning opportunities in university-level education.

Considering the huge emerging benefits, the popularity of using mobile devices among students in developed and developing countries has been growing quite exponentially. This has emerged because of the fact that most of the students have owned these mobile devices in recent times. At this point, Bangladesh, despite being a developing country, is following the same notion. It is found that majority of university students in Bangladesh are in possession of their own mobile phones or other forms of mobile devices simply due to technology being easily available. The other reason for taking in mobile devices is to reduce the cost of Internet and to compare it to that of the last few years. All these aspects bring a huge opportunity to use mobile devices in Bangladeshi universities.

In addition to that, it is realized that there has been very little research in the world (including Bangladesh) that explores the students' experiences of using mobile devices in their learning. The majority of the previous research on m-learning mainly focused on identifying different factors of using mobile devices in education (Al-Fahad 2009; Park et al. 2012); how mobile devices can facilitate student learning (Rogers et al. 2010) or the students' perceptions on mobile learning (Hashim et al. 2014; Kafyulilo 2014) that were based on either a mixture of quantitative and qualitative paradigm or other methodologies including surveys. However, none was found to have been conducted with the use of phenomenography as their theoretical and methodological perspectives. Whereas, a significant number of prior research investigated students' conceptions of learning by using phenomenography and subsequent studies showed evidence for contributing improvement of student learning (Duarte 2007; Eklund-Myrskog 1998; Ellis et al. 2008; Vermunt and Vermetten 2004; Virtanen and Lindblom-Ylänne 2010). Considering this gap (theoretical and methodological), it is urgent to conduct research on students' experiences of learning through these devices. These experiences will be crucial for formulating instructional strategies (pedagogy) that will assist the teachers in properly facilitating the teaching and learning practices. In order to fill up this emerging gap, the main purpose of this study was to identify the qualitatively different ways of experiencing the role of mobile devices in the

Bangladesh University students' learning practices. In order to achieve this purpose, the following research question was used to guide the study:

- What are the qualitatively different ways in which university students understand the role of mobile devices in their learning practices?

8.3 Methodology

This study was qualitative based and was carried out using qualitative research methodology. It was conducted using phenomenography as its theoretical and methodological framework. Phenomenography is a research methodology that is used to qualitatively differentiate ways in which different people experience, understand, and conceive a phenomena (Marton 1981). The main purpose of phenomenography is the description of the various experiences and conceptions that people have for a specific phenomenon (Khan 2014; Marton and Booth 1997). Phenomenographically, a conception is considered to be the way in which one is seeing or understanding something, or in other words, comprehending the exact meaning of something to a specific individual (Sin 2010). In this context, it can therefore be said that conceptions are always expected to be different when various people are involved. Therefore, phenomenography was used in this study to identify the different ways of students' conceptions of m-learning in university education. The final outcomes of this research were revealed as the "categories of description."

8.3.1 Sample

Each student who was selected for this study, was considered to have experience of using any mobile or handheld devices like smartphones, tablets, iPads, and iPod in their learning for at least 6 months. It was required to have the minimum level of experience toward the phenomenon and creating variations (getting participants' in-depth awareness) while taking the interviews. However, the degree of experiences among different respondents and the type of handheld devices they use were not necessarily the same and were tolerated to vary from one respondent to another. In total, a sample of three students from the University of Bangladesh were recruited by using purposive sampling technique. The main characteristics of the students who participated were:

- *Disciplines*: students were selected from two disciplines, one from electrical and two from computer science.
- *Institutions*: students were invited from an engineering university.
- *Study level*: two from postgraduation and one from undergraduation.

- *Experience of m-learning*: 2–4 years.
- *Language*: fluent in English.
- *Gender*: three male students.

8.3.2 Data Collection

In this study, the major tool that was used for collecting data was the phenomenographic interviews (Åkerlind 2005; Barnard et al. 1999; Bruce et al. 2004; Harris 2011; Limbu and Markauskaite 2015). In the method of investigating the students' conceptions on m-learning in university education, interviewees were asked to share their reflections on the role of mobile devices in their learning as well as how these devices could be useful in their learning. A semi-structured in-depth interview protocol was used to gather data and each interview lasted for about 40–50 min. Initially, the participants were asked about "what aspect," for instance, *what does m-learning mean to you*? In order to get a much deeper understanding, the follow up questions were asked. For example, *could you explain this further*?

8.3.3 Data Analysis

The interviews were recorded by an audio recorder and each interview was listened several times (Åkerlind 2005; González 2009; Limbu and Markauskaite 2015). The audio-recorded data was transcribed verbatim. This process was then followed by reading the transcripts many times in order to get a deep insight of the various experiences received from the participants (Åkerlind 2005; Limbu and Markauskaite 2015). At this stage, similarities and differences from each transcript were recognized and later followed the preliminary categories, which was then checked with transcripts. The final outcome spaces were confirmed based on back and forth discussions with the research members. No category was identified without supporting the quotations from the transcripts.

8.4 Results

The results revealed four different categories of description:

- Category A: storing learning materials
- Category B: accessing information and knowledge
- Category C: effective learning tool
- Category D: effective tool for collaboration

The detailed elaboration of each of those categories is followed by the most appropriate quotations obtained from the interview transcripts. Some identification numbers were used at the end of every quotation to help the researchers keep track of the ones that have been used and to keep the interviewees anonymous in the study.

8.4.1 Categories of Description

8.4.1.1 Category A: Storing Learning Materials

In Category A, mobile learning is viewed as a way of getting various learning materials from different sources and storing them in these handheld devices for further use as required. In this way, students will be able to get their learning materials from different sources and store them in their mobile and later they can access them. For instance, if a teacher gives a lecture using PowerPoint presentations, students can easily download those presentations from the sharing device (teacher usually uploads that presentation for the students) by using their mobile devices and save them. In that way, they have wider scope for keeping learning materials safe. With reference to this argument, some participants stated that:

> May be I came in a little bit late, but my friends took notes and I don't have that much time to copy and write everything, so I just get my phone, take a snap of the notes and then when I go back to my room … Then I just read them direct. [B3]
>
> … You will get everything like PDF that you can put in your phone, you can even download many books in your phone and pictures also. [B2]

Besides in this category, a mobile device is also seen as a recoding tool for future learning. For example, the participating students mentioned that with their mobile devices, they can record the lecture live during classes so that the teachers' explanations will be used later during their free time:

> You can even make records. You can record the lectures … [B2]

8.4.1.2 Category B: Accessing Information and Knowledge

Category B represents the view that mobile devices facilitate the access to information and knowledge that are important in their learning. First, this perceived ease of access to information was expressed in various ways. For example, the use of free online and offline dictionaries that may have been installed in the phone, as expressed:

> I installed a dictionary application. In case I get a word that I don't understand, I use the dictionary on my mobile phone then I can know the meaning of that word. [B1]

Alternatively, participants also discussed that they can get the meaning directly from the Internet in case they get a terminology that is new or ambiguous for them:

> I can access the internet like google search in case I get a terminology that I don't understand. I can search the internet and use it… [B1]

Second, in this category, the mobile device is perceived as a way to access knowledge. It is seen that students use different search tools such as Google, Google scholar, and the likes by using their mobile devices to gain related knowledge that provide them more explanations and clarifications about a specific topic on the Internet.

> Then maybe another thing mobile devices, there are a variety and a vast number of apps, educational apps. So I can just go to google play, search and then I can get a very long list… [B3]

8.4.1.3 Category C: Effective Learning Tool

In this category, learning with the help of mobile devices is perceived as an effective learning tool. This effectiveness is perceived mainly through criteria such as time-saving, cost as well as mobility. M-learning allows students to access a vast variety of information and knowledge within the shortest time possible. It is viewed that learning is much quicker in m-learning than it could be in the traditional or other learning methods:

> There, I will be wasting time writing everything down. But I just go direct, read, understand then memorize. So it saves some time while revising. [B3]
>
> If I just take a snap, it will take like a second but if my colleague decides to draw it in his book, it will take him like 20 min. So in such a way, it saves time to me. [B3]

In this category, m-learning is also perceived to be cost effective. Although it involves an initial cost to buy a mobile device but in the long run it saves students' money:

> Then another thing [is that], it saves money. In which way? For example if a teachers gives us a slide which has like 56 pages, it means if I print it will be costly. But if I just copy the slide to my phone, I think in that way, it saves me some TAKA [Bangladeshi Currency]. [B3]

Additionally, this category viewed mobile device as a means of mobility in student learning. University students in this category perceived m-learning as the learning that occurs anytime and anywhere that students want. For example, the participants stated that in most cases, they can move with these devices anywhere they go. It enables them to access to whatever they want to learn at their convenient time.

> It always depends but the major point is that it's mobile. The mobility aspect. It's like wherever I go I have my mobile phone… [B3]
>
> Mobile learning, I understand it by using some devices which you hold in your hands and can have access to it anywhere and anytime for your use in learning. [B1]

In brief, in this category the use of mobile devices is seen as a time-saving cost effective and portable devices for enhancing student learning in university.

8.4.1.4 Category D: Effective Tool for Collaboration

In this category, m-learning is viewed as an essential means for collaboration. For example, phone calls from mobile phone or Skype could be used for direct communication, text message from mobile or email could be used for sending information to enhance their learning. Some of the participants mentioned that their mobile devices enable them to communicate with their teachers, supervisors, colleagues as well as senior students in case they are in need of some assistance.

> During that time, our teacher was not in the campus. Even he was not in Bangladesh. He gave us his Skype and I used one time to ask him one question.... I practice most of the problems, I got some difficulties. So I sent a message to the teacher through Skype, he answered me and I got the answered, I practiced and it worked. [B2]
>
> Also having communication with the teacher because I can easily consult the teacher through the email for more clarification. [B1]

In addition to that, Category D presents another understanding of using mobile devices in students' learning which is direct (synchronous) and indirect (asynchronous) collaboration among student and teacher and/or student and student. In this point, students are seen to use different social media such as Facebook, LinkedIn, and WhatsApp for stated collaboration. For instance, university students in many cases used Facebook for collaborating with their supervisors when they face any difficulties with regards to their projects, theses and so on.

> For example in this semester, I have a supervisor for my thesis. So in case I have a query and he is not around, I just log into Facebook then I ask him via Facebook then he replies. [B3]
>
> Collaboration is also seen while students work in a group. Students generally use their mobile devices to get in touch with their colleagues (peer groups) to complete their group works such as assignments, solving problems, group discussion. One of the participant stated this in his response:
>
> ... But remember you have to work on the assignment in time. So I may do something, maybe my part, first of all maybe we can divide the assignment. So I do my part, maybe go to Facebook, send him what I have done, when he is at home. When he reads through he also maybe sends me *his. So by the time he comes back to school...* [B3]

8.5 Discussion and Conclusions

Before discussing the results, we would like to state the limitations of this study. The participants were recruited from one university in Bangladesh and were small in number. However, a sample of three is not an unusual practice in phenomenographic research approach. For example, Forster (2013) interviewed three professionals from nursing practice about their conceptions of information literacy. Moreover, the results depend on the setting or the context of each study; therefore, these results may not be generalizable for other contexts. However, the aim of phenomenographic research approach is not to provide generalizable results rather its focus is on a particular phenomenon that needs to be investigated deeply.

Turning to discussion, the findings are limited in scope in relation to previous phenomenographic studies, because students' experiences of m-learning are a new area of investigation. However, the results of this study could be interpreted in a wider context. The results revealed four qualitatively different ways of seeing mobile devices in student learning: storing learning materials; accessing information and knowledge; effective learning tool; and effective tool for collaboration. The four categories are placed from lower level to higher level understanding. Therefore, the four categories are broadly divided into two orientations: *fragmented orientation* (Category A and B) in which the mobile devices are considered as a way to store and access information in student learning. Students do not consider mobile devices for constructing their knowledge or solving their problem or engaging collaborative learning. It mainly focuses on students' surface level of learning. In contrast, *cohesive orientation* (Category C and D), in which the mobile devices are viewed as a means to develop students' understanding, to construct their own knowledge, and to engage them in collaborative learning. It is mainly involved with deep level of learning. These findings are broadly consistent in previous phenomenographic studies (Biggs and Tang 2011; Eklund-Myrskog 1998; Ellis et al. 2008; Ellis et al. 2006; Lucas 2001). Generally, these studies reported students' conceptions of learning in different context and were broadly placed into deep and surface level of learning. Nevertheless, the results provide emerging conceptions of m-learning.

As m-learning becomes a growing concern in the teaching and learning practice of a developing country, the role of using mobile in student learning is becoming a major focus of research initiatives (Kafyulilo 2014; Rogers et al. 2010). It is suggested then that the findings of this study could be used to inform these initiatives, as this study provides a second order experience (the findings derived from participants who had experiences of m-learning) of the investigated phenomenon. In recognition of the significance of these findings, this research provides different ways of using mobile devices in student learning, which is a potential input for improving teaching practice. For example, it may help teachers to create different teaching approaches that will match students' learning approaches, which will guide university students to make maximum use of mobile devices in their learning. The emerging results also contribute the improvement of professional development program. In addition to that, policy makers and curriculum developers could get empirical evidence about students' experiences so that they can develop a curriculum that will encourage and promote the use of mobile devices in the university education. Previous research reported that students' conceptions of learning are linked with teachers' conceptions of teaching (Prosser and Trigwell 1999). Therefore, future research is proposed to investigate Bangladesh University teachers' conceptions of m-learning. The main aim of conducting such future study is to find out the relationships between students' conceptions of m-learning and teachers' conceptions of m-learning. It is important to acknowledge that our study reports a preliminary exploration of using mobile devices in student learning, thereby suggesting a future investigation with a broader sample from more than one university. It is also suggested to explore its analysis in different dimensions to understand the investigated phenomenon in a more conclusive manner.

Acknowledgments The authors wish to thank Bob Fox, Professor, School of Education, UNSW for his useful suggestions.

References

Åkerlind, G. S. (2005). Learning about phenomenography: interviewing, data analysis and the qualitative research paradigm. In J. Bowden & P. Green (Eds.), *Doing developmental phenomenography* (pp. 63–74). Melbourne: RMIT University Press.

Al-Fahad, F. N. (2009). *Students' attitudes and perceptions towards the effectiveness of mobile learning in King Saud University, Saudi Arabia*. Online Submission, *8*(2).

Barnard, A., McCosker, H., & Gerber, R. (1999). Phenomenography: A qualitative research approach for exploring understanding in health care. *Qualitative Health Research, 9*(2), 212–226.

Biggs, J., & Tang, C. (2011). *Teaching for quality learning at university: What the student does*. UK: McGraw-Hill Education.

Bruce, C., Buckingham, L., Hynd, J., McMahon, C., Roggenkamp, M., & Stoodley, I. (2004). Ways of experiencing the act of learning to program: A phenomenographic study of introductory programming students at university. *Journal of Information Technology Education: Research, 3*(1), 145–160.

Duarte, A. (2007). Conceptions of learning and approaches to learning in Portuguese students. *Higher Education, 54*(6), 781–794. doi:10.1007/s10734-006-9023-7.

Eklund-Myrskog, G. (1998). Students' conceptions of learning in different educational contexts. *Higher Education, 35*(3), 299–316. doi:10.1023/a:1003145613005.

Ellis, R. A., Goodyear, P., Prosser, M., & O'Hara, A. (2006). How and what university students learn through online and face-to-face discussion: conceptions, intentions and approaches. *Journal of Computer Assisted learning, 22*(4), 244–256. doi:10.1111/j.1365-2729.2006.00173.x.

Ellis, R. A., Goodyear, P., Calvo, R. A., & Prosser, M. (2008). Engineering students' conceptions of and approaches to learning through discussions in face-to-face and online contexts. *Learning and Instruction, 18*(3), 267–282.

Forster, M. (2013). A phenomenographic investigation into information literacy in nursing practice —Preliminary findings and methodological issues. *Nurse Education Today, 33*(10), 1237–1241.

González, C. (2009). Conceptions of, and approaches to, teaching online: a study of lecturers teaching postgraduate distance courses. *Higher Education, 57*(3), 299–314. doi:10.1007/s10734-008-9145-1.

Hammond, M. (2014). Introducing ICT in schools in England: Rationale and consequences. *British Journal of Educational Technology, 45*(2), 191–201.

Hammond, M., Reynolds, L., & Ingram, J. (2011). How and why do student teachers use ICT? *Journal of Computer Assisted learning, 27*(3), 191–203.

Harris, L. R. (2011). Phenomenographic perspectives on the structure of conceptions: The origins, purposes, strengths, and limitations of the what/how and referential/structural frameworks. *Educational Research Review, 6*(2), 109–124. doi:10.1016/j.edurev.2011.01.002.

Hashim, K. F., Tan, F. B., & Rashid, A. (2014). Adult learners' intention to adopt mobile learning: A motivational perspective. *British Journal of Educational Technology, 46*(2). doi: 10.1111/bjet.12148.

Kafyulilo, A. (2014). Access, use and perceptions of teachers and students towards mobile phones as a tool for teaching and learning in Tanzania. *Education and Information Technologies, 19* (1), 115–127.

Karim, M. A. (2010). *Digital Bangladesh for Good governance*. Paper presented at the Bangladesh Development Forum 2010, Bangabandhu International Conference Centre. Retrieved from http://www.lcgbangladesh.org/bdf-2010/BG_Paper/BDF2010_SessionVI.pdf.

Khan, S. H. (2014). Phenomenography: a qualitative research methodology in Bangladesh. *International Journal on new trends in education and their implication, 5*(2), 34–43.

Koehler, M. J., & Mishra, P. (2005). What happens when teachers design educational technology? The development of technological pedagogical content knowledge. *Journal of Educational Computing Research, 32*(2), 131–152.

Limbu, L., & Markauskaite, L. (2015). How do learners experience joint writing: University students' conceptions of online collaborative writing tasks and environments. *Computers & Education, 82*, 393–408.

Lucas, U. (2001). Deep and surface approaches to learning within introductory accounting: A phenomenographic study. *Accounting Education, 10*(2), 161–184.

Marton, F. (1981). Phenomenography-describing conceptions of the world around us. *Instructional Science, 10*(2), 177–200.

Marton, F., & Booth, S. (1997). *Learning and awareness*. New Jersey: Lawrence Erlbaum Associates.

Noor-Ul-Amin, S. (2013). An effective use of ICT for education and learning by drawing on worldwide knowledge, research, and experience: ICT as a change agent for education. *Scholarly Journal of Education, 2*(4), 38–45.

Ozdamli, F., & Cavus, N. (2011). Basic elements and characteristics of mobile learning. *Procedia-Social and Behavioral Sciences, 28*, 937–942.

Ozdamli, F., & Uzunboylu, H. (2014). M-learning adequacy and perceptions of students and teachers in secondary schools. *British Journal of Educational Technology.*

Park, S. Y., Nam, M. W., & Cha, S. B. (2012). University students' behavioral intention to use mobile learning: Evaluating the technology acceptance model. *British Journal of Educational Technology, 43*(4), 592–605.

Potyrala, K. (2001). *ICT tools in biology education.*

Prosser, M., & Trigwell, K. (1999). *Understanding learning and teaching: The experience in higher education*. Buckingham: Society for Research into Higher Education & Open University Press.

Rogers, Y., Connelly, K., Hazlewood, W., & Tedesco, L. (2010). Enhancing learning: a study of how mobile devices can facilitate sensemaking. *Personal and Ubiquitous Computing, 14*(2), 111–124.

Serin, O. (2012). Mobile learning perceptions of the prospective teachers (Turkish Republic of Northern Cyprus sampling). *Turkish Online Journal of Educational Technology-TOJET, 11*(3), 222–233.

Sha, L., Looi, C.-K., Chen, W., Seow, P., & Wong, L.-H. (2012). Recognizing and measuring self-regulated learning in a mobile learning environment. *Computers in Human Behavior, 28* (2), 718–728.

Sin, S. (2010). Considerations of quality in phenomenographic research. *International Journal of Qualitative Methods, 9*(4), 305–319.

Vermunt, J., & Vermetten, Y. (2004). Patterns in student learning: relationships between learning strategies, conceptions of learning, and learning orientations. *Educational Psychology Review, 16*(4), 359–384. doi:10.1007/s10648-004-0005-y.

Virtanen, V., & Lindblom-Ylänne, S. (2010). University students' and teachers' conceptions of teaching and learning in the biosciences. *Instructional Science, 38*(4), 355–370. doi:10.1007/s11251-008-9088-z.

Author Biographies

Md. Shahadat Hossain Khan completed his Ph.D. from the University of Sydney, Australia. He has been working as an assistant professor of the department of Technical and Vocational Education (TVE), at the Islamic University of Technology (IUT), Bangladesh, since 2006. He has a wide experience in ICT-enhanced teaching and learning at national and international levels. He has wider expertise on TPCK Framework, m-learning in tertiary level, professional development particularly focus on scholarship in teaching (student-centered teaching, ICT-integration). Currently he is researching on Open Source Technology (OST) to integrate it into teaching and learning context in higher education.

Benadjih Oiriddine Abdou has completed Masters of Science in Technical Education (MScTE), Islamic University of Technology (IUT), Dhaka, Bangladesh. He has been working as a Teaching assistant of the department of Technical and Vocational Education (TVE), at the Islamic University of Technology (IUT), Dhaka Bangladesh, since June 2015. He is currently working on widening his knowledge on phenomenography and m-learning in university education.

Che Kum Clement Ph.D. is Chairman of the Department of Technical and Vocational Education, Islamic University of Technology (IUT) Board Bazar, Gazipur, Bangladesh, a subsidiary organ of the Organization of Islamic Cooperation comprising 57 countries. Before joining IUT he has been working for the Cameroon's Ministry of National Education as senior management staff in charge of Vocational schools' management at the Directorate of Secondary Technical and Vocational Education, Republic of Cameroon. Che Kum's research interest is on Curriculum development in TVE, ICT in teaching-learning, competency based TVE and higher education and social change.

Chapter 9
Mobile Learning, Student Concerns and Attitudes

Zoran Putnik

Abstract In this paper, examples of the current methods of employment in mobile learning, both in schools and in independent projects is presented and commented. After that, statistical data collected in a survey of students is given, concerning student's opinions about the use of mobile services in mLearning, reasons for such a situation with, and attitudes about the future possible use. This data is accompanied with the ideas about the proper method of application of mobile learning not only in formal environment in schools, but also in an informal environment. Suggestion of this paper is to develop and organize mLearning as an addition, enhancement and supplement to both classic classroom and already accepted "classic" eLearning, because of ample additional possibilities for use of mobile technology in any type of education.

9.1 Introduction

Possible changes in learning theories, due to the developments in information and communication technologies, attract globally increasing interest internationally. An intention to meet the needs of knowledge-based economy and society, and provide high-quality living, initiated adjustments of technology-based learning—from computer assisted, through computer-based education, then over web-based learning, finally touching elements of mobile learning.

Reason for this situation is quite obvious. Declaration that "Today's exciting opportunities require innovative thinking, practical know-how and tremendous depth and breadth of expertise" in SRI International (2013), describes in plain words that the two main drivers for such transformation, which are not a surprise, are: demand and supply.

Z. Putnik (✉)
The University of Novi Sad, Novi Sad, Serbia
e-mail: putnik@dmi.uns.ac.rs

D. Churchill et al. (eds.), *Mobile Learning Design*, Lecture Notes in Educational Technology, DOI 10.1007/978-981-10-0027-0_9

Demand because of rapid obsolescence of knowledge, need for a specific piece of information and understanding at a specific moment, desire to satisfy learning needs of distributed employees in a profitable way, and enable anytime access to lifelong learning. Similarly, supply since advances in the digital field enables construction of multimedia and interactive learning objects, and since Internet access, both wireless and mobile, became standard at work and at home, because technology and educational standards became more able to facilitate production of reusable and compatible learning resources.

An easy conclusion comes from this. Knowledge-based society has new requirements which finally could be gratified for both general education, and on-time and in-place training. We are finally able to create settings for well-informed and flexible people, erudite ready to be continuously (re)trained and (re)educated. Not only in order to stay competitive as a workforce, but also for their personal improvement, fulfilment and satisfaction. And, as noted in Homan and Wood (2003), "… with wireless phones and handheld devices, the relationship between the device and its owner becomes one-to-one, always on, always there, location aware and personalized"—exactly what we need for just-in-time education. In addition to that—why stop only at education? As said in Traxler (2007), "… we have to recognize that mobile, personal, and wireless devices are now radically transforming societal notions of discourse and knowledge, and are responsible for new forms of art, employment, language, commerce, deprivation, and crime, as well as learning".

While most of the current papers discussing points about mobile learning try to compare it with some other form of learning, or discuss technological issues of it, this paper has different intention. By showing collected answers to a survey about current and possible future use of mLearning at the Department of Mathematics and Informatics, Faculty of Science, University of Novi Sad, Serbia, our idea is to show the possibilities to use mLearning as an addition to all other forms of teaching and learning, starting from classic classroom teaching, and going over well established practice of eLearning at our Department that is in much more detail explained in Putnik et al. (2014), Zdravkova et al. (2012), and Ivanović et al. (2010). In a wider sense, the idea is to let students select time, place and amount of learning, using technology that comes so naturally to them who are digital natives—mobile technology.

9.2 State of the Art

Common contemporary learning model includes ideas of learning content management systems (LCMSs). Reusable learning objects, digital learning activities and available educational software, all of those are ready, created by subject experts and instructional designers, with the help of interested learners fascinated to get involved in this field, as discussed in Putnik (2014). Created repositories of learning objects are available to be accessed and searched, to allow users to share and retrieve needed learning resources, according to their individual learning objectives. However, such organization is still rather traditional in its nature. Interested learners access common repository of learning objects, with the purpose of acquiring "body of knowledge", created so that it can be assessed later. What is missing here are situation and circumstances, conversation, dialogue and exchange of opinions.

Such or similar claim is repeated in many research papers. For example, in Homan and Wood (2003) authors say that "With increased popular access to information and knowledge anywhere, anytime, the role of education, perhaps especially formal education, is challenged and the relationship between education, society, and technology are now more dynamic than ever". Similarly, in Vavoula and Karagiannidis (2005) authors claim "The emergence of the knowledge society poses new requirements for education and training: the knowledge-based economy requires a flexible, very well-trained workforce; and the citizens of the information society need to be continuously (re)trained in order to remain competitive…"

There are a lot of definitions of mobile learning, but even the most informal one will satisfy a point we find important to make. A large line of authors describe mobile learning as learning that takes place using mobile and wireless devices, such as personal digital assistants, tablet computers, or smart mobile phones. Such a definition makes a distinction between mobile and other forms of eLearning considering specific type of equipment used. However, in our opinion, mobile learning should be defined with the emphasis on something else, allowing for mobile learning from the point of view of learner! Mobility in such a sense enables learning to take place everywhere and in any situation. For example, doctors can recheck their medical knowledge while doing the hospital rounds, computer experts can update their expertise while fixing a software bug, person can improve its' language skill while travelling abroad. Of course, as claimed in Motiwalla (2007) among other papers, after reading and researching into attempts of mLearning usage "… we know from these studies that mLearning approach must complement an existing learning environment, developers must understand limitations of mobile devices and use them for appropriate learning pedagogies…"

Generally speaking, the idea is not new—serious projects and evaluations of mobile learning are performed for more than a decade now. What follows are some of the available results. In an introductory survey course in sociology described in McConatha and Praul (2007), an opportunity to use mLearning product developed by HotLava Software was offered. About 40 % of the students selected to use this product, and access data via their personal devices. Their responses were collected and analyzed, and their performance was compared to the outcomes of those students who chose not to use mLearning tool. The conclusion was that students using mLearning software "… demonstrated a higher level of knowledge on the subject matter covered in the course, when compared to students choosing not to use the tools".

Another example is given in Zanela Saccol et al. (2010), where a more specific question was investigated. The actual possibilities of mLearning were researched for the development of individual competences and for collaboration in the organizational setting. After developing a mobile virtual learning environment called

COMTEXT, the analysis was conducted about the training of IT professionals. This analysis did not give such a complimenting result for mLearning. Or, to quote the paper's conclusion "… learners showed interest and excitement for the innovation characteristic of mLearning … However, excitement turns into frustration when mobile and wireless technological limitations are faced, as well as the mobile device ergonomic limitations". Since in our research, IT professionals are in question, the result is of a larger interest, because we are also dealing with (future) IT professionals, students of Computer Science at our Department.

In Fayyoumi et al. (2013) a situation and students satisfaction with mLearning was studied in Arab countries. The results are very favourable, which probably might be to a certain level connected with the facts that the research is (a) very recent, which covers use of even smarter smart phones, and (b) conducted in rich countries, which ensures that students do have necessary technical equipment. Either way, data obtained within a survey shows that 70 % of the students agreed that "learning skills are enhanced through mLearning", 60 % claimed that "mobile examination is useful", while fascinating 96 % of students agreed that "mLearning is very useful and is very helpful for those students who live at remote areas and cannot attend the university daily". This last fact should not be connected only to "living at remote areas", since mLearning has the same effect on students who are employed, thus being limited by time and cannot attend university lectures because of other obligations.

Finally, visiting the other part of the world through (Organista-Sandoval and Serrano-Santoyo 2014), we can read about the situation and opinions on mLearning in Mexico. Being the most recent study, it is not a big surprise to find that about 97 % of teachers and students have "some kind of cell phone or smart-phone", and that about one of every four interactions with the mobile device has a concrete educational purpose. While authors complain that "… in general the educational use of the cell phone is mainly aimed to establish communication between the students and to access information via Internet", we find this fact exactly in line with our ideas!

Smart phones and ability to access Internet at all times, which in turn includes access to teaching and learning resources, is what gives mLearning its greatest value. This, even without creation of any particular mLearning educational systems, or repositories of obligatory teaching materials, facilitate the most of necessary environment for just-in-time and just-in-place learning. At the same time, presented examples of projects using advantages and possibilities of mobile technologies show that mobile learning is, for some time now, able to take a step forward from experimental pilot projects, towards institutionalized implementations. An example of such use of mLearning is given in Corbeil and Valdes-Corbeil (2007) where it is been said "Instructional uses: Students can download audio and video lectures and podcasts to their smart phones. They can play audio, video, and Flash movies; display and edit text documents; access e-mail and Web contents; send IM and text messages; and use the phone for mass storage".

All of the mentioned examples show, one way or another, that mLearning can be an excellent extension and complement of formal learning and eLearning, particularly in a sense of situation dependent, lifelong, or just-in-time learning. And this is exactly the point we are trying to investigate in this paper and show that even in a lesser economically developed country such as Serbia, student concerns and attitudes towards mobile learning are rather similar to student opinions all over the world.

While considering such a prospect, we must also make a comment on possible problems in application of mobile learning in practice. Pretty much the same inconveniences are mentioned in a lot of papers dealing with mLearning, so we will just try to abstract them from for example Organista-Sandoval and Serrano-Santoyo (2014), Frydenberg (2007), or Litchfield et al. (2007). The first things usually mentioned are:

- Limited resources compared to desktop technologies;
- Problems with compatibility between devices;
- Different operating systems and applications, or simply
- Fast progress and frequent innovations in the development of mobile technologies.

Limited resources, same as with other technologies, are becoming less of a problem with the growth and evolution of mobile devices. Still, while the problem of speed or storage size for example will be or already are overcome, in our opinion, one of the limitations will stay longer, and that is the size of the screen. Even with the advancements in precision and screen resolution of mobile devices, the size itself will for the most of users be everlasting problem. Yet, if we accept that as a fact, work harder on the design of suitable user interface, and not limit ourselves only to mobile technology for learning; this should not be a devastating fact.

Compatibility problem, as time passes, becomes less and less important also. Mobile phone producers are trying to either make machines that are standardized and similar to others, or to enable use of the same/similar operating system, which helps overcoming the issue. Same situation as the one we encountered with desktop technology is on the verge to happen with mobile devices. Device itself is one thing, while the operating system is something else, not bound to the device, left for user to select. This way, combined problems of compatibility and different operating systems and applications help solving each other. Companies that produce games and utilities for mobile devices showed us that it is possible to cover wide range of mobile platforms, so we are sure that this will be the case with mobile learning also.

Finally, we mentioned problem of "fast progress and frequent innovations in the development of mobile technologies". While this can be a challenge and can cause certain difficulties, we should keep in mind that it's not a technology that is in the centre of teaching, but methods. Of course, let us also not forget that "digital natives" can cope with any technological advancement much easier than us "digital immigrants" can, as nicely explained in Prensky (2001). And, since they not only like digital technologies, but use them as an integral part of their lives, if schools do not join, students will feel even greater separation between school and life.

9.3 The Present Study

Going further in our research, besides analyzing general issues, obstacles, and potential of mLearning, we tried to dig deeper into specific situation at our country, and in particular at our institution. According to Wikipedia (2015) Serbia is economically speaking not a highly developed country, suffering consequences of wars, breakdown of the country of Yugoslavia, bombing and economic sanctions, just to mention a few things. Still, before all of the these things, Serbia and Yugoslavia in general, was one of the most developed countries of East Europe with much more developed connections with the Western Europe and USA than any of other so-called "communist" countries. The question we were investigating was are there some traces of that left, and if people of Serbia, and specifically young people, are ready to accept the challenges of technological development in all areas of life and in particular changes and developments in the area of education. We tried to research into the situation considering the mobile technology, phones, notebooks, tablets and similar tools and equipment, but limited ourselves to the situation in education, and students' attitudes and standpoints in this area.

9.3.1 Current Situation in Serbia in General

Traditionally and historically, we can claim that Serbia is the country rather fond of phones. Only 7 years after Alexander Graham Bell patented the telephone in 1876, the first phone conversation was conducted in Serbia in 1883 between the Geography Department of the Ministry of Military affairs and army barracks in charge of engineering in Belgrade (Trninić 2013). Not long after that, in 1899. the public phone traffic was started in Serbia using inductor phone central, while in 1902. a new "Siemens and Halske" central with 1000 new phone numbers was installed in Belgrade, because of the growing needs of interested public.

Checking the "World Factbook", we can also notice that Serbia, country of a little more than seven million inhabitants, uses (data is from 2012) 2.98 million of main lines, and 9.138 million of mobile phones. There are also about 4.107 millions of Internet users (data is from 2009, from CIA "The World Factbook" for 2012.). Data available at "100 People: A World Portrait" mentions that 75 out of 100 people in the world owns a cell phone, and 30 are Internet users. Comparing to situation in Serbia, we can notice that speaking of Internet Serbia is ahead of the world average, while considering the mobile phones it is at the forefront of average by far!

9.3.2 Specific Situation at the Department of Mathematics and Informatics, Novi Sad, Serbia

Before presenting the data and results we collected, let us repeat in a few words a basic idea behind the chapter.

Current research in the area of mLearning usually presents different results about successful use of mLearning at Universities, but also in various informal educational projects. At the same time, we are at the point where some portion of our teaching is delivered in some form reachable by mobile learning facilities, intentionally or not. While originally it means that we are aware that some people will interact and communicate with our teaching resources using mobile devices, further it should lead us towards decision on how to publish and distribute this information and these resources.

Creation of learning activities and teaching material suitable for mobile learning should be governed by the ideas connected and guided to learning, not to technology —same as is the case with the implementation for any other technology-based learning. Use of mobile devices is not the purpose, objective, or sole goal—it is a medium, an instrument to enable activities that otherwise were not possible, to increase usability of those that had drawbacks because of technological reasons, all in all to increase benefits for learners. As a consequence, in our opinion it is definite that the use of mobile technologies is suitable only for the part of learning activities, while other parts are still better supported by some other types of technologies. And, we are satisfied with that, because we confirm to the stand that "learning can't be managed, but can and should be facilitated" (Ivanović et al. 2014).

9.4 Methodology

9.4.1 Instrument

Study is focused on the attitudes and views of undergraduate students about the possibilities of mobile learning. It is a quantitative research, and we developed a short questionnaire to collect the data needed. As one of the common possibilities with this type of study, we decided to use structure of close-ended, Likert scale five-point measure survey. For each question, students were asked to give opinion ranging from 1 = strongly disagree, to 5 = strongly agree. The questionnaire was distributed to students via e-mail, and it took only several minutes to complete, since the questions covered only the basic opinions. This in turn leads to a high response rate.

Survey was conducted on two occasions, with two generations of students of the Department of Mathematics and Informatics, Faculty of Science, University of Novi Sad. We narrowed our survey only to students of the Computer Science direction.

Students were informed that, since data collection went through e-mail, their answers will not be anonymous, so that they are allowed to refuse to answer the survey. Great majority of students declared that they do not mind answering publicly, actually that they wish that their opinion is heard and taken into account. While it may be contrary to some other institutions or countries, this type of behaviour is recognized at our Department earlier, as presented in Ivanović et al. (2013).

Survey took about 2 weeks each year, and first time covered 178 students, while the second time there were 138 students communicated. Not all of the students answered and completed the survey, so altogether we collected 198 surveys for the analysis.

Major descriptive statistics is presented in Table 9.1. The most of the respondents, as can be noticed, have some type of mobile phone, but this time we didn't investigated further into the type of the phone/tablet/e-book they use. Still such a majority of students using mobile phones, together with their opinion about possibilities, with extremely rare persons being strictly against mLearning, shows that there is a large space for improvement of the use of mLearning at our Department.

Considering the age of the respondents, it can be noticed that the most of them are "older" students, students of final year of bachelor studies (3.) final year of diploma studies (4.) or students of master studies. Does this mean that we can take their opinions more seriously, we will not assess.

Also, we must make a comment about "frequency of use" of mobile phones. While we honoured answers of "I don't use mobile phones" type and counted six of those, we noticed that those same persons later answered that they sometimes read they mail using their phone. We interviewed one of them in person and at least with

Table 9.1 Descriptive statistics of the respondents

Item	Options	Number of people (total is 198)	Percentage (%)
Gender	Male	129	65.15
	Female	69	34.85
Mobile device	Phone	198	100
	No phone	0	0
Year of study	1. year	5	2.53
	2. year	11	5.56
	3. year	73	36.87
	4. year	52	26.26
	Master studies	57	28.79
Use frequency	Does not use	6	3.03
	Calls and sms's, rarely other functions	77	38.89
	Calls, sms's, often other functions	75	37.88
	"Everything" available at the Department (calls, sms's, e-mail, instant messaging with LMS, Wiki, Forums, learning resources, access to LMS, contact with lecturers)	40	20.20

him, thing was cleared-up. “I don’t use” refers to his beliefs, to the fact that he does not like mobile phones, mainly for the privacy issues. Still, as a person of a modern age and profession, he is aware that mobile phones are a necessity, he has one, and having it he also uses, unwillingly, the most of its functions.

Of course, this fact throws a slight shadow on our statistics, since some of the numbers do not fit with answers to other questions, but we consider this a normal and usual matter with such a large survey.

Numbers and percentages of students using mobile phones for other functions here probably does not represent only their wishes, but mainly availability of possibilities for use of mobile phones and mobile services of the LMS we employed. The complete statistics will shed much more light on the topic.

Table 9.2 shows the results of students’ answers regarding mobile phones from the viewpoint of their use in education. It can be separated into two groups of questions, the first one dealing with how our students use mobile phones at the faculty and for which (educational) purposes.

The second group of questions tries to recognize the reasons for the situation, and identify the causes and motives for the state of the art at the Department. Finally, the third “group” contains only a single question, and tries to find out whether our students are willing to use mobile phones in their studies more than they do now.

There are 12 items discussed, with the distribution of opinions presented, and the mean score given in order to describe the strength of the item. There are only two mean scores higher than three, showing that only reading e-mail as a service is accepted at the moment. The highest grade after that one gets willingness of our students to use mobile phones more in their education, having a mean value of exactly 3.

The lowest mean score is gained for “I do not use mobile phone” question, showing that our students are accustomed to mobile phones and use them in life outside of the faculty. In our opinion, this suggests that with adding more abilities for clever use of mobile phones for education, there is a chance for introduction of mLearning at our Department.

The next two lowest mean scores are in connection of use of mobile phones for access to forums and wikis available within our LMS, and for communication with lecturers. Both of those low scores are easily explainable. Forums and wikis within LMS are at our Department used for the obligatory assignments. Consequently, that requires reading of posts of other students in forums, and of additions and changes of wikis from other team members. Finally, it requires text typing, sometimes a lot of text, perhaps even addition of some drawings, which is much more difficult through the mobile phones.

The other low mean score is even easier to explain. Communication with the lecturer by mobile phones requires a prerequisite that the lecturer agrees to something like that, which is generally speaking not very likely. Namely, having dozens or even hundreds of students looking for help/opinion/assistance at possible weird hours is a good reason not to agree to that kind of communication.

Table 9.2 Percentage distribution of opinions about mobile phones in education

Item	Strongly disagree (%)	Disagree (%)	Slightly agree (%)	Agree (%)	Strongly agree (%)	Mean
What do I use?						
I do not use mobile phone	180	7	3	2	6	1.22
I use mobile phone only for calls and sms's	71	28	32	31	36	2.66
Sometimes I use mobile phone for reading e-mail	48	18	27	26	78	3.35
I use mobile phone to receive notifications from faculty LMS	99	8	25	25	40	2.49
I use mobile phone to communicate with lecturers	138	24	21	5	10	1.61
I use mobile phone for LMS forums and wiki	140	22	21	4	10	1.59
I use mobile phone to download learning resources from LMS	104	24	22	18	30	2.22
Mobile phone is a usual part of my studies, and I use it for everything	80	34	41	17	23	2.33
Why don't I use it?						
I do not use mobile phone at the faculty because of the high price of Internet access	121	19	30	10	18	1.91
I do not use mobile phone at the faculty because my phone is not good enough	104	24	20	12	36	2.24
I do not use mobile phone at the faculty because that is wrong, and I concentrate better on written material	53	23	44	42	32	2.88
Do I want to?						
I would like to have a possibility to use my mobile phone more for my studies	44	29	53	27	45	3.00

That the above is true is easily visible if we check the mean scores for other 3 possible uses of mobile phones in connection with the LMS, which are higher almost by a whole grade on the average. Namely, possibility to download learning resources, even with the recognized problem of small screen of mobile phones, is graded higher and used more often. Even higher is a mean score for "receiving notifications from lecturers". The most promising point is still the fact that even with a lot of individual services with a mean score relatively low, bellow two, mean score for the assessment if mobile phones are "regular part of studies" is higher than that, showing that with some changes in the approach of lecturers, mobile phones can be used to a much greater benefit in education at our Department.

As mentioned, the final big question is why do not our students use mobile phones in their education more that they do? The first two possibilities that come to mind and that would easily explain the situation were in our opinion:

- It's too expensive, or
- My mobile phone is not strong enough.

The survey gives a definite "No" answer for the first possibility. Almost 2/3 of the respondents "strongly disagree" with the opinion that access to Internet needed for mobile phone use is expensive, while only 9.09 % of students strongly students agrees with that opinion.

The second opinion also has "No" answer almost to the same degree. More than a half of students strongly disagree that their phone is not strong enough, while mere 18.18 % see that as a problem. Considering the trends and developments in the field, it is only natural to expect that with time, number of those with weak mobile phones can only drop down.

So, the situation might be explained by the answers to the third question within that section of a survey—is it good to use mobile phones for learning? Are we better suited or accustomed to use written material? Opinions about this question are highly divided, and the distribution is very balanced. This probably can also mean that we should re-phrase our definition of mLearning and think of it more as the meeting point of mobile devices and eLearning pedagogy. Not dependent anymore on use of computer laboratories, students can work on their knowledge at their homes, on field trips, or even while travelling to those field trips.

There is only one question with more balanced answers amongst students. The final question "Do you want to use mobile phones in your education more than you do currently?" has almost perfect balance between answers:

- Strongly agree versus strongly disagree: 45 : 44;
- Agree versus disagree: 27 : 29;
- Slightly agree (or we can asses it as "I'm-not-sure"): 53.

In our opinion, these answers show hidden fears behind it. Will use of mobile phones put some more pressure on them? Require some additional work? Or will it relax their studies and allow them to learn whenever they want and wherever they want?

And this is the key question and the key point of our research, giving us appropriate idea. We do not want to suggest introduction of mLearning into our studies as an obligatory form, as a system that will require shopping for the expensive and powerful equipment, learning of use of complicated applications, or ruining ones eye health by trying to read and study on a small screen. The idea is to organize mLearning as a welcomed supplement, as an ability to download, read, listen, or watch learning resources when it is convenient for a user, but with a plenty of other possibilities and types of learning materials. That will in our opinion attract more students to use mobile phones in education, and connect the best of all worlds and types of learning. The power of mLearning is highly increased by enhancing existing blended learning courses with otherwise weakly existing features such as notifications, easy communication services, access to discussion/interaction services, or personalized agents in non-productive, “dead” times.

It is easy to recognize that both undergraduate and master students are not satisfied with the mobile learning and the fact that it is not offered currently at our Department. Still, collected data proves that almost all of them are both equipped for its use, and are using some of the services they chose. So, if we offer more services, and yet do not force their use, we expect much better opinions of our students considering mobile learning.

9.5 Conclusion

Lifelong learning, ubiquitous learning, learning anytime, by anyone in anyplace, we can pick any of those buzzwords, and still end up with the more-or-less similar concept. Today’ economy is knowledge based, and it requires well-educated and flexible personnel, personas prepared to be continuously re-educated and re-trained, in order to be and stay competitive with the others. Also, the rapid development of learning theories and methodologies, together with the advances in information and communication “machinery”, creates prospects for satisfying these needs, and enables abandoning of (only) the traditional learning models. Incredible growth of mobile and wireless technologies allows incorporation of learning into everyday surroundings.

At the Department of Mathematics and Informatics of University of Novi Sad, we created and distributed a survey to several generations of students of computer science study direction. Except for some minimal number of students who declare that they do not use mobile phones, the most of the other students’ opinions and attitudes are almost unanimous. Collected answers to a survey about mLearning just slightly simplified, show that our students are technologically equipped for it, are accustomed to use of mobile phones, are not using services of mobile phones too much in education, but, the most importantly, are willing to use them more. It is simply our job to give them a better chance to do so.

This study is definitely limited not only in a sense of number of students surveyed, but more importantly in a type of students analyzed. Results we gained for students of computer science study direction may not be the same as for students of some other directions, especially for some human or social sciences, we expect. Still, as it was our hypothesis, students of computer science at the Department of Mathematics and Informatics of University of Novi Sad showed very similar opinions and attitudes to their colleagues of the same study direction all over the world reported in research papers. Being the area of fast development, mLearning requires constant insight into the views and beliefs of students, so an obligatory future direction of this research should be repeated and more detailed survey each several years, or even more often. Advances in mobile technology and software applications for mobile phones and tablets, lowering of prices for mobile services, and constant unification and standardization attempts for information formats, require regular alertness of educators in order to make the best use of possibilities offered. Another possible direction of study that can be very interesting and perhaps give some different results is to extend the study to students that are not so much connected to technology by their interests and study direction.

Consequently, let us hereof forget about the assessments of mobile learning, comparisons with eLearning, distance learning, or even classic classroom teaching, and let us take what's usable from it. As cleverly noted long time ago in Tough (1979), "… when the person's central concern is a task or decision, he will not be very interested in learning a complete body of subject matter. Instead, he will want just the knowledge and skill that will be useful to him in dealing with the particular responsibility of the moment". To provide for such people, we should not stick just to a single, traditional learning model based on the concept of one tutor, helping students to acquire in-advance-defined knowledge, and later assessing and measuring their success. We should give students a chance to choose their time and amount of learning, select a problem or part of it to concentrate on, and present them with enough learning resources that will satisfy any learning style and philosophy. Mobile learning will never and should not replace either other types of eLearning approaches or classroom teaching in our opinion. Yet, if applied properly, it can complement and append value to existing learning models and practice.

To apply mLearning properly, we must also consider and answer several wider questions. How should a university lecturer plan hers/his activities to help students accept mobile learning as a natural extension of other activities? What type of resources and digital activities should be obligatory, what should be additionally available? Should some of the resources become strictly mobile, or should there always be a stable and classic variant of everything? Should lecturers wait for the official recognition of the need for mLearning at their institution, or should they act as enthusiasts and start offering services and resources in mobile forms by themselves? The most of these questions are not only philosophical, they invoke also some very practical, sensible, realistic, and useful conclusions, since dealing with the proper development of any type of learning resources requires a great deal of effort, and careful planning and realization.

Developments in information and communication technology, and particularly in wireless and mobile technologies, can help us go away from traditional learning models, because nowadays learning can be easily carried, brought or even implanted into everyday environment. What makes mLearning thrilling is the fact that even though most of the individual features contained in current mobile devices are around for years, bringing all of them together in one small, powerful, and always available device is new. Joining the features, functionalities, and ability to go online ensures adoption of such devices even by the most unwilling users.

References

Corbeil, J. R., Valdes-Corbeil, M. E. (2007). *Are You Ready for Mobile Learning.* Retrieved from the EDUCAUSE review website: http://www.educause.edu/ero/article/are-you-ready-mobile-learning.

Fayyoumi, A., Mohammad, H., & Faris, H. (2013). Mobile based learning and examination: students and instructors perception from different Arab countries. *Journal of Software Engineering and Applications, 6*, 662–669. doi:10.4236/jsea.2013.612079.

Frydenberg, M. E. (2007). Teaching and learning information technology with mobile devices. *IADIS International Journal of Internet /WWW, 4*(2), 34–45.

Homan, S., & Wood, K. (2003). Taming the mega-picture: Wireless quizzing. *Syllabus Magazine,* 7–8 October 2003.

Ivanović, M., Putnik, Z., & Budimac, Z. (2010). eLearning in Serbia: The state and development of e-Learning in Serbia. In *Cases on challenges facing e-learning and national development: institutional studies and practice* (Anadolu University-2010, pp. 773–797). Eskisehir, Turkey.

Ivanović, M., Putnik, Z., Komlenov, Ž., Welzer, T., Hölbl, M., & Schweighofer, T. (2013). Usability and privacy aspects of moodle—students' and teachers' perspective. *Informatica, An International Journal of Computing and Informatics, 37*(3), 221–230.

Ivanović, M., Budimac, Z., Putnik, Z., & Komlenov, Ž. (2014). Encouraging teamwork, Web 2.0 and social networking elements in distance learning. In *Overcoming challenges in software engineering education: delivering non-technical knowledge and skills.* doi: 10.4018/978-1-4666-5800-4.ch005, (pp. 26–45). Hershey, USA: IGI Global.

Litchfield, A., Dyson, L. E., Lawrence, E., Zmijewska, A. (2007). *Directions for m-learning research to enhance active learning*. Paper presented at the ascilite Singapore ICT: Providing choices for learners and learning, pp. 587–596, Singapore.

McConatha, D., & Praul, M. (2007). *Mobile learning in the classroom: An empirical assessment of a new tool for students and teachers.* Paper presented at the Washington Interactive Technologies Conference, Virginia, USA.

Motiwalla, L. F. (2007). Mobile learning: A framework and evaluation. *Computers & Education, 49*, 581–596. doi:10.1016/j.compedu.2005.10.01.

Organista-Sandoval, J., & Serrano-Santoyo, A. (2014). Appropriation and educational uses of mobile phones by students and teachers at a public university in Mexico. *Creative Education, 5,* 1053–1063. doi: 10.4236/ce.2014.512119.

Prensky, M. (2001). Digital natives, digital immigrants. *On the Horizon*, *9*(5) (MCB University Press). doi: 10.1108/10748120110424816.

Putnik, Z. (2014). *Contributions to the development of eLearning—Possibilities for conversion of learning activities and materials into electronic form*, (Doctoral Thesis, University of Novi Sad, Serbia) (in Serbian). doi: 10.2298/NS20140214PUTNIK.

Putnik, Z., Ivanović, M., & Budimac, Z. (2014). *Web 2.0 in educational practice—a decade of experience*. Paper presented at the international conference on e-Learning'14, Tenerife, Spain, (pp. 194–200).

SRI International. (2013). *Client Sponsored Research & Development.* Retrieved from http://www.sri.com/engage/client-sponsored-research-development.

Tough, A. (1979). The adult's learning projects: A fresh approach to theory and practice in adult learning, 2nd edn. Toronto: Ontario Institute for Studies in Education. Pfeiffer & Co. (April 1979).

Traxler, J. (2007). Defining, discussing, and evaluating mobile learning: The moving finger writes and having writ…. *International Review of Research in Open and Distance Learning*, *8*(2), (Special issue—Mobile learning).

Trninić, Ž. (2013, June). Jubilees—One hundred and thirty years of telephones in Serbia, JAT New Review.

Vavoula, G., & Karagiannidis, C. (2005). Designing mobile learning experiences. *Advances in Informatics. Lecture Notes in Computer Science*, *3746*, 534–544. Berlin: Springer.

Wikipedia. (2015). *Economy of Serbia*. Retrieved from https://en.wikipedia.org/wiki/Economy_of_Serbia.

Zanela Saccol, A., Reinhard, N., Barbosa, J. L. V., & Schlemmer, E. (2010). m-Learning (mobile learning) in practice: A training experience with IT professionals. *Journal of Information Systems, 7*(2), 261–280.

Zdravkova, K., Ivanović, M., & Putnik, Z. (2012). Experience of integrating Web 2.0 technologies. *Educational Technology Research & Development, 60*(2), 361–381. doi:10.1007/s11423-011-9228-z.

Author Biography

Zoran Putnik is an assistant professor at the Faculty of Science, Department of Mathematics and Informatics, University of Novi Sad, Serbia. He received doctoral degree (teaching computer science) in 2014. His research interest is currently focused on e-learning and virtual learning environments, and software engineering. He has published around 140 articles in proceedings of international conferences and journals, and written several university textbooks. He participated within 20 international and national projects, was a member of organizing committees of 17 conferences, and is a Managing Editor of Computer Science and Information Systems (ComSIS) Journal.

Chapter 10
Access Moodle Using Mobile Phones: Student Usage and Perceptions

Xiao Hu, Leon Chi Un Lei, Jinbao Li, Nathalie Iseli-Chan, Felix L.C. Siu and Samuel Kai Wah Chu

Abstract This study investigated how often students used mobile phone to access various activities on Moodle. A survey on self-reported usage was filled by 252 university students in courses offered by four different faculties at the University of Hong Kong. Follow-up interviews were conducted to solicit students' perceptions on mobile access to Moodle and the underlying reasons. The results show significant differences in students' usage of various Moodle activities via mobile phones. Students' responses also suggest that mobile access to Moodle is a necessary complement to computer access but its limitation on usability and reliability may have restricted its potential in enhancing teaching and learning.

X. Hu (✉) · L.C.U. Lei · J. Li · N. Iseli-Chan ·
FelixL.C. Siu · S.K.W. Chu
The University of Hong Kong, Hong Kong, SAR China
e-mail: xiaoxhu@hku.hk

L.C.U. Lei
e-mail: culei@hku.hk

J. Li
e-mail: jinbao@hku.hk

N. Iseli-Chan
e-mail: iselicn@hku.hk

FelixL.C. Siu
e-mail: flcsiu@hku.hk

S.K.W. Chu
e-mail: samchu@hku.hk

D. Churchill et al. (eds.), *Mobile Learning Design*, Lecture Notes in Educational Technology, DOI 10.1007/978-981-10-0027-0_10

10.1 Introduction

The learning management system (LMS), Moodle, has been adopted by many higher education institutions around the world. To date, Moodle has been registered in more than 1800 sites over 120 countries, and is available in more than 60 languages (Kennedy 2004). Despite the increasing use of Moodle, concern has been expressed as to how Moodle is being used (Carvalho et al. 2011).

With the rapidly increasing use of handheld mobile devices among staff and students in higher education, it has become more and more common for them to access teaching and learning-related information and services using mobile devices (Peters 2009). A 2011 survey on mobile services in academic libraries in Hong Kong and Singapore reveals that the possession rate of mobile devices was 93.4 % among Hong Kong college students, and 61.9 % of them used smartphones to access the Internet (Ang et al. 2012). It is not uncommon to see university students use smartphones to access learning resources on Moodle and other LMSs. However, how frequent students use Moodle via mobile phones to carry out different Moodle activities and the possible reasons behind such usage patterns have rarely been formally investigated. The current research aims at filling this gap by examining students' frequencies of mobile access to Moodle for different activities and exploring possible reasons behind the usage patterns.

10.2 Related Work

10.2.1 Use of LMS

Research has been conducted to describe and analyse the use of LMS in higher education. Francis and Raftery (2005) defined three levels of LMS usage. The first level is for depositing materials and distributing information; the second is for enhancing teaching and learning by using various tools in LMS for communication, collaboration, assessment and quiz tests. The third and highest level is for supporting fully fledged online courses where most learning takes place on the LMS. It is indicated that even though an e-learning platform is available, the institutions might not make full use of it (Nichols 2008). Carvalho and her colleagues (2011) surveyed around 15,000 students for their use of two LMSs, Blackboard and Moodle. They found that for the majority of students, the use of the LMSs was still in the lowest level, that is for accessing learning materials and course announcements. Only some of them used LMSs for sending emails or taking quiz tests. The course forum, course chat room and virtual classroom were among the least used functionalities.

On the other hand, the importance of learning through social interaction and collaboration has been confirmed repeatedly (Tu and Corry 2003). Interaction plays a crucial role in academic success and persistence (Shea et al. 2006), and it is believed that knowledge construction begins when a student has engaged in a collaborative activity, because knowledge is created in situation (Chavez 2011). Therefore, educators increasingly make efforts to bring the use of LMS to a higher level that involves more interaction and collaboration among students.

It has been shown that technology usage patterns could vary across users' experience and information technology competency. For example, a study on organizational IT implementation (Venkatesh and Bala 2008) derived positive relations between usage experience and technology usage. Research has also been conducted to investigate whether Computer Self-efficacy (CSE) is related to users' perceived usefulness of e-learning (Hayashi et al. 2004). Different genders may also affect technology usage patterns. Ong and Lai (2006) concluded that males and females perceived e-learning differently, which influenced their behavioural intention to use e-learning. Meanwhile, Horvata et al. (2013) claimed that males and females were equally satisfied with Moodle quality characteristics. In terms of students' perception on Moodle, Kennedy's (2005) study on Hong Kong students' attitudes towards Moodle found that students liked the convenient accessibility of learning resources and the flexibility of organizing online materials on the Moodle pages. Carvalho et al. (2011) also found that students mainly perceived helping to find necessary information as the most useful function of Moodle. As such, previous studies considering users' experience, IT competency and genders have focused on technology usage patterns and perception towards e-learning, leaving the context of mobile access to e-learning less researched.

10.2.2 Mobile Learning

Mobile learning is thought to enhance opportunities for building a learning community, interaction and collaboration among students (Donaldson 2011). Çavus et al. (2008) investigated students' opinions of mobile learning by surveying 317 undergraduate students. They found that students' learning greatly benefited from using emails, forums and chat via mobile devices, and mobile learning was deemed effective by students during their communication with other students and instructors. In their study, there was no statistically significant difference in mobile learning across departments, gender or nationality. In terms of mobile access to LMS, researchers are divided as to their opinions towards the integration of LMS into mobile learning. For instance, Çavus (2011) presented the benefits of mobile learning using LMS and advocated that the integration of LMS into mobile learning would be a trend of learning platform in the future learning environment. On the other hand, Kouninef et al. (2012) brought up some constraints of mobile learning in using mobile technologies, including the small screen size and other device limitations. Besides, Ssekakubo et al. (2013) found that mobile phones were students' least used electronic devices for accessing LMS services and the main reason was the inadequate design of the LMSs for mobile interaction.

Although a number of studies investigated how frequent students used Moodle to accomplish various activities, there seemed a lack of a direct study relating the frequency of mobile access of Moodle to the background of users. In this study, we therefore attempt to identify students' Moodle usage patterns across different disciplines, their Moodle experience, and IT competency levels and between genders. Specifically, the research questions this study aims to answer are: (1) how would the frequencies of mobile access to Moodle activities vary across students in different disciplines, with different Moodle experience, IT competency levels, and between genders? (2) why do students use (or not) mobile access to Moodle.

10.3 Methodology

10.3.1 The LMS and the Courses

Moodle (version 2.6) was used in all the courses included in this study at the University of Hong Kong. Although there is a mobile app for Moodle, it cannot be integrated into the Moodle installation due to information security implementations in the University. Alternatively, the Moodle installation provides a Mobile Theme, which is a custom-designed display for smartphone browser screens. When users use smartphones to access Moodle, the Moodle server can detect the access device and then automatically display the Mobile Theme. Students can use the Mobile Theme to view course content pages, submit assignments and access a number of the Moodle functions including Forum, Choice, Feedback, Quiz, URL and Wiki.

Seven courses of four instructors were selected for this study. The instructors were in four different disciplines, Education, Engineering, Social Sciences and Humanities and Arts. The four instructors used Moodle in different levels and styles. The instructor from Social Sciences used Moodle as a repository of teaching materials and a platform for making course announcements. Besides uploading teaching materials, the instructor from Education also used discussion forums for student–student and student–instructor interaction. Links to external web sites were also put on Moodle of this course. As for the course in Engineering, the instructor used Moodle as a platform where students can read/download learning materials, submit assignments, take quizzes, conduct group projects and receive feedback from the instructor. The instructor from Humanities and Arts used Moodle to host learning materials, send announcements and messages to students, answer questions students raised, as well as Wiki and Glossary activities where students could post course-related information they collected off-class. The Engineering course was a Common Core course that could be taken by any year 1 and year 2 students across the University. As the class size was big, there were six teaching assistants in this

Table 10.1 Distribution of Moodle activities across courses

Moodle activities	Education	Social science	Engineering	Humanities and arts[a]			
	Course 1	Course 2	Course 3	Course 4	Course 5	Course 6	Course 7
Accessing resources	69	48	62	30	9	58	68
Submitting assignments (assignment, Turnitin assignment)	2	0	12	0	0	0	0
Taking tests (quiz, questionnaire)	2	0	15	0	0	0	0
Interaction (discussion forums, feedback, chatroom, choice)	9	0	3	3	1	6	0
Collaboration (wiki, glossary)	5	0	1	4	0	4	0
Total	87	48	93	37	10	68	68

Notes [a]The instructor in Humanities and Arts taught four courses

course. The Education course was a Master level course and the other courses were on the undergraduate level. Table 10.1 summarizes the number of various Moodle activities across these courses.

10.3.2 Participants and Procedure

This study adopts a mixed method with survey and interview data collected and analyzed.

10.3.2.1 The Survey

The survey was conducted in the last class of the courses. 389 students from the seven courses were invited to participate in the survey. 253 students in total responded to the questionnaire with valid answers (65 % response rate). The responses were collected partially online (n = 142) and partially on paper (n = 111). Table 10.2 presents the demographic distribution of the participants.

10.3.2.2 The Interview

After the survey data were collected, emails were sent to 80 survey respondents (20 from each discipline) to invite them to the follow-up interviews. Twelve of them

Table 10.2 Demographic information of questionnaire respondents

	N	Gender		Moodle experience[a]		IT competency[b]	
		Male	Female				
Education	17	3	14	1.71	1	2.88	3
Social science	57	25	32	2.41	3	2.93	3
Engineering	125	91	34	2.16	2	2.74	3
Arts	54	15	39	2.93	3	3.19	3
All	253	134	119	2.35	3	2.89	3

Notes [a]Ratings of "Moodle experience" are based on a 4-point Likert-type scale: 1—"less than 3 months", 2—"months to less than 1 year", 3—"1 year to less than 2 years", and 4—"2 years or more"

[b]Ratings of "IT competency" are based on a 5-point Likert-type scale: 1—"not competent", 2—"of little competency", 3—"somewhat competent", 4—"competent" and 5—"very competent"

accepted the invitation and participated in the interviews (3 in the Education course, 3 Social Sciences, 5 in Engineering and 1 in Humanities and Arts). The interviews were conducted partially face-to-face (n = 2) and partially through phone (n = 10). After the interviews, each interviewee was paid 30HKD for their participation.

10.3.3 Instruments

A questionnaire asking about the experience of using Moodle of the selected courses (Appendix 1) was used for collecting quantitative data. It included two parts: demographic information and frequency of course Moodle use. Part 1 asked for basic demographic information as well as their experience with Moodle and self-perceived IT competency level; Part 2 asked about the frequencies of using different categories of Moodle activities with variables in a 7-point Likert scale: ranging from 1 (never) to 7 (several times a day). A semi-structured interview protocol was designed to collect interview data. The main questions are shown in the following:

- What did you usually do when you access Moodle of this course via mobile phones? Why and why not using it for other purposes?
- Did you have any difficulties in using Moodle of this course using either computers or mobile phones? If yes, what were they?
- What kinds of supports do you think would be helpful when you encountered difficulties in using mobile phone to access Moodle of this course?

Table 10.3 Descriptive statistics of frequency of using Moodle via mobile phones

Moodle activities	N	Minimum	Maximum	Mean	Std. Deviation
Accessing resources	252	1	7	3.70	1.526
Submitting assignments	251	1	7	2.22	1.553
Taking tests	252	1	7	2.30	1.567
Interaction	251	1	7	2.06	1.457
Collaboration	252	1	7	2.08	1.508

Notes Ratings are based on a 7-point Likert-type scale: 1—"never", 2—"Once a month or less", 3—"Once every 2 weeks", and 4—"1–2 times a week", 5—"3–6 times a week", 6—"Once every day", 7—"Several times a day"

Table 10.4 Statistics of frequency of using Moodle via mobile phones across disciplines

Moodle activities		Humanities and Arts	Education	Social Science	Engineering	Sig. Kruskal–Wallis
Accessing resources	N	54	17	56	125	0.002**
	Mean	3.35	3.06	3.39	**4.08**	
	Median	4.00	4.00	4.00	4.00	
Submitting assignments	N	53	17	56	125	0.000**
	Mean	1.38	1.53	1.50	**2.99**	
	Median	1.00	1.00	1.00	3.00	
Taking tests	N	53	17	56	125	0.000**
	Mean	1.41	1.00	1.50	**3.22**	
	Median	1.00	1.00	1.00	4.00	
Interaction	N	53	17	55	125	0.000**
	Mean	1.69	1.35	1.62	**2.52**	
	Median	1.00	1.00	1.00	2.00	
Collaboration	N	54	17	56	125	0.000*
	Mean	1.43	1.24	1.55	**2.71**	
	Median	1.00	1.00	1.00	2.00	

Notes Ratings are based on a 7-point Likert-type scale: 1—"never", 2—"Once a month or less", 3—"Once every 2 weeks", and 4—"1–2 times a week", 5—"3–6 times a week", 6—"Once every day", 7—"Several times a day"
**Indicates significance at $p < 0.01$ level

10.4 Results

10.4.1 Questionnaire Responses

Table 10.3 shows the statistics of student self-reported usage of Moodle via mobile phones. Access to learning materials was the most frequent activity while interacting with instructors and other students was the least frequent. It is noteworthy that students' responses varied from "never" to "several times a day" in all activity categories.

Statistics across different disciplines are presented in Table 10.4. Students in the Engineering course reported the highest frequency across all Moodle activities accessed via mobile phones among all participating students. As the data are in ordinal scale, the non-parametric Kruskal–Wallis test is used to compare the frequencies across courses. The significance levels (p values) are reported in Table 10.4. Statistically significant differences were found in all five categories of activities: accessing resources submitting assignments, taking tests, interaction and collaboration.

Experience of using Moodle may have affected students' usage of Moodle via mobile access. Kruskal–Wallis tests revealed that students with different Moodle

Table 10.5 Descriptive statistics of frequency of using Moodle via mobile phones across experience of using Moodle

Moodle activities	less than 3 months		3 months to less than 1 year		1 year to less than 2 years		2 years or more		Sig. Kruskal–Wallis
	N	Mean	N	Mean	N	Mean	N	Mean	
Accessing resources	86	3.65	35	3.94	85	3.86	45	3.29	0.164
Submitting assignments	86	2.48	35	2.29	85	2.15	44	1.82	0.155
Taking tests	86	2.51	35	2.37	85	2.40	45	1.67	0.020*
Interaction	85	2.13	35	2.29	85	2.14	45	1.64	0.069
Collaboration	86	2.19	35	2.29	85	2.22	45	1.47	0.015*

Notes Ratings are based on a 7-point Likert-type scale: 1—"never", 2—"Once a month or less", 3—"Once every 2 weeks", and 4—"1–2 times a week", 5—"3–6 times a week", 6—"Once every day", 7—"Several times a day"

*Indicates significance at $p < 0.05$ level

experience reported significantly different usage frequency in taking tests and collaboration ($p < 0.05$, Table 10.5). Follow-up pair-wise tests revealed that, for both activity categories, students with "2 years' or more" experience with Moodle actually reported lower frequencies than those with "less than 3 months'" or "1 year to less than 2 years'" experience ($p = 0.02 \sim 0.04$). There was no significant difference between other pairs of experience values.

Besides, difference in the frequency of using Moodle via mobile phones across IT competency was also analysed. Table 10.6 indicates a statistically significant difference of access frequencies in interaction and collaboration activities ($p < 0.05$). For interaction, a follow-up pair-wise test found that students who rated themselves as "not competent" reported significantly more frequent access than those who rated themselves as "somewhat competent" ($p = 0.02$) or "competent" ($p = 0.03$). For collaboration, students who rated themselves as "not competent" reported significantly more frequent access than those who rated themselves as "competent" ($p = 0.04$). There was no significant difference between other pairs of IT competency values.

The study also compares the difference of reported usage frequency between genders, and the statistics and results of Mann–Whitney tests are shown in Table 10.7. There are statistically significant differences in all activity categories but accessing resources. Results showed that male students displayed a higher self-reported frequency in Moodle activities including submitting assignments, taking tests, interacting and collaborating with one another.

Table 10.6 Descriptive statistics of frequency of using Moodle via mobile phones across IT competency

Moodle activities	Not competent		Of little competency		Somewhat competent		Competent		Very competent		Sig. Kruskal–Wallis
	N	Mean	N	Mean	N	Mean	N	Mean	N	Mean	
Accessing resources	29	3.83	54	3.91	96	3.75	59	3.44	12	3.58	0.5
Submitting assignments	29	2.83	53	2.21	96	2.08	59	2.19	12	2.25	0.145
Taking tests	29	2.76	54	2.31	96	2.17	59	2.27	12	2.50	0.26
Interaction	29	2.79	54	2.09	96	1.96	59	1.93	11	1.82	0.018*
Collaboration	29	2.69	54	2.22	96	1.97	59	1.90	12	1.92	0.032*

Notes Ratings are based on a 7-point Likert-type scale: 1—"never", 2—"Once a month or less", 3—"Once every 2 weeks", and 4—"1–2 times a week", 5—"3–6 times a week", 6—"Once every day", 7—"Several times a day"

*Indicates significance at $p < 0.05$ level

Table 10.7 Difference of frequency of using Moodle via mobile phones between genders

Moodle activities	Male		Female		Sig. Mann–Whitney
	N	Mean	N	Mean	
Accessing resources	133	3.79	119	3.61	0.341
Submitting assignments	118	**2.55**	118	1.85	0.002**
Taking tests	133	**2.74**	119	1.80	0.000**
Interaction	132	**2.31**	119	1.79	0.040*
Collaboration	133	**2.38**	119	1.75	0.003**

Notes Ratings are based on a 7-point Likert-type scale: 1—"never", 2—"Once a month or less", 3—"Once every 2 weeks", and 4—"1–2 times a week", 5—"3–6 times a week", 6—"Once every day", 7—"Several times a day"

*Indicates significance at $p < 0.05$ level

**Indicates significance at $p < 0.01$ level

10.4.2 Themes from Interviews

Table 10.8 presents several representative quotes from students' responses to the questions on how and why they used mobile access to Moodle to carry out the corresponding activities.

All interviewed students answered that they used mobile phones to access Moodle of their courses, because using mobile phones allowed them to access Moodle at any place and any time. They could read learning materials and important information such as assignment deadlines when no computer or Wi-Fi connection was available. Mobile access also enabled them to read announcements, comments and feedback as soon as they were available online. Students' tendency in using Moodle for resource depository and information retrieval in this study demonstrated consistency with previous studies on students' perception on Moodle (Kennedy 2005; Carvalho et al. 2011). The students from the Engineering course

Table 10.8 Representative quotes from the interviews

Moodle activities	Students' responses
Accessing resources	Student A: *I would download some information [into my mobile phone's storage]. Whenever or wherever I want to view it, using mobile phones comes in more convenient*
Submitting assignments	Student B: *If I need to submit assignments, I would not use my mobile, as the files are not in my mobile phone*
Taking tests	Student C: *When I have turned off my computer, but I still want to change the answers [of the previously completed quiz]. I would then quickly use it [mobile phone]. It is more convenient this way*
Interaction	Student D: *We preferred face-to-face discussion [to online interactions with groupmates]*
Collaboration	Student E: *[I] Never [used mobile access to Moodle for collaborative projects]. There are too many words in the project. So it is difficult to read on a small device*

(n = 5) also mentioned that they used mobile phones in class to access Moodle because one of the course requirements was to complete a short quiz within 4 h after each class. Therefore, when the students did not bring their laptop to class, they would use mobile phones to finish the quizzes.

However, students also indicated that using mobile phones was not a preferred method to access Moodle. Most of them referred to usability issues such as small screens and an awkward keyboard. As a result, they would only be comfortable to conduct simple and low-stake tasks using mobile access. It was a common theme among students that the Mobile Theme of Moodle was inconvenient. To start a Moodle session on mobile phones, they needed to launch a browser window/tab, type in the URL, and log into the system. As the session expired after a short period of idle time, students had to log in again virtually at each time of access. Besides, the display of Moodle course pages on mobile phones was mentioned quite often during the interviews. All the course pages contained a rich amount of information. While the texts on the course pages were well displayed on computer screens, with proper headings and indentions, the format could become cluttered on the screen of mobile phones. Moreover, a majority of interviewed students preferred face-to-face discussion when working on collaborative projects, rather than using mobile phones for online interaction. Mobile interaction was only a choice when group members could not gather at the same time. Last but not least, several students mentioned that they did not know how to upload files to Moodle from their mobile phones or to find files to be downloaded from Moodle.

10.5 Discussion

Both the survey and interview data indicated that students used mobile phones to access Moodle for learning materials much more often than for other activities (Table 10.3), which indicates that the use of mobile access to Moodle was still at the lowest level as suggested in Francis and Raftery (2005). One possible reason is that the usability limitations of mobile access discouraged the students from using it for complicated tasks (e.g., Wiki edits, discussion posts) or activities that were deemed not urgent. Such usability limitations echo with the view of Kouninef et al. (2012) on the constraints of mobile learning using mobile technologies. In addition, depositing learning materials is the most widely used function of Moodle across all courses in this study, and there were much fewer Moodle activities related to interaction and collaboration across these courses (Table 10.1). Also, a possible reason for limited interpersonal interaction on Moodle is its inadequate design for mobile interaction (Ssekakubo et al. 2013), though usability issues like small screens seem to outweigh this technical inadequacy.

The distribution of Moodle activities shown in Table 10.1 could partially explain the significant differences on students' self-reported Moodle usages via mobile phones presented in Table 10.4. For accessing resources, a pair-wise test following the Kruscal–Wallis test reveals that the only significant difference ($p = 0.02$) lied in between the Engineering course and the courses in Humanities and Arts where much fewer learning resources were hosted in two of the courses. The Moodle of the Engineering course had substantially more assignments and test activities than others, and this could justify why the frequencies of using these activities reported in this course were significantly higher than those of all other courses ($p < 0.01$). In addition, the quizzes in the Engineering course were designed in small sizes, with 3–5 multiple choices questions in each, and students reflected that they felt comfortable to access those quizzes via mobile phones since they only spent a little time to complete and did not involve much typing on the keyboard. On the grounds that students held positive attitudes towards accessing short quizzes via mobile access to Moodle, a recommendation for promoting mobile access to Moodle is that instructors could make adjustments to the course design by adding in-class online short quizzes as an additional assessment task, so that it would be desirable for students to access Moodle via mobile phones during class time or soon after classes.

For interaction and collaboration activities, even though the Engineering course had fewer activities in these two categories compared to other courses, the reported usage frequencies via mobile phones were still significantly higher than those in other courses (Table 10.3). This result suggests that creation of Moodle activities that are designed for interaction and collaboration does not necessarily result in more frequent access to those activities via mobile phones. Students from the Engineering course reported that they felt there was a learning community built on the course Moodle (Shea et al. 2006). There were a variety of learning activities that involved interactions and collaborations, including a group project, a group presentation and peer assessments (inter- and intragroups) (Lei et al. 2013). In addition, the instructor and teaching assistants responded to students' posts in a timely manner. These may all have contributed to the stronger motivations of the students in accessing the course Moodle via mobile phones. Accordingly, it is recommended that interactive and collaborative learning activities like peer assessments and group projects should be implemented in the future course design, as students could opt for contributing to these tasks using mobile access to Moodle at their convenience. Besides, instructors and teaching assistants need to be more responsive and more active in facilitating student interactive activities.

Interestingly, the results also revealed that students who have used Moodle for a shorter period of time tended to use mobile access more often to take tests and collaborate on Moodle than those who have used Moodle for 2 years and more (Table 10.5). In addition, students with low self-perceived IT competency used more mobile access to Moodle for interaction and collaboration activities (Table 10.6). These seem to contradict with many studies where experience and IT competency are positively associated with technology usage (Venkatesh and Bala 2008). Such findings could potentially supplement previous research on the relationship between IT competency and technology usage patterns. We conjecture that the statistics might have been dominated by the students in the Engineering course who rated higher usage frequencies and lower Moodle experience and IT competency than other students. However, this would need further analysis to be confirmed.

The study also found male students used mobile access significantly more often than female students in all listed Moodle activity categories except for resource access. During the interviews, some female students complained about the complexity of some Moodle activities and expressed the need of instructional help on using those activities. Such gender differences have also been found in other studies on gender difference in educational technology (e.g., Heemskerk and Dam 2009), as well as in studies on the significance of gender differences in users' perceived usefulness of e-learning (e.g., Ong and Lai 2006). An implication is that providing instructions on how to use Moodle activities, especially with mobile access, would be helpful for users of both genders and would reduce the feeling of complexity to female students. On another note, student gender distributions vary a lot across the courses and the Engineering course was the only one with much more male than female students (Table 10.2). Therefore, it is possible that the observed gender difference may be partially affected by the higher ratings among students in the Engineering course.

10.6 Conclusion and Future Work

This study compared the usage patterns of Moodle for different activities via mobile phones among college students enrolled in courses across four disciplines, and analysed the reasons behind these usage patterns.

In general, students in this study did not prefer using their mobile phones to access Moodle, due to the limitations of mobile access on usability and reliability. However, most of them indeed used mobile phones to access Moodle when it was necessary. In terms of Moodle activities, it was found that students preferred carrying out easy and low-stake Moodle tasks on their mobile phones, such as accessing learning materials. The students expressed the need for a more user-friendly mobile access. In comparing survey responses from students across the courses, it was found that good pedagogical design could at least partially mitigate the limitations of mobile access and encourage students to use Moodle more often for activities involving interaction and collaboration.

A possible limitation of this study is that the data collection was limited to a single university (HKU) in Hong Kong. Since different universities might employ their own LMSs in different ways, resulting in variant perceptions with and opinions on the LMS, the conclusion made in this study might not be generalizable to all universities. Follow-up studies can expand the sample by recruiting participants from different universities and in different regions. Another limitation is that the findings of this study are solely based on self-reported data from participants, which might be subject to the difference in students' own perception. Future studies could rely on objective data sources such as the usage patterns as reflected in the LMS system logs.

This study focused on examining the activity-specific usage patterns of Moodle via mobile access, while paying relatively less attention to students' opinions on mobile access to LMS, such as perceived usefulness. Future work will include the analysis of students' perceptions on the usefulness of mobile access to Moodle and the factors that might affect their perceptions. Also, in forthcoming studies, LMS system logs will be collected for further analysis of students' behaviour on Moodle.

Acknowledgments We would like to thank the students who participated in this study. This study is partially supported by a Teaching Development Grant in the University of Hong Kong.

Appendix 1: Questionnaire

Part 1: Demographic information

What is your gender?
How old are you?
Where did you spend most of your life?
How long have you used Moodle?
Have you ever used any other learning management systems?
What is your IT (information technology) competency level?

Part 2: Frequency of using different Moodle functions

I used Moodle of this course via mobile phones to access learning materials (e.g., slides, notes, readings, assignments)
I used Moodle of this course via mobile phones for submitting assignments.
I used Moodle of this course via mobile phones for taking tests/quizzes/exams.
I used Moodle of this course via mobile phones for interacting with instructors/classmates (e.g., replying to posts, sending messages, chatting, etc.).
I used Moodle of this course via mobile phones for collaborating with classmates (e.g., editing wikis, contributing to glossary, discussing group projects, etc.).

References

Ang, S., Chia, YB., Chan, I., Leung, K., Li, K., & Ku, K.M. (2012). The survey on mobile library services in hong kong and singapore academic libraries (pp. 1–53). Retrieved from http://hdl.handle.net/10722/152520.

Carvalho, A., Areal, N., & Silva, J. (2011). Students' perceptions of Blackboard and Moodle in a Portuguese university. *British Journal of Educational Technology, 42*(5), 824–841. doi:10.1111/j.1467-8535.2010.01097.x.

Çavus, N., Bicen, H., & Akçil, U. (2008). The opinions of information technology students on using mobile learning. In *Proceedings of the 08 International Conferences on Educational Sciences*. Magosa, North Cyprus.

Çavus, N. (2011). Investigating mobile devices and LMS integration in higher education: Student perspectives. *Procedia Computer Science, 3*, 1469–1474.

Chavez, J. F. (2011). *A knowledge management tool for collaborative learning A case study using a wiki (doctoral dissertation)*. New Mexico: The University of New Mexico Albuquerque.

Donaldson, R. L. (2011). *Student acceptance of mobile learning*. The Florida State University College of Communication & Information. Retrieved from http://diginole.lib.fsu.edu/etd/716/.

Francis, R., & Raftery, J. (2005). Blended learning landscapes. *Brookes eJournal of Learning and Teaching, 1*(3), 1–5.

Hayashi, A., Chen, C., Ryan, T., & Wu, J. (2004). The role of social presence and moderating role of computer self efficacy in predicting the continuance usage of e-learning systems. *Journal of Information Systems Education, 15*, 139–154.

Heemskerk, I., & Dam, G. T. (2009). Gender inclusiveness in educational technology and learning experiences of girls and boys. *Journal of Research on Technology in Education, 7*(3), 253–276. Retrieved from http://www.tandfonline.com/doi/abs/10.1080/15391523.2009.10782531.

Horvat, A., Dobrota, M., Krsmanovic, M., & Cudanov, M. (2013). Student perception of Moodle learning management system: a satisfaction and significance analysis. *Interactive Learning Environments*, (ahead-of-print) (pp. 1–13).

Kennedy, D. M. (2004). Challenges in evaluating Hong Kong students' perceptions of Moodle Moodle: An open-source learning management system. In *The potential of LMSs in teacher education*, (July pp. 327–336).

Kennedy, D. M. (2005). Challenges in evaluating Hong Kong students' perceptions of Moodle. In Proceedings of Australasian Society for Computers in Learning in Tertiary Education (ascilite) (pp. 327–336).

Kouninef, B., Tlemsani, R., Rerbal, S. M., & Lotfi, A. (2012). Developing a mobile learning approach in platform LMS INTTIC. *Information Technology Journal, 11*(8), 1131–1137.

Lei, C. U., Wan, K., & Man, K. L. (2013). Facilitating teaching CPS and technology-based content with learning management systems. In *Proceedings of the International MultiConference of Engineers and Computer Scientists* (Vol. 2).

Nichols, M. (2008). Institutional perspectives: The challenges of e-learning diffusion. *British Journal of Educational Technology, 39*(4), 598–609. doi:10.1111/j.1467-8535.2007.00761.x.

Ong, C. S., & Lai, J. Y. (2006). Gender differences in perceptions and relationships among dominants of e-learning acceptance. *Computers in Human Behavior, 22*(5), 816–829.

Peters, K. (2009). M-learning: Positioning educators for a mobile, connected future. In M. Ally (Ed.), *Mobile learning: Transforming the delivery of education and training*. Vancouver: Marquis Book Printing

Ssekakubo, G., Suleman, H., & Marsden, G. (2013). Designing mobile LMS interfaces: learners' expectations and experiences. *Interactive Technology and Smart Education, 10*(2), 147–167.

Shea, P., Sau Li, C., & Pickett, A. (2006). A study of teaching presence and student sense of learning community in fully online and web-enhanced college courses. *Internet & Higher Education, 9*(3), 175–190.

Tu, C., & Corry, M. (2003). Building active online interaction via a collaborative learning community. *Computers in the Schools, 20*(3), 51–59.

Venkatesh, V., & Bala, H. (2008). Technology acceptance model 3 and a research agenda on interventions. *Decision Sciences, 39*(2), 273–315.

Author Biographies

Xiao Hu an Assistant Professor in the Division of Information and Technology Studies in the Faculty of Education of the University of Hong Kong. Her research interests include learning analytics, applied data/text mining, and information retrieval. She is leading several projects on using learning analytics to improve teaching and learning, and has co-organized Learning Analytics Summer Institute, LASI-Hong Kong in 2013 and 2014. Dr. Hu holds a Ph.D. degree in Library and Information Science and a Master's degree in Computer Science, both from the University of Illinois at Urbana-Champaign.

Leon Chi Un Lei received his B.Eng. (first class honors) and Ph.D. in Electrical and Electronics Engineering from the University of Hong Kong (HKU) in 2006 and 2011, respectively. He is now an E-learning Technologist in the Technology-Enriched Learning Initiative, HKU. He is also an Honorary Assistant Professor in the Department of Electrical and Electronics Engineering, HKU. His current research interests include learning analytics, MOOC and smart learning environment. He was awarded with the Best Paper Award in IEEE TALE 2013 and IAENG IMECS 2012, Best Student Paper Award in IAENG IMECS 2010 and IAENG IMECS 2007.

Jinbao Li an Assistant Professor in the Department of Geography at the University of Hong Kong. He received his Ph.D. degree in Earth and Environmental Sciences from Columbia University in 2009. His research centers on long-term variations in El Niño, the most powerful inter-annual climate force on Earth, and the Asian monsoon, the most crucial climate system to the Asian community. He carries out diagnostic studies using both instrumental and proxy records to gain insights into long-term climate dynamics and their impacts on society. He teaches courses related to climate change, ecosystem and contemporary global environmental issues.

Nathalie Iseli-Chan has been teaching French as a foreign Language for over 10 years, 7 of which with the support of web 2.0 technologies. Since 2007 she has been developing and devising Moodle based activities, including self-directed, self-assessed and cross level activities, aiming at bridging the gap between language learning and teaching, and building online learner communities. She joined the School of Modern Languages and Cultures 5 years ago. Her research interests include language learning and teaching strategies as well as the pedagogical application of new technologies to learning and teaching French as a foreign language.

Felix L.C. Siu has a solid background in Computing and Science Education with many years teaching experience in primary, secondary and tertiary education. He is senior lecturer in the Division of Information and Technology Studies, Faculty of Education at The University of Hong Kong. His research interests and experiences are diverse, covering areas such as multimedia in education, information literacy, information management, information and communication technology in education, and mobile & ubiquitous technology in Education. He has published, as author or co-author, many academic papers on the use of ICT in education.

Samuel Kai Wah Chu the Head of Division of Information and Technology Studies and an Associate Professor at the Faculty of Education, The University of Hong Kong. He is also the Deputy Director of Centre for Information Technology in Education in the Faculty. His publications include articles, books and key journals in the areas of IT in education, information and library science, school librarianship, academic librarianship and knowledge management. He has been involved in over 50 research/project grants and is also the recipient of his Faculty's Outstanding Researcher Award in 2013.

Author Biographies

[illegible]

[illegible]

[illegible]

[illegible]

[illegible]

[illegible]

[illegible]

Part III
Mobile Learning Analytics

Chapter 11
An Effective Cloud-Based Simulator Facilitating Learning Analytics on Mobile Devices

Vincent Tam, Alex Yi, Di Xu and Edmund Y. Lam

Abstract Learning analytics is targeted to better understand each learner's interests and unique characteristics in learning so as to build a personalized learning environment. However, many learning analytics techniques may require relatively intensive computation, thus inappropriate for any mobile application. In this paper, we propose an interactive and personalized e-learning system named the COMPAD+ simulator facilitated by an intelligent learning analytics algorithm running on the cloud server to quickly estimate the learner's areas of interests on the simulation results as based on his/her initial inputs and then flexibly generate the simulation details as appropriate. To protect the data privacy of each individual learner, the personal data is stored in the learner profile under each password-protected account on the cloud server with all the intermediate simulation data to be erased after each learning task. Up to our understanding, this work represents the first attempt to successfully develop a flexible, interest-based, and platform-independent simulator directly run on any mobile device facilitated by an efficient learning analytics algorithm working on the cloud server. To demonstrate its feasibility, a prototype of our cloud-based COMPAD+ e-learning system is built and carefully evaluated on various mobile devices. Clearly, there are many promising directions in terms of both pedagogical and technological impacts to extend and enhance our interest-based COMPAD+ simulation platform for future e-learning systems.

V. Tam (✉) · A. Yi · D. Xu · E.Y. Lam
The University of Hong Kong, Hong Kong SAR, People's Republic of China
e-mail: vtam@eee.hku.hk

A. Yi
e-mail: alexyi1124@gmail.com

D. Xu
e-mail: h1395716@connect.hku.hk

E.Y. Lam
e-mail: elam@eee.hku.hk

D. Churchill et al. (eds.), *Mobile Learning Design*, Lecture Notes in Educational Technology, DOI 10.1007/978-981-10-0027-0_11

11.1 Introduction

In recent years, cloud computing technologies have been frequently used to improve the competitiveness, efficiency, and reliability of daily operations or services in many enterprises or government units through providing extra computational resources and/or data storage available on the underlying public or private cloud platforms. Nevertheless, there are relatively less work targeted to investigate on how cloud computing (Velicanu et al. 2013) may enhance students' understanding of a specific subject in e-learning systems. Through careful observations on a first year course on Computer Systems in the Faculty of Engineering, the University of Hong Kong over the past few years, we found that many Engineering students encountered various difficulties in understanding some essential concepts in computer systems, especially the program execution and the underlying data transfer among the various devices/registers in a specific computer system. Intrinsically, these concepts are abstract and often involve a complex knowledge structure, and therefore are difficult to understand. Furthermore, most existing simulators for computer systems are text-based and mainly focused on showing the final results after program execution without clearly showing the intermediate steps and key "operations," and particularly the essential components/concepts involved in such operations. In many cases, the ultimate results are simply presented to students without knowing how such results are generated. Undoubtedly, several existing simulators can only provide a limited set of debugging functions such as monitoring the values of selected registers at a certain computation step during the program execution. Yet without knowing which components, or specifically which internal registers, are actually involved in the process of computation, it is totally impossible and meaningless to use such debugging functions for monitoring the changes of values on all the registers in order to better understand the behavior of program execution in the specific computer system. In a recent research project awarded by Microsoft Research Asia (MSRA), we successfully built a very interactive and user-friendly simulator, namely the COMPAD+ simulator as an extended version of the original "learning PAD for COMputer systems" (COMPAD) on a cloud computing platform, which will quickly enhance students' learning of essential concepts related to computer systems through the live animation of program execution on their areas/topics of interests for a specific computer architecture. Through analyzing the students' selected topics of interests in computer systems recorded on the back-end cloud server, the course instructor will have a better picture about which topics are possibly more attractive to his/her own students. On top of it, the detailed information including the list of instructions/operations issued, the simulation speed used, and the progress of simulation made by individual students can provide an excellent data repository for thorough investigations of learning analytics that can be flexibly done by any course instructor or programme director with the aid of some quickly developed cloud sever programs or scripts.

The original COMPAD simulator (Fung et al. 2010) was developed as a standalone e-learning application to run on the Microsoft .NET platform on any desktop or notebook computers. Yet, our cloud-based COMPAD+ simulator is carefully designed and recently built on the Windows Azure Cloud platform (Windows Azure Development Team 2015) to provide "anytime and anywhere" simulation services for our students revising important concepts on computer systems through web browsers running on their own mobile devices such as smartphones or tablet PCs. It is worth noting that the client application of our COMPAD+ simulator is platform-independent to be run on any operating systems including the Android, iOS, Linux, Mac OS, Microsoft Windows, Unix and many others through the web interface. In addition, the design of our COMPAD+ simulator is so generic that it can be readily implemented on any cloud computing platform, and easily integrated with other existing e-learning systems. Furthermore, the underlying cloud computing platform may quickly provide additional computational resource to boost the performance of our COMPAD+ simulator so as to tackle any complex simulation task while the cloud storage will provide a large repository to carefully organize all the simulation models in our designated e-learning system. More importantly, through adopting the IEEE learning object metadata standard to represent each key concept/component of various computer systems as "reactive" models of learning objects, our cloud-based COMPAD+ simulator may easily facilitate the sharing and reuse of relevant concepts for future e-learning applications. To demonstrate the feasibility of our proposal, we carefully consider the system design and build a prototype implementation of our cloud-based COMPAD+ simulator with a thorough evaluation plan to investigate on how novice or experienced learners may benefit from our interactive simulator in various ways. Clearly, there are many interesting directions including the plausible uses of the COMPAD+ simulator to evaluate students' performance for learning analytics (Hong et al. 2005) such as the learning path optimization method (Tam et al. 2012) for our future investigation.

This chapter is organized as follows: Sect. 11.2 reviews the previous work on applying cloud computing technologies to improve e-learning systems (Wong and Looi 2009), particularly their overall system architectures. Section 11.3 details the system design of our enhanced COMPAD+ simulator on any cloud computing platform to enhance students' understanding of a specific subject. Section 11.4 provides an empirical evaluation of our implemented prototype with a thorough evaluation plan. Last but not least, some concluding remarks of this interesting work are given in Sect. 11.5.

11.2 Previous Work

Most previous work (Fernández et al. 2012; Masud and Huang 2012; Pocatilu 2010) solely focus on improving the "efficiency" of existing e-learning systems with cloud computing technologies through restructuring the overall infrastructures or system architectures of the e-learning systems on cloud computing platforms. Fernàndez et al. (2012) give a clear overview on the current state of the structure of cloud computing for e-learning applications, and also detail the most common infrastructures that have been developed for such e-learning systems. Masud and Huang (2012) carefully consider an e-learning system architecture based on cloud computing. Lastly, Pocatilu et al. (2010) try to measure the positive impact of using cloud computing architectures upon the development of e-learning systems by advancing a set of cloud computing efficiency metrics for enhanced process control of e-learning system implementation. The long-term overall efficiency of the cloud computing usage in e-learning systems is also evaluated in their studies.

Generally speaking, there is rarely any work that tries to investigate on how cloud computing technologies can be used to enhance both "efficiency" and "effectiveness" of e-learning systems to facilitate students' learning in a specific subject, and more importantly the potential of integrating some appropriate learning analytics technique(s) to better understand students' learning difficulties and progress in relevant topics. Therefore, this motivates us to initiate a research project on building cloud-based and interactive e-learning systems as supported by the Microsoft Research Asia to carefully investigate on how cloud computing techniques can enhance students' understanding in various computer systems through "efficient" and "effective" simulations displayed on their mobile devices while facilitating the possible integration of learning analytics technique(s) to analyze the unique learning characteristics of each individual learner through the detailed log files collected at the back-end cloud server.

In addition, Tam et al. (2012) study the learning path optimization (LPO) method for learning analytics on the next-generation e-learning systems and propose a rule-based evolutionary algorithm (Mitchell 1998) to obtain the optimized learning paths. The reference learning paths are essentially some specific sequences of learning concepts/topics (Chen et al. 2008, Chuang and Shen 2008) as recommended by experts in the underlying subject domains. After being extracted from the reference paths, a set of precedence rules (Tsang 1993) will be used as the evaluation criteria during the process of LPO in the corresponding course. The quality of the generated learning paths (Xu et al. 2012) are determined by what extent the generated learning path violates the precedence rules. Along this direction, with our cloud-based and enhanced COMPAD+ simulator, users' interests will become a new factor to be considered in the LPO so as to generate learning paths of a better quality, thus with fewer violations of precedence rules with respect to the experts' reference paths. An optimization scheme based on both reference paths and users' interests will be proposed and carefully considered in the subsequent sections.

11.3 System Design of Our Cloud-Based COMPAD+ Simulator

Figure 11.1 shows an overview of the system design of our enhanced COMPAD+ simulator on the Windows Azure Cloud platform. The system offers a series of interactive and user-friendly simulation services through the web interface which

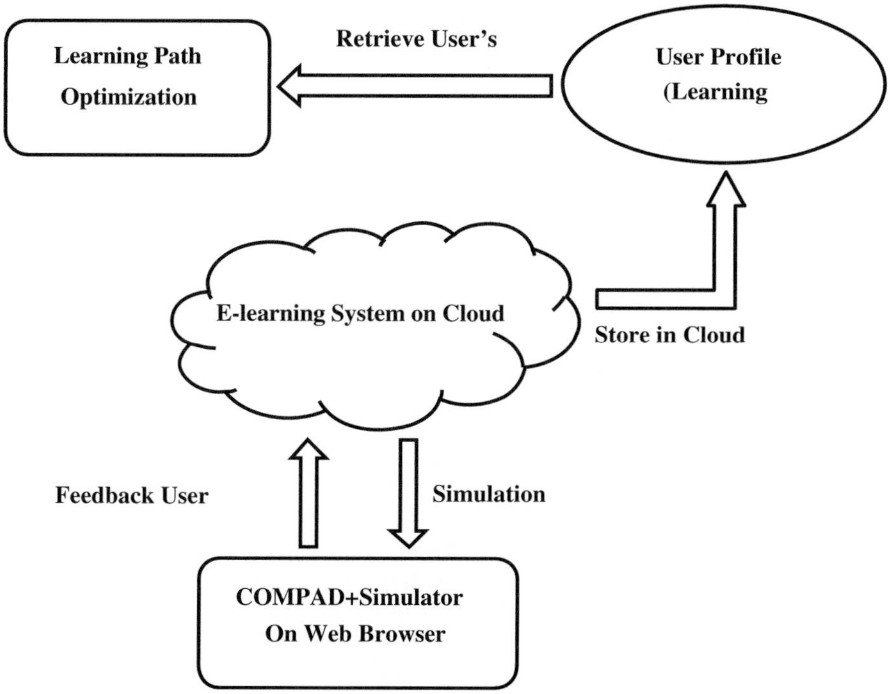

Fig. 11.1 The system design of our cloud-based COMPAD+ simulator

shows the intermediate steps and also the ultimate simulation results to learners while collecting the individual learners' profiles including their learning interests and feedback back to the Windows Azure cloud databases for performing learning analytics by the back-end cloud server program. With the learners' learning interests, the COMPAD+ simulator is able to provide more adaptive and personalized interfaces that can be dynamically changed for each individual learner as based on his/her own learning interests. Besides, each individual's learning interests can be considered as a new criterion during the process of learning path optimization (LPO). In this way, the ultimate learning paths generated by the LPO method have carefully considered the experts' valuable inputs together with each individual's learning interests. This will help to more precisely capture each learner's unique learning characteristics, thus providing more personalized learning path(s) to suit the individual learner's requirements.

Our cloud-based COMPAD+ simulator will be consisted of two major parts, namely the adaptive user interface as the front-end of the simulator and the improved learning path optimization algorithm with the consideration of each individual's learning interests as a new learning analytics technique computed by the back-end cloud server. The details about these two major subsystems will be explained in the following subsections.

11.3.1 The Adaptive User Interface of the COMPAD+ Simulator

To facilitate the generation of adaptive user interface in order to suit the learner's interests/needs, after each learner successfully logs into our cloud-based COMPAD+ simulator, (s)he will need to have a ranking of his/her topics of interests in relevant subject area to be stored in the individual's learner profile. For instance, when a learner specifies in his/her profile that the topic of "Binary Arithmetic" as the most interesting topic among all relevant topics in Computer Systems, the cloud-based COMPAD+ simulator will tend to focus more on the detail of the binary arithmetic operations such as binary additions or subtractions during program execution, and therefore "adapt" the generated user interface according to suit this individual's learning interest. On the other hand, when another learner expresses in his/her learner's profile that the topic of "Memory" is the most interesting, the simulation details and generated user interface will be adapted to detail the memory map and stored values inside each memory cell of the underlying computer system.

Table 11.1 shows a listing of 11 topics of interests in the subject area of Computer Systems for ranking by each individual learner in our COMPAD+ simulator. For each topic, a user may give a ranking score from 1 to 10. When two or more topics receive the same ranking score, they will be treated as equally important, and therefore with the relevant simulation details to be displayed in all subsequent simulations until the learner makes any change to the ranking of topics in his/her learner's profile. After the learner gives a ranking of all predefined topics in relevant subject, the ranking scores will be uploaded and stored in the Windows Azure SQL databases as the learners' profiles for later retrieval and processing. Other than being used for generating adaptive user interface to suit each individual's needs, the stored ranking scores of the involved concepts can also be used for learning analytics that will be explained in greater detail in Sect. 11.3.2.

Table 11.1 Topics in computer systems for ranking in our COMPAD+ simulator

Topic number	The involved concept
1	Assembler, linker and loader
2	Binary arithmetic
3	Computer organization
4	Digital circuits
5	Exceptions
6	Memory
7	Microcomputer structure and operation
8	Number systems
9	Physical input/output, interrupt and DMA
10	Programming techniques
11	The 68HC11 microprocessors

11.3.2 The Improved Learning Path Optimization Algorithm for Learning Analytics

Based on the original learning path optimization (LPO) method described in (Tam et al. 2012), we propose in this work a newly revised LPO method that carefully considers both the experts' opinions and the individual learner's interests during the process of finding the optimized learning paths for learning analytics to be performed at the back-end cloud server. In particular, each individual's learning interest will be considered as a new criterion during the optimization process of our revised LPO method. Essentially, both the original and revised LPO methods make use of an evolutionary algorithm to simulate the natural selection process of biological evolutions. Inside our LPO methods, the evolutionary algorithm will produce the first generation of random learning sequence and then iteratively evolve them into learning paths of better quality, that will fit most of the precedence constraints/rules as extracted from the experts' reference paths, through the mutation and crossover operators as occurring in the natural selection process until the whole population is converged to a (sub)-optimal solution or the predefined maximum number of iterations exceeded.

The initial learning paths will be put into a list called "CurrentGeneration" with which the evolutionary algorithm (Affenzeller et al. 2009) will use its *fitness()* function to compute the fitness value of each generated learning path so as to refine its solution quality through the mutation and/or crossover operator. In the original LPO method for learning analytics, the first criterion used to compute the fitness value is mainly the difference between the generated path and the reference path(s). For instance, in our previous prototype implementation of the original LPO method, we considered two reference paths as extracted from two different sources of references (Miller 2004; Tocci and Ambrosio 2003) The two reference paths are shown as below.

$$\begin{aligned} &\text{Reference Path 1}: < 8,2,4,5,6,3,1,7,11,10,9 > \\ &\text{Reference Path 2}: < 8,4,5,6,3,7,11,1,10,2,9 > \end{aligned}$$

where each number in the list denotes the topic number as defined in Table 11.1 in the previous Sect. 11.3.1. For example, the number 8 represents the Topic 8 for "Number Systems" in Table 11.1.

In order to compute the difference between the generated paths and the reference paths as proposed in (Tam et al. 2012), precedence rules need to be extracted from the relevant learning paths first. Considering the above reference path 1 <8,2,4,5,6,3,1,7,11,10,9> as an example, the extracted precedence rules are: (8, 2), (2, 4), (4, 5), (5, 6) and so on. The first precedence rule (8, 2) means that Topic 2 for "Binary Arithmetic" should be learned only after finishing the Topic 8 for "Number Systems." Suppose this specific precedence rule (8,2) is violated in the newly generated learning path as <6,3,2,10,4,8,5,1,7,11,9> in which Topic 2 is placed in front of Topic 8 instead of being placed after it as required by the precedence rule.

Accordingly, the number of violated paths will be incremented by 1. Besides, to measure the extent to which this rule is violated, the index offset is defined as the absolute difference between the positional indices of the two involved topics, i.e., $|5-2|=3$ in this particular case with the positional index starting at 0. The fitness value of each generated learning path is then defined as the product of the number of violated paths and the summation of all the involved index offsets, i.e.,

$$fitness\_value = no\_of\_violated\_paths \times \sum_{i=0}^{k} index\_offset\_rule[i]$$

where k is the total number of rules extracted from any particular reference path.

However, in our enhanced COMPAD+ simulator, each individual's learning interests is also included as a new parameter for a more thorough consideration in the new fitness value to be defined as follows:

$$new\_fitness\_value = w1 \times no\_of\_violated\_paths \times \sum_{i=0}^{k} index\_offset\_rule[i] + w2 \times \sum_{j=0}^{n} index\_offset\_concept[j]$$

where n is the total number of involved concepts for ranking by each individual learner. And $w1$ and $w2$ are the two arbitrarily assigned weightage values to reflect the relative importance of minimizing the violation of reference paths as compared to that of minimizing the violation of the learner's interests as specified in his/her profile. For instance, in our subsequent test cases, we set both $w1$ and $w2$ as 0.5 and 0.5, respectively, thus implying that the relative importance of minimizing the violation of reference paths is the same as that of minimizing the violation of the learner's interests in the overall optimization process of our enhanced COMPAD+ simulator.

11.4 Our Empirical Evaluation

To demonstrate the effectiveness of our cloud-based and enhanced COMPAD+ simulator (Tam et al. 2013) to facilitate students' learning through interest-based and interactive simulation together with learning path optimization as detailed in Sect. 11.3, an empirical evaluation of our prototype of the cloud-based COMPAD+ simulator is conducted and clearly explained in the following subsections. The interactive simulation services are implemented with the JavaScript, Hypertext Markup Language (HTML), and Document Object Model (DOM) on the Microsoft Windows Azure Cloud platform. With the HTML DOM, the JavaScript can access and change all the elements of an HTML document dynamically generated by our cloud-based simulation engine. On the other hand, the enhanced learning path optimizer is implemented as a computing service routine in C#, which can be flexibly and independently run on the cloud server to perform learning analytics for any individual learner based on the collected log file(s).

11.4.1 The Adaptive User Interface of the COMPAD+ Simulator

When a learner tries to access our cloud-based COMPAD+ simulator through any HTML5-supported web browser, the learner needs to log into his/her registered user account so as to retrieve the learner's profile. After a successful login, the learner will be prompted to rank all the relevant concepts/topics based on his/her own interests in a subject area as shown in Fig. 11.2.

After completing the concept ranking, a personalized user interface will be dynamically generated by our cloud-based COMPAD+ engine as based on the learner's interests expressed in his/her concept ranking. From this point onward, all

Ranking of Learner's Interests

Please rank concepts below, then click **Submit** to submit ranking.

Concepts	Score
Number Systems	1
Binary Arithmetic	10
Digital Circuits	3
Computer Organization	9
Microcomputer Structure & Operation	6
The 68HC11 Microprocessors	6
Programming Techniques	7
Physical Input/Output, Interrupt & DMA	2
Memory	6
Exceptions	1
Assembler, Linker & Loader	2

SUBMIT

Fig. 11.2 The user interface of our cloud-based COMPAD+ simulator for ranking concepts

the simulation data, including the animation scripts and displayed values, as generated by the back-end simulation engine will be focused on the learner's interests. Then, the simulation engine will send those simulation data to the web browser with which the preloaded JavaScript program will process the data to generate the final simulation results for display. Figure 11.3 shows the personalized user interface of our COMPAD+ simulator focusing on the learner's interest on the topic of "Binary Arithmetic" in Computer Systems. Since the learner is interested in "Binary Arithmetic," the values on the Accumulator A (ACCA) and Data Register (DR) as used for binary arithmetic operations in the Arithmetic Logic Unit (ALU) are clearly displayed in the concerned user interface to suit the learner's interest.

To better evaluate the prototype implementation of our COMPAD+ simulator, 10 students were invited to use and compare the various learning features and adaptive user interface of our simulator. After experimenting with our COMPAD+ simulator, the students were asked to answer three different questions so as to collect their feedback. Below is a question for students to compare the organization of the general user interface as compared to that of the adaptive user interface of the COMPAD+ simulator to facilitate their learning.

Question 1: Do you think that the layout of adaptive user interface for "Binary Arithmetic" is better organized than the general one to help you more focused on the "Binary Arithmetic"?

After comparing the adaptive and general user interfaces, a total of 8 (out of 10) students "Agree" or "Strongly Agree" the layout of the adaptive user interface for "Binary Arithmetic" is better organized than that of the general user interface to focus on the "Binary Arithmetic." Only two students express "Neutral" to this question, thus thinking that the two interfaces have no difference in helping them

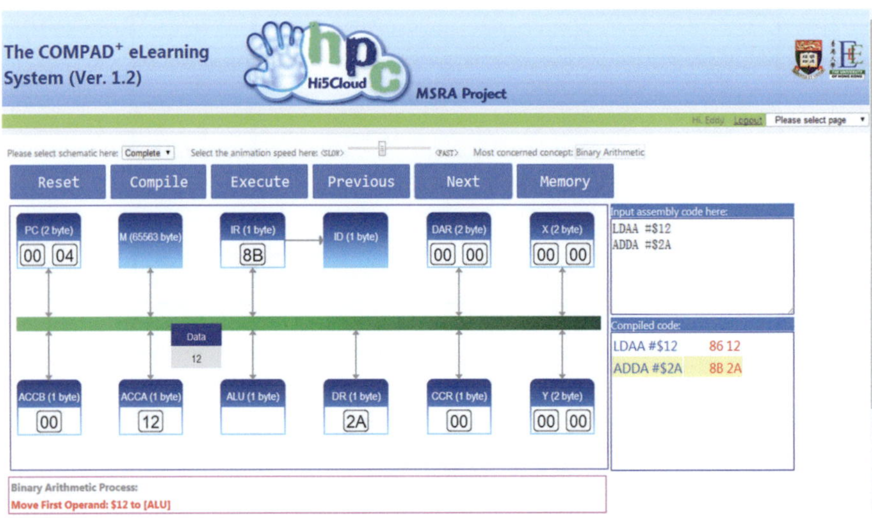

Fig. 11.3 The adapted user interface of our COMPAD+ simulator for "binary arithmetic"

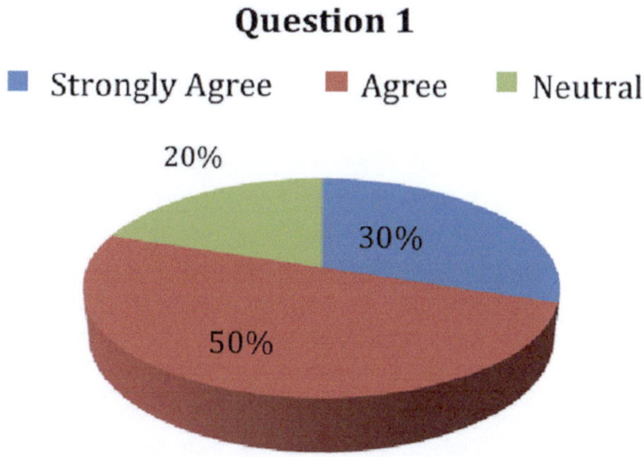

Fig. 11.4 The students' survey results of Question 1 about our COMPAD+ simulator

focused on the "Binary Arithmetic" of the concerned computer system. The distribution of the students' survey results on Question 1 is shown as below (Fig. 11.4).

11.4.2 The Improved Learning Path Optimization Algorithm for Learning Analytics

To evaluate the effectiveness of our proposed learning path optimization schemes, a prototype implementation of the evolutionary algorithm is implemented in C#. In all subsequent test cases, the following two reference paths are considered with the topic number as defined in Sect. 11.3.1 in the subject of Computer Systems.

$$\text{Reference Path 1}: <8, 2, 4, 5, 6, 3, 1, 7, 11, 10, 9>$$
$$\text{Reference Path 2}: <8, 4, 5, 6, 3, 7, 11, 1, 10, 2, 9>$$

The Evaluation Scheme 1 for learning path optimization considers the above reference paths only whereas the Evaluation Scheme 2 considers both the two reference paths and a specific learner's concept ranking as the topics of interest as below.

$$\text{Concept Ranking}: <3, 2, 8, 5, 9, 4, 6, 11, 7, 10, 1>$$

The population size of the evolutionary algorithm is set to 100 with the top 10 % of the whole population selected as the optimized learning paths for further improvements. The probabilities of crossover, mutation, and generating new random paths are 0.75, 0.5, and 0.15, respectively. Table 11.2 shows the computed results with the average and best fitness values obtained from the whole population over 10, 50, and 100 successive runs.

Table 11.2 Results obtained by the two proposed evaluation schemes for learning path optimization

Number of generation	Evaluation scheme 1: considering reference paths only	Evaluation scheme 2: considering both reference paths and user's interest
10	Average violated distance: 66.10	Average violated distance: 83.27
	Average integrated score : 120.61	Average integrated score : 101.49
	Best path: <8,2,4,5,6,7,3,11,1,9,10>	Best path: <3,8,2,5,9,4,6,7,11,1,10>
	Violated distance of best path: 12.8	Violated distance of best path: 57.6
	Integrated score of best path: 57.9	Integrated score of best path: 36.8
50	Average violated distance: 47.43	Average violated distance: 57.26
	Average integrated score: 100.17	Average integrated score: 70.27
	Best path: <8,4,2,5,6,3,7,11,1,10,9>	Best path: <3,8,2,4,5,9,6,1,7,11,10>
	Violated distance of best path: 6.4	Violated distance of best path: 32.4
	Integrated score of best path: 56.7	Integrated score of best path: 33.7
100	Average violated distance: 43.05	Average violated distance: 57.09
	Average integrated score: 90.24	Average integrated score: 69.15
	Best path: <8,2,4,5,6,3,7,11,1,10,9>	Best path: <3,8,2,4,5,9,6,7,11,1,10>
	Violated distance of best path: 4.4	Violated distance of best path: 34.6
	Integrated score of best path: 50.7	Integrated score of best path: 32.8

The first column of Table 11.2 gives the number of generations used in the evolutionary algorithm; the second column shows the results obtained by the Evaluation Scheme 1 which considers the two reference paths only whereas the third column shows the results of the Evaluation Scheme 2 considering both reference paths and the specific learner's interests. From the results obtained, it is obvious that the average of violated distances of the Evaluation Scheme 1 and also the corresponding average of integrated scores of the Evaluation Scheme 2 are decreasing while increasing of the number of generations employed by the evolutionary algorithm. This is due to the fact that both schemes optimize the learning paths based on their predefined criteria.

In addition, when comparing the obtained results across the two evaluation schemes in Table 11.2, it can be found that the learning paths obtained in the Evaluation Scheme 1 tend to have better quality than those returned by the Evaluation Scheme 2 based on the criterion of the violated distances of the two reference paths. This is because adding the specific learner's interests into the criteria of the Evaluation Scheme 2 will diverge the focus of the search, thus degrading the learning paths' quality based on the criterion of violated distances of the two reference paths. However, with a fairer and more objective comparison on the integrated scores, the learning paths obtained in the Evaluation Scheme 2 definitely return learning paths with better quality than those returned by the Evaluation Scheme 1 since both criteria of violated distances to the reference paths and the learner's interests are more thoroughly considered in the Evaluation Scheme 2 as compared to that of the Evaluation Scheme 1.

11.5 Concluding Remarks

In this chapter, we successfully integrated the cloud computing technologies, personalized learning through the learners' profiles to specify their topics of interests and learning path optimization into our interactive COMPAD+ simulator for learners to quickly focus on their interested topic(s) in the dynamic simulations of computer systems on mobile devices. Our proposed framework of the enhanced COMPAD+ simulator is so generic that it can be readily applicable to other subject areas, and easily integrated into other existing e-learning systems. To demonstrate the effectiveness of our proposal, a prototype of our cloud-based COMPAD+ simulator was implemented using the JavaScript and HTML DOM method on the Microsoft Windows Azure Cloud platform for which some initial and positive students' feedback was collected. Furthermore, an enhanced optimization scheme considering both experts' inputs and the learner's interests is devised to provide a more thorough learning analytics on the back-end cloud server. All in all, this chapter reports our ongoing work that has initiated many interesting directions for future investigation including the detailed analysis on pedagogical impacts of using the cloud-based COMPAD+ simulator to facilitate students' understanding of computer systems, the possible integration with other existing e-learning systems and the plausible uses of the cloud-based COMPAD+ simulator together with adaptive and online quizzes to evaluate students' performance for more detailed learning analytics for future investigation.

Acknowledgments The authors are deeply grateful to Dr. Daniel Churchill, Dr. Kinshuk, and Professor Yi Shang for their fruitful discussions and valuable inputs on learning analytics and mobile learning.

References

Affenzeller, M., Winkler, S., Wagner, S., & Beham, A. (2009). *Genetic algorithms and genetic programming: Modern concepts and practical applications* (pp. 1–23). Boca Raton: Chapman & Hall/CRC.

Chen, C.-M., Peng, C.-J., & Shiue, J.-Y. (2008). Ontology-based concept map for planning personalized learning path. In *Proceedings of the IEEE Cybernetics and Intelligent Systems*, Chengdu, November 2008, pp. 1337–1342.

Chuang, H. M., & Shen, C. C. (2008). A study on the relationship among learning path, learning style, and e-learning performance. In *Proceedings of the 7th International Conference on Machine Learning and Cybernetics*, Kunming, July 2008, pp. 2481–2486.

Fernández, A., Peralta, D., Herrera, F., & Benítez, J. (2012). An overview of e-learning in cloud computing. *Workshop on LTEC, 2012*, 35–46.

Fung, S., Tam, V., & Lam, E. Y. (2010). Improving an interactive simulator for computer systems with learning objects. In: *Proceedings of the IEEE 2nd International Conference on Education Technology and Computer (ICETC 2010)* (Vol. 3, pp. 16–20). Shanghai, China: IEEE Computer Society Press, June 22–24 2010.

Hong, C., Chen, C., & Chang, M. (2005). Personalized learning path generation approach for web-based learning. In *4th WSEAS International Conference on E-ACTIVITIES*, Miami, Florida, USA, November 2005, pp. 62–68.

Masud, M. A. H., & Huang, X. (2012). An e-learning system architecture based on cloud computing. *World Academy of Science, Engineering and Technology 2012*, vol. 62, pp. 74–78.
Miller, G. H. (2004). *Microcomputer engineering* (3rd ed.) Upper Saddle River, N.J.: Pearson Education/Prentice Hall.
Mitchell, M. (1998). *An introduction to genetic algorithms* (pp. 2–26). Cambridge: MIT Press.
Pocatilu, P., Alecu, F., & Vetrici, M. (2010). Measuring the efficiency of cloud computing for e-learning systems. *WSEAS Transactions on Computers, 9*, 42–51. ISSN: 1109-2750.
Tam, V., Lam, E. Y., & Fung, S. (2012). Toward a complete e-learning system framework for semantic analysis, concept clustering and learning path optimization. In *Proceedings of the 12th IEEE International Conference on Advanced Learning Technologies (ICALT 2012)* (pp. 592–596). Rome, Italy: The IEEE Computer Society Press, July 4–6 2012.
Tam, V., Yi, A., & Lam, E. Y. (2013). Building an interactive simulator on a cloud computing platform to enhance students' understanding of computer systems. In *Conference Proceedings of the 13th IEEE International Conference on Advanced Learning Technologies (ICALT 2013)* (pp. 154–155). Beijing, China: The IEEE Computer Society Press, July 15–18 2013.
Tocci, R. J., & Ambrosio, F. J. (2003). *Microprocessors and microcomputers: hardware and software* (6th ed.). Upper Saddle River: Prentice Hall.
Tsang, E. (1993) *Foundations of constraint satisfaction* (pp. 305–310). London: Academic Press.
Velicanu, A., Lungu, I., Diaconita, V., & Nisioiu, C. (2013). Cloud e-learning. In *Conference proceedings of "eLearning and Software for Education" (eLSE),* Issue 02, pp. 380–385.
Windows Azure Development Team. (2015). *Microsoft azure: cloud computing platform and services*. Retrieved March 31, 2015, from http://azure.microsoft.com/en-us/.
Wong, L., & Looi, C. (2009). Adaptable learning pathway generation with ant colony optimization. *Educational Technology and Society, 12*(3), 309–326.
Xu, D., Wang, Z., Chen, K., & Huang, W. (2012). Personalized Learning Path Recommender Based on User Profile Using Social Tags. In *2012 Fifth International Symposium on Computational Intelligence and Design (ISCID)* (Vol. 1, pp.511–514). Hangzhou, October 2012.

Authors Biography

Vincent Tam completed his Ph.D. in computer science from the University of Melbourne, Australia in 1998. Dr. Tam was the winner of Innovative Teaching Award (2000) in the School of Computing, National University of Singapore, and also the Best Teacher Award (2011) in the Faculty of Engineering, University of Hong Kong. He is a Principal Lecturer and Honorary Associate Professor in the University of Hong Kong, the part-time deputy director of an EMB-funded project on e-learning packages, and also the principal investigators of various teaching development projects.

Alex Yi received the M. Eng. degree in electrical and electronic engineering from the University of Hong Kong. He is now a Senior Research Assistant in the E-learning Pedagogical Support Unit (EPSU) at the University of Hong Kong.

Di Xu obtained the M. Eng. degree in electrical and electronic engineering from the University of Hong Kong in the summer of 2014. He is now working as an engineer in the mainland China.

Edmund Y. Lam received the B.S., M.S., and Ph.D. degrees in electrical engineering from Stanford University. He is now a Professor in Electrical and Electronic Engineering at the University of Hong Kong, and the Director of its Computer Engineering program.

Chapter 12
A New Wave of Innovation Using Mobile Learning Analytics for Flipped Classroom

Gary K.W. Wong

Abstract Flipped classroom is designed to enrich students' learning experience through active learning activities in the classroom. To prepare the students for these active learning activities, the teachers typically provide pre-recorded video lectures and various computer-mediated learning activities for the students to go through online before the lessons. When students meet with the teachers face-to-face, they are engaged with interactive and collaborative learning tasks. This flipped learning is further facilitated by mobile technology as the students can access these learning materials anytime anywhere on their mobile devices within and outside the classroom. This chapter describes a conceptual model and an initiative of using mobile learning analytics to understand the learners' behaviours inside and outside the classroom under flipped learning approach. Empirical data on the students' perceptions of this initiative is presented as well to supplement the analysis. Issues and implications for designing flipped learning with mobile technology and learning analytics are discussed.

12.1 Introduction

Flipped learning approach is a new way of promoting and supporting active learning (Jensen et al. 2015; Lage et al. 2000; Roach 2014), in which in-person classroom lectures are "flipped" with other learning activities at home. This approach offers one possible solution as a way to realize the student-centred pedagogy and the benefits of active learning, which primarily focuses on bringing activities, promoting student engagement in class, and encapsulating the idea of "learning-by-doing" in the pedagogy (Wong and Cheung 2015). With the emerging mobile technology, students are able to bring their mobile devices anywhere at any time to access the learning materials including videos through the learning

G.K.W. Wong (✉)
The Hong Kong Institute of Education, Hong Kong, SAR China
e-mail: wongkawai@ied.edu.hk

D. Churchill et al. (eds.), *Mobile Learning Design*, Lecture Notes in Educational Technology, DOI 10.1007/978-981-10-0027-0_12

management platform (e.g. Moodle, Schoology), which extends the flipped classroom to mobile platform for more collaborative learning outside the school (Kiger et al. 2012; Wong 2014).

The flipped learning model relies on students' preparation outside the class and in-class interaction (Herreid and Schiller 2013). Yet, this brings challenges for teachers to assess and evaluate their students' learning progress. One way to resolve this issue is through learning analytics, which often refers to the collection, analysis, and reporting of data about learners in their learning context using the techniques of data mining (Ali et al. 2012; West 2012). Learning analytics provides a possible new way of looking into these data and is an emerging research area in educational technology to assist formative assessment (Kumar et al. 2015; Ma et al. 2015; Tempelaar et al. 2014). Besides, the learning analytics has been suggested for an extension to mobile platforms in order to analyse the learning process of students when mobile devices are adopted (Shoukry et al. 2014). However, this area of study is very limited and challenging, and how to take the advantage of the learning analytics is also not well addressed in literature to cooperate with mobile learning experience and even teaching pedagogy (Fulantelli et al. 2015; Persico and Pozzi 2015). In this chapter, it aims to investigate how mobile learning analytics is perceived by the learners in their learning process, which can provide insights to educators in order to understand further the learners' behaviours outside the classroom in flipped mobile learning when it is implemented. Our goal is to understand the students' perceptions on mobile learning analytics so that learning design and formative assessment are more effective in higher education by generating timely and informative feedback for learners (Yorke 2003).

This chapter describes our conceptual framework and an initiative of using learning analytics to predict the learners' behaviours inside and outside the classroom under flipped learning approach with the support of mobile technology in their learning. Issues and implications for designing flipped learning with mobile technology and learning analytics are discussed based on the responses of the learners. More importantly, the implication of how teachers effectively use mobile learning analytics to enhance the quality of formative assessment will be shared.

12.2 Conceptual Framework

In this section, our goal is to develop a unified conceptual framework based on the four components, which support the flipped classroom for active learning, namely *mobile learning, formative assessment, instant feedback* and *learning analytics* as shown in Fig. 12.1. Indeed, flipped classroom for active learning can be done by allocating more time in interactive classroom for providing learners with true mobile learning experience, which supports learners in shifting between contexts directly during the class meeting time (Pegrum 2014), where formative assessment for instant feedback can be accomplished by the learning analytics technology (Boud and Falchikov 2006). To facilitate our discussion in this chapter, the

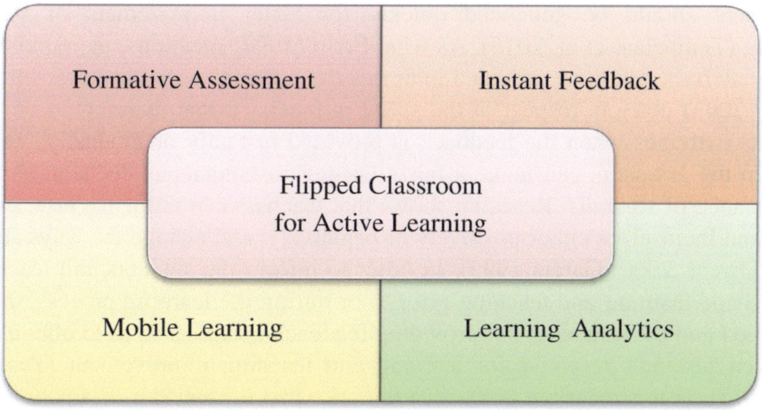

Fig. 12.1 Four major components to formulate the flipped classroom for active learning: Mobile learning, formative assessment, instant feedback and learning analytics

formative assessment and feedback, mobile learning analytics and flipped learning approach will be discussed separately, and a unified conceptual model is presented as the foundation of the learning design in higher education.

12.2.1 Formative Assessment and Feedback

Formative assessment becomes an important component in higher education (Nicol and Macfarlane-Dick 2006; Yorke 2003). It offers both students and teachers a way to improve the learning and pedagogical design, and it sets forth the direction where learning can be present. Sandler (1998) defines this type of assessment as an evaluation which it is specifically intended by the assessors to offer their feedback on students' performance in order to improve and accelerate students learning. Nicol and Macfarlane-Dick (2006) extend this notion to a more central argument by which the formative assessment and feedback are provided to empower students in higher education as self-regulated learners. Although the definition of formative assessment is fuzzy and may be oversimplified, Yorke (2003) suggested that this assessment could be done formally or informally as long as it can contribute to student learning through the provision of information about performance as valuable feedback. The effectiveness of formative assessment relies on the quality of the feedback obtained by learners (Yorke 2003). Researches have shown the value of formative assessment and feedback to contribute to the important parts of the meaningful and effective learning process (Boud and Falchikov 2006; Butler and Winne 1995; Lehmann et al. 2014).

The relationship of formative assessment and feedback is inevitably close and the feedback in return to inform the students concerning the performance assessment should be generated quickly for better improvement of students' learning (Tempelaar et al. 2015). As what Eraut (1994) mentions, the rapidity with which feedback has to be provided indicates the different ways the assessors work. Yorke (2003) notes in respect of the speed of feedback that there could be in two opposite extremes when the feedback is provided instantly or gradually, where in between the assessors can make a fairly but not instantaneous decision about the performance of students. Research shows that teachers can influence how students feel about themselves either positively or negatively, and change the ways students learn (Dweck 2000; Garcia 1995). In order to inform the students and teachers to better shape learning and teaching prior to or during the learning process, the time (i.e. speed and the frequency) of providing feedback is crucial so as to offer a timely and even instantly response for learning and teaching improvement (Tempelaar et al. 2015). In this regard, it is sensible to notice that formative assessment together with the feedback depends on the appropriateness of time and appropriateness in providing feedback.

By all means, it is crucial that the feedback is generated as the outcome of formative assessment to regulate the performance of students and motivate their learning in the progress so that students can become self-regulated learners (Dweck 2000; Irons 2007). Self-regulated learning is indeed the key element in active learning and requires learners to engage recursively in a cycle of cognitive activities when they attempt to complete a given task, where effective learners in this approach rely on feedback (Butler 2002).

12.2.2 Mobile Learning Analytics

With the prevalence and capability of mobile devices (i.e. smartphone and tablet devices) through wireless access to the Internet to obtain bundles of digital learning resources, the educators have called for redesign of pedagogies and practices in instruction and operations to realize the mobile learning opportunities (Brown and Green 2010; Herro et al. 2013; Kiger et al. 2012). It has been observed that educators are taking the advantage of this mobile technology to promote a new way of learning with the objectives to, for instance, improve student engagement, enhance peer interaction and collaboration, obtain feedback from both inside and outside the classroom, support mobile communications and extend the place and time of learning and exploring the world, even analytic tools to enhance learning (Shuler 2009; Kiger et al. 2012). Thus, mobile technology opens many new doors for further investigations in both learning design and pedagogies when devices, learners and the learning experience are all mobile (Pegrum 2014).

In the meantime, with the help of computers and advanced data analysis in this digital age, feedback from the assessment can be given instantly with prediction as well (Tempelaar et al. 2015). Data mining techniques are designed commonly for collecting large scale of data, extracting actionable patterns and obtaining insightful knowledge (De Liddo et al. 2011; Gundecha and Liu 2012). Using these data mining techniques to effectively analyse the interaction data, personal data, systems information and academic information collected from the learning management systems (LMSs), educators can better understand the deep thinking of students during the learning process, as well as capture and visualize their behavioural intention or motivation in learning (Mazza and Milani 2004; Romero et al. 2008). This information will facilitate the assessment of students through a systematic and real-time approach, to identify effective pedagogic changes for particular students (West 2012) and to guide them through the learning process with the ultimate goal of optimizing their learning outcome (Ferguson 2012).

Moreover, with the help of mobile technology, learning analytics can be extended to cover the mobile data analysis for learning on mobile platforms known as *mobile learning analytics* (Aljohani and Davis 2012; Fulantelli et al. 2013), where the learning process of students outside the classroom can be analysed and extended when mobile devices are adopted (Shoukry et al. 2014). The literature discusses a few examples of how learning analytics can be used to assess or evaluate the students' participation in mobile learning:

- A low access rate of online materials even mobile phones with wireless connection may indicate dropout tendency, disengagement or the need for special help by individual students (Deschacht and Goeman 2015).
- Textual mining technique can analyse how much the online posts to discussion forum with their mobile devices are correlated with other peers (Baker and Inventado 2014).
- The video watching behaviours using mobile phones may reveal information about how students learn while watching the video, e.g. how often they rewind or forwards a video lecture and the frequency of students pause at certain scenes (Giannakos et al. 2015).
- Learning analytics can capture how active students learn through mobile devices and their mobile social behaviours with other classmates as a new trend for mobile analytics apps (Chen et al. 2012).

Despite the benefits with mobility extension, analysing mobile data is challenging because of the nature of mobility leading to complicated and non-deterministic social behaviours and interactions, which will require a new way to investigate and solve the issue (Ferguson 2012). At the present, this area of study is very limited and challenging, and how to take the advantage of the learning analytics to cooperate with mobile learning experience, formative assessment and teaching pedagogy is also not well addressed in literature (Aljohani and Davis 2013; Fulantelli et al. 2015). In terms of technical feasibility, mobile mining has been an active research area in data management (e.g. Lu et al. 2011; Musolesi 2014; Nagalakshmi and Sumathi 2014; Zheng et al. 2008). Yet, these techniques are not directly designed for learning behavioural analysis; it is inevitable that educators and computer scientists should work together to provide more pedagogical solutions to extend the existing mobile mining mechanisms to learning management systems. The study of pedagogical and technological challenges in mobile learning analytics is beyond our scope in this chapter.

12.2.3 Flipped Learning Approach

The major objective of the flipped learning is to introduce a new teaching arrangement in which didactic lectures are moved outside of the face-to-face teaching sessions to allow more time for *active learning* in the classroom (Bergmann and Sams 2012; Bishop and Verleger 2013). Active learning involves engaging students in an activity that "forces them to reflect upon ideas and how they are using those ideas" (Collins and O'Brien 2011, p. 5). This reflection, according to Bonwell and Eison (1991), is associated with higher order thinking tasks including analysis, synthesis and evaluation. There is empirical evidence suggesting that the flipped learning model leads to better students' engagement and learning outcomes (Bishop and Verleger 2013). One recent study also suggests that flipped learning model can help increase the learning motivation of students and balance their cognitive load (Abeysekera and Dawson 2015).

In practice, swapping in-class learning with out-of-class learning as compared to the traditional model operationalizes the notion of flipped learning (Jinlei et al. 2012). Before the teaching sessions, students usually participate in out-of-class learning by watching video lectures and completing pre-lesson assignments at home or even on mobile devices (Fulton 2012; McDonald and Smith 2013). This rearrangement frees up more time for in-class active learning activities in the face-to-face sessions since contents have already been delivered in prior (Roehl et al. 2013). This is in contrary to the traditional teaching model, in which the in-class learning component is dominated by didactic lectures, whereas the students have little chance to actively reflect on the contents until they do the homework or revision at home (Wong and Cheung 2015).

There is plenty of empirical literature on how practitioners implement the flipped learning model in their classroom (e.g. Davies et al. 2013; Herreid and Schiller 2013; Milman 2012). A similar approach can be found in several other cases with different subjects and yet positive influences on students' learning such as engineering, business, teacher's education, and medical education (Findlay-Thompson and Mombourquette 2014; Mason et al. 2013; Sharma et al. 2015). With the advanced mobile technology and wireless networking infrastructure, the active and mobile learning experiencing both in-class and out-of-class can be extended anywhere at any time, such as using mobile devices to access the video lectures and completing in-class learning activities using their devices (Kong 2014; Roehl et al. 2013; Smith and McDonald 2013). Therefore, flipped classroom can be integrated with mobile and learning analytics to provide a complete learning experience to support at different stages of learning. Yet, no significant work has been conducted on the impact of learning analytics in mobile-enabled flipped classroom.

12.2.4 Unified Conceptual Model for Active Learning

To formulate a unified conceptual model for active learning, it is attempted to integrate the models on learning design and analytics with formative assessment into our *4-step Flipped Learning Model* (Wong and Cheung 2015). First of all, learning design, defined by Lockyer et al. (2013) as "the sequence of learning tasks, resources, and supports that a teacher constructs for students over part of, or the entire, academic semester" (p. 1442), provides a complete picture with a series of premeditated pedagogical actions for learning and teaching. Figure 12.2 shows the basic model to conceptualize the relationship among the three common categories, namely *resources*, *tasks* and *support mechanisms (R–T–S)*. As suggested, the resources can be files, diagrams, questions, web links, pre-readings, and videos particularly in flipped learning approach. Students can access these materials in order to complete the expected tasks at different times of learning stages. In the meantime, the support mechanisms should be designed to assist the learners to complete the assigned tasks, where peers and instructors can both serve as the support. Learning analytics stipulates a passive approach of collecting and organizing information on how students interact with learning resources, each other and their teachers (Lockyer et al. 2013), which can be aligned with the design of learning.

Several generic models in learning analytics exist (Greller and Drachsler 2012; Xing et al. 2015), and the classification from Lockyer, Heathcote and Dawson framework (2013) has been adopted. Two broad categories of analytic applications can be classified, which are *checkpoint analytics* and *process analytics*. According to the definition, *checkpoint analytics* is considered as the "snapshot data that indicate a student has met the prerequisites for learning by accessing the relevant resources of the learning design" (p. 1448). One example would be the record indicating the login time and frequency of students in learning management system; *process analytics*, in the meanwhile, is defined as those data which can "provide direct insight into learner information processing and knowledge application within the tasks that the students completes as part of a learning design" (p. 1448). For example, the textual analysis on the coherence of students' comments collected from a discussion forum as a part of the assigned tasks can bring insights about how deep the students engage in the discussion and the association with the topic.

In terms of the formative assessment, based on Shum and Crick's learning dispositions model (2012), meaningful data being collected and analysed by the

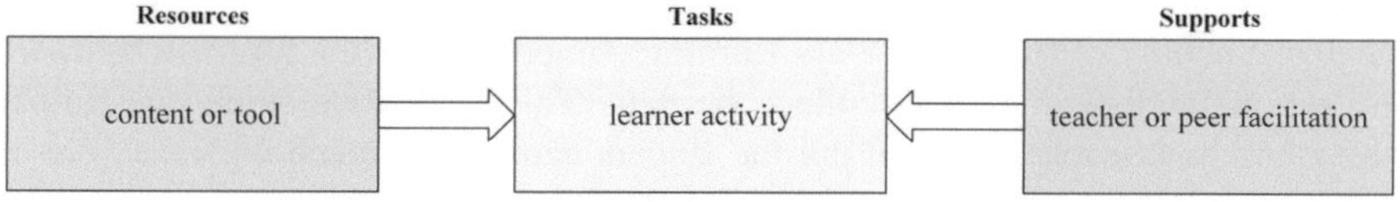

Fig. 12.2 A learning design with three common categories (resources, tasks and supports) (Lockyer et al. 2013)

system should reflect the relatively enduring tendency of learners behaving in a certain way, which can be linked up to motivation, affect and valuing. According to the argument behind the analytics model, some cognitive behaviours typically reveals higher order information about one's state or intentions for instance, intentionally self-disclosing 'metadata' about themselves, having the knowledge or expectation that it may be sensed and possibly acted on by people or machines, known to or hidden from them (Shum and Crick 2012). With the collection of this type of data, feedback can be provided to students to better shape teaching and enhance learning during the process when the intentions of students are captured purposefully and knowingly.

Figure 12.3 shows the unified conceptual framework, which captures the flow of flipped learning approach with the support of mobile learning analytics and the types of data for formative assessment. In the 4-step flipped learning model (Wong and Cheung 2015), the R–T–S fits into different stages where the analytics can facilitate the formative assessment about the learning behaviours. Note that mobile and non-mobile learning can be introduced seamlessly to enhance the experience wherever appropriate. With this unification of the models, this conceptual framework is the guiding model for the design of authentic flipped learning experience.

12.3 Research Design and Methodology

12.3.1 Research Objectives

While there is an increasing interest of research in flipped classroom to study the learning motivation, learning design with interactive activities, and effectiveness of teaching and learning with video lecturing (e.g. Bishop and Verleger 2013; Chorianopoulos et al. 2014; Mirriahi and Dawson 2013; Wong and Chu 2014), no initiative has been done to study the perception of learners and their change of learning behaviours when mobile learning experience and analytics are implemented. In this study, the main objective is to offer the empirical data and analysis on the students' perceptions to mobile learning analytics with the flipped learning approach, where the perceptual constructs are aligned to the learning dispositions of three different types in Shum and Crick's model suggested by Tempelaar et al. (2015) learning motivation and engagement, learning styles and attitude.

To investigate the use of learning analytics in flipped classroom, the author conducted a case study to analyse the learning analytics collected from two undergraduate courses that used the flipped learning approach. This section documents the setting of the out-of-class learning components of the flipped classroom and the methodology used to collect the data. This qualitative study can further strengthen how teachers should put the student-centred approach with the help of learning analytics as well as the design of the future analytics tools.

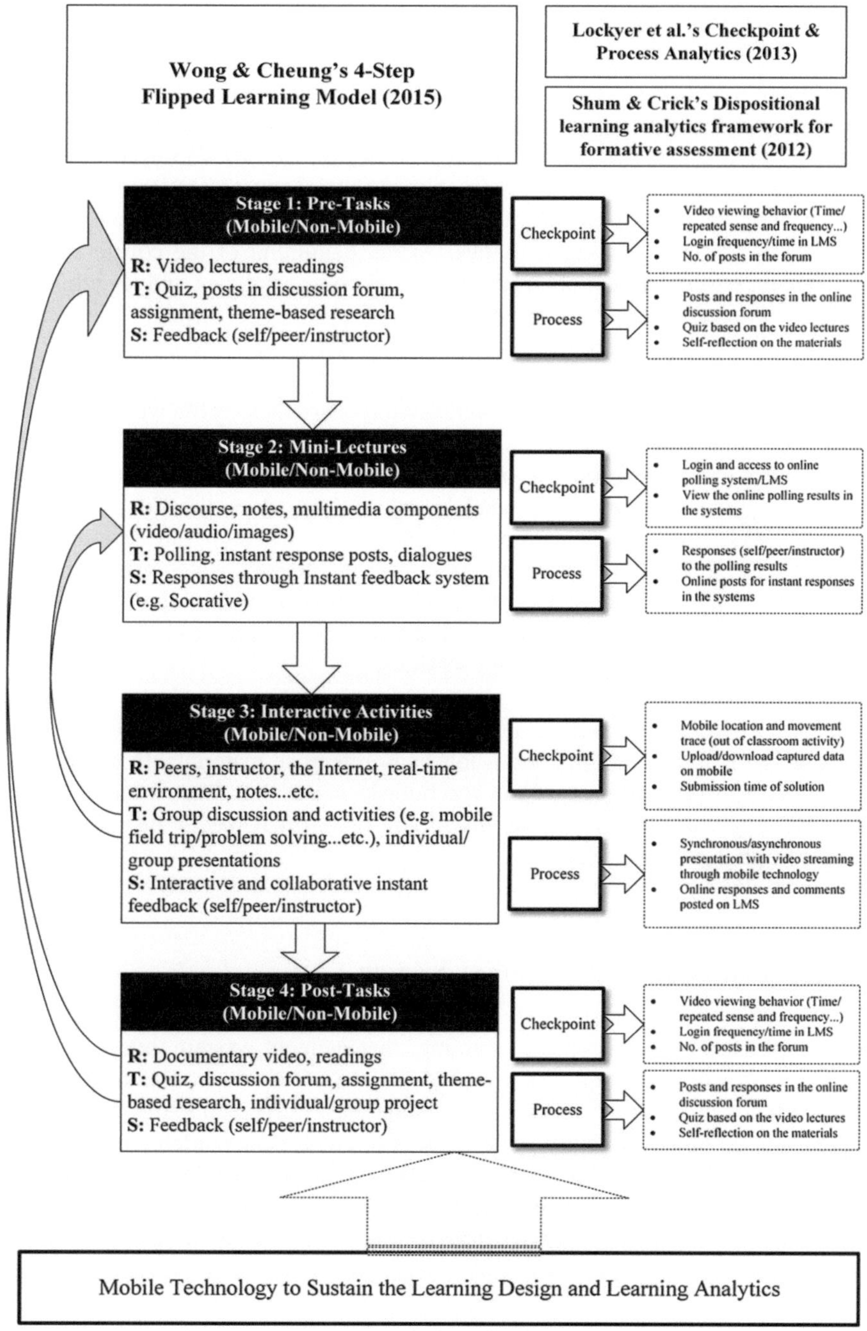

Fig. 12.3 Conceptual model of active learning in flipped classroom supported by mobile learning analytics

12.3.2 Context of Study

Two undergraduate courses (mathematics education and general education) taught in Hong Kong by the author were selected for the study. One of the courses was taught in the summer semester and fall semester of the 2014/15 academic years, while the other was taught in the spring semester of the same academic year. Both courses adopted the flipped learning model consisting of out-of-class learning followed by in-class learning activities. At the beginning of the course delivery, these students were briefed about the role of learning analytics and its implementation as a reference to their mobile learning experience through their mobile devices in this flipped learning and teaching approach.

The out-of-class learning components consisted of video lectures that were either prepared by the teacher or adopted from existing resources on the web. The videos prepared by the teachers were segmented into short clips (5–8 min) and each focused on a particular sub-topic of the lesson. The students were asked to watch the videos regularly before the lessons. In one of the courses, the videos taught the concepts and procedures required to complete in-class lab exercise and problem-solving activities. In the other course, the students watched the videos to gain the foundational knowledge of the topics followed by online discussion and peer reviews. Afterwards, they prepared individual presentations on the topics to be delivered in the face-to-face teaching sessions.

To facilitate the students to access the videos, the links to the videos were posted to Schoology, a cloud-based LMS service that was free of charge to use. Also, available on Schoology were course notes and reference materials. In addition, the teacher set up discussion threads in the built-in discussion forums and asked the students to participate in online discussion with their peers or submit their pre-lesson preparation works.

12.3.3 Learning Management System

Schoology is a mobile-friendly learning management system adopted in the teaching and learning of the courses. Students can access the course materials and take part in the learning activities via its mobile view or the Schoology app installed in their mobile phone. Using mobile devices to engage the students in the flipped classroom could extend their learning experience outside the classroom and afford the reflective engagement (Kong and Song 2015; Smith et al. 2015). Schoology offers various educational-based features such as document and multimedia sharing, discussion forum, announcements and updates, assignment and assessment, polling, mobile native apps version and most importantly the built-in analytic tools. YouTube was chosen to host the videos because it offers more features targeting to video sharing, such as high storage volume, various selections on video playback quality, completely available on various mobile operating system platforms and targeted video learning analytics. In the current study, YouTube videos were embedded through adding their links to the Schoology. Students could then link to these videos from Schoology and watch them anytime anywhere.

12.3.4 Participants

In the academic year 2014/15, a total of 114 students from the author's courses were taught using flipped learning model; 53 students were from year 3 of the mathematics education programme; 61 students were from general education courses ranging from year 1 to year 4 in various majors. All of them experienced the whole process of flipped learning where both in-class and out-of-class activities were logged and analysed either by the Schoology, YouTube studio (lecture video host) or by mobile instant feedback system (e.g. Socrative). Their biographic backgrounds were not collected explicitly in this study. This study adopted convenient sampling (Creswell 2009) to select the students from the author's teaching courses to participate in the study, and these students registered the sections of the author on the self-decisional basis as there were other sections/courses available to them in the semesters.

12.3.5 Instruments and Procedure

Learning analytics in Schoology and YouTube was used for data collection. In Schoology, there are a few built-in analytics such as Course, User, Assignment, Discussion and Links. For example, Schoology can show the total hits per day in the course on Course Analytics. The User Analytics captures the "Last logged in", "Last course access", "Total time in course" and "Number posts". The Assignment Analytics reports the total views on the assignment description. The Discussion Analytics shows the number of posts in each discussion forum. The Links Analytics provides the count of clicks to the URL provided by the instructors.

While Schoology provides useful access statistics to course materials in general, it does not offer any analytics to monitor the video watching activity because the latter were hosted in YouTube and added to Schoology as embedded hyperlinks. It was therefore necessary to rely on the tools in YouTube to observe access rate of video lectures. Particularly, the Analytics available in the Creator Studio of YouTube offers some sophisticated features to capture the activities of users when they watch these video lectures. For example, it tells the number views in overall, estimated minutes watched, number of Likes/Dislikes, number of comments, and even their demographics (e.g. playback location and gender). Statistics in each individual video are available as well.

Table 12.1 Open-ended questionnaire about the student's perceptions in learning analytics

Question #	Questions
1	How does the learning analytics affect your learning motivation and learning methods in this course? Why?
2	Do you think the use of learning analytics by the teacher will be helpful to your learning? Why or why not?
3	What is your feeling about the teacher's use of learning analytics to analyse the process of your learning?
4	How will it change your learning if the learning analytics data is available to you as students instead of only being available to teachers?
5	What other analytics data should be introduced to help understand better the student learning?

The study was also supplemented by questionnaire data on the student's perceptions of the learning analytics. Only the students from one of the general education courses were sampled to participate in this questionnaire since their dynamic background can provide with various insights in respect to their different major studies. The questionnaire consisted of the following open-ended questions distributed to the students in one of the courses in the study (see Table 12.1).

The questionnaire was made using online Google Form available through embedding to the Schoology. The students were invited via the Schoology where they could indicate their voluntariness in the questionnaire. These students were from different major departments, and their biographic data were not collected, as they are not of concern in this study. All student participants were assured of the confidentiality of their identity in data reporting based on the consent form they submitted before the beginning of the questionnaire. Each participant was expected to complete the questionnaire in approximately 10 min. The data collection took 2 weeks and their submission is anonymous.

12.3.6 Data Analysis

Since the learning analytics tools in the system provide the analysis with visualization as the presentation, no further data analysis in their participation and interactions was required in respect to students' perceptions. Concerning the questionnaire, the questions are all open-ended and therefore the answers are qualitative and unstructured. The responses were mainly in Cantonese while some were written in English. The author translated them into English verified by other colleagues before analysing it qualitatively using the iterative coding process in Creswell (2002) to identify the categories, themes and patterns that emerged from the data. There were 36 students from the spring semester 2014/15 being invited to participate in the questionnaire; 19 students responded completely to the questionnaire (57.8 % response rate). These students were assigned sequence numbers of 1–19. Selected excerpts are translated to English (wherever necessary) and tabulated in Table 12.3. Each excerpt is marked with a [#n] in the end where n is the sequence number of the student who submitted the questionnaire with responses.

12.4 Results and Analysis

This section presents the selected learning analysis about the students' participations and learning behaviours as the examples of how learning analytics was collected and studied. The analysis and implication of students' perception on these learning analytics implemented in the course are also presented and discussed.

12.4.1 Access to Course Notes and Learning Activities

The following online viewable analytics data were collected from the course during fall semester 2014/15. The visualization of the learning analytics concerning the behaviour of students outside of the classroom was observed periodically on the weekly basis. The different types of statistical analysis shown in the following figures highlight the participation and learning process of the class and the individual students. Figures 12.4 and 12.5 show the students' frequency of access to the materials in Schoology and their activeness in participating all the tasks assigned

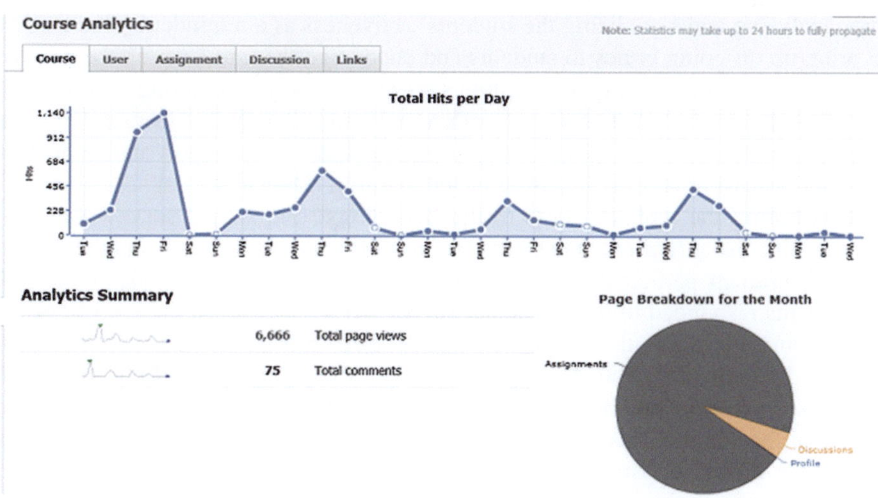

Fig. 12.4 Course analytics with total hits per day and overall summary

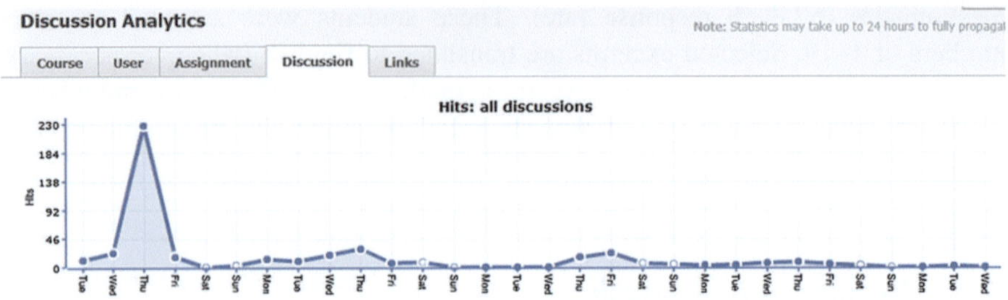

Discussion	Posts
Discussion 1: Principles and Theories of IT and Mathematics Education	18
Discussion 2: Grouping	12
Discussion 3: Self Reflection on Lesson Plan Design	22
Discussion 4: Simple GeoGebra Showcase	16
Discussion 5: Reflection on Technology in Mathematics Education	0
Angry Bird Showcase	0
Total	68

Fig. 12.5 Discussion analytics and summary of post frequency

outside the classroom. Specifically, Fig. 12.4 shows that students usually had a high peak participation rate in Schoology around the time for face-to-face lessons. A similar pattern is observed from other activities such as online discussion (see Fig. 12.5).

Indeed, this learning analytics could provide instant insights to teachers about how active students are participating in the learning process, and if they are making progress on a daily basis (see Fig. 12.6). For example, emails were regularly sent after analysing and visualizing the students' activeness as a reminder by the teacher to bring up on-going issues to students and encourage them to keep up the pace of learning. In flipped learning, the analytics serve as the key-monitoring scheme with visualization so that teachers can understand more about the learners in their mobile-supported learning especially when the LMS can support native mobile apps for accessing and participating in the learning (Dyckhoff et al. 2012).

Concerning the mobile access to the Schoology, one simple survey was conducted with the group of students at the end of each semester to understand the medium used in participating the flipped learning. A total of 70 out of the 114 participants responded to this particular survey (61.4 % response rate). 93 % of the respondents indicated that they have used their mobile devices to access to Schoology for at least once. Among them, 46 % of them used their iPhone/iPad and 45 % used Android mobile phones to access to Schoology. Table 12.2 shows the results of the survey. This may indicate that mobile learning experience could be a part of the flipped classroom approach learning and teaching.

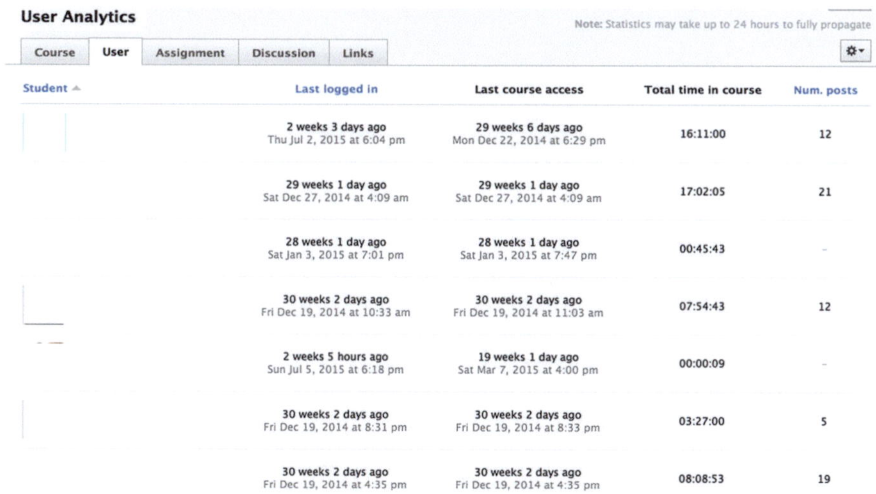
User Analytics

Note: Statistics may take up to 24 hours to fully propagate

Course | User | Assignment | Discussion | Links

Student	Last logged in	Last course access	Total time in course	Num. posts
	2 weeks 3 days ago Thu Jul 2, 2015 at 6:04 pm	29 weeks 6 days ago Mon Dec 22, 2014 at 6:29 pm	16:11:00	12
	29 weeks 1 day ago Sat Dec 27, 2014 at 4:09 am	29 weeks 1 day ago Sat Dec 27, 2014 at 4:09 am	17:02:05	21
	28 weeks 1 day ago Sat Jan 3, 2015 at 7:01 pm	28 weeks 1 day ago Sat Jan 3, 2015 at 7:47 pm	00:45:43	-
	30 weeks 2 days ago Fri Dec 19, 2014 at 10:33 am	30 weeks 2 days ago Fri Dec 19, 2014 at 11:03 am	07:54:43	12
	2 weeks 5 hours ago Sun Jul 5, 2015 at 6:18 pm	19 weeks 1 day ago Sat Mar 7, 2015 at 4:00 pm	00:00:09	-
	30 weeks 2 days ago Fri Dec 19, 2014 at 8:31 pm	30 weeks 2 days ago Fri Dec 19, 2014 at 8:33 pm	03:27:00	5
	30 weeks 2 days ago Fri Dec 19, 2014 at 4:35 pm	30 weeks 2 days ago Fri Dec 19, 2014 at 4:35 pm	08:08:53	19

Fig. 12.6 User analytics indicating the login information

Table 12.2 Mobile learning with access to Schoology (n = 70)

Main purpose of accessing to Schoology with mobile devices	Counts	Percentage %
Watching video lectures	26	16
Posting comments in discussion forums	38	23
Read updates and announcement	48	29
Access and read additional learning resources	17	10
Send and receive Schoology inbox messages with instructor	6	4
Download course materials (course outline, assignment…etc.)	30	18
Others	2	1

12.4.2 Access to Video Lectures

Figures 12.7, 12.8 and 12.9 show some examples of general statistic about viewing frequency and estimated minute watched by the students available in YouTube. By aligning the YouTube statistics with Schoology statistics by date, it shows that students could be focusing on learning through watching video before attempting to complete assignments or respond to comments in Schoology. This demonstrates that students accessed to video lectures as a preparation for further learning activities including face-to-face classroom participation, and the teacher could conduct follow-up activities to evaluate their understanding of the video contents.

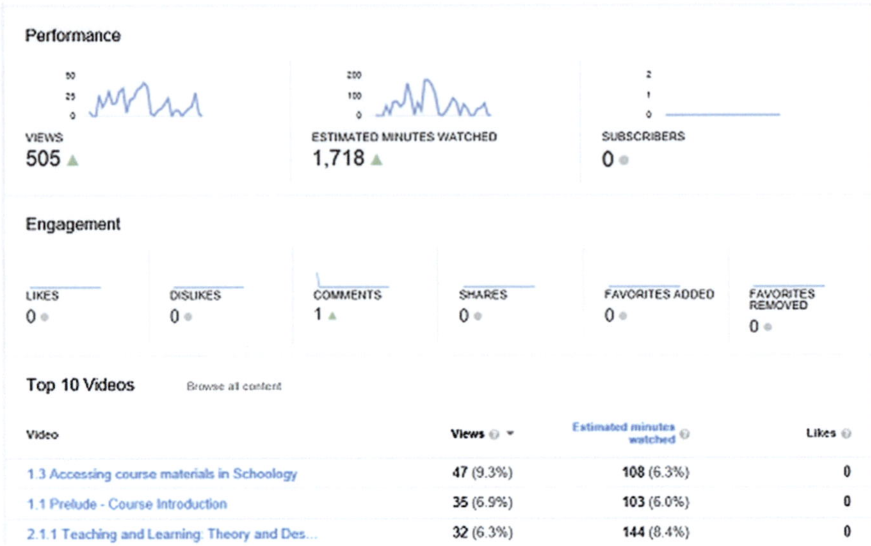

Fig. 12.7 Example of summary page of video watching analytics in YouTube

Fig. 12.8 Example of number of view versus date in YouTube

12.4.3 Access Through Mobile Platforms

In addition, Figs. 12.10 and 12.11 show the statistics related to mobile learning in particular. From the analysis we can see that roles of mobile devices served as complimentary extension to the stationary learning. The statistic shows that 83 % of

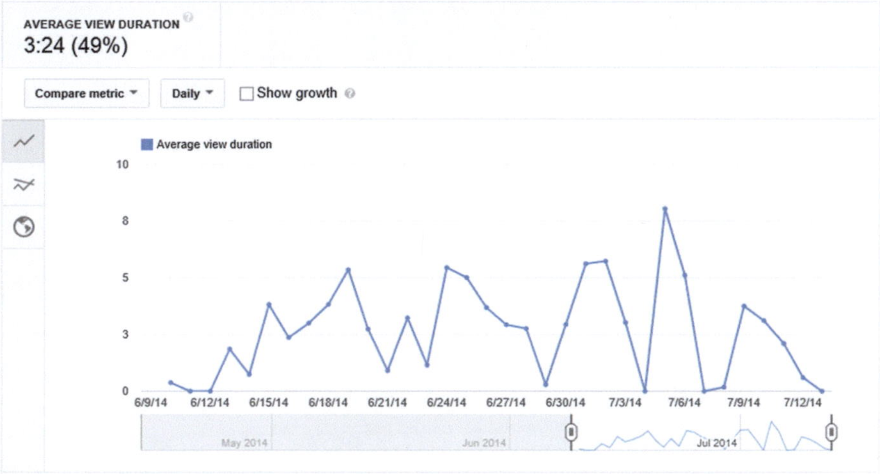

Fig. 12.9 Example of average view duration (minutes) versus date in YouTube

the views were through Windows/Macintosh, which is assumed to be relatively stationary than iOS and Android. This may be due to the screen size and watching experience outdoor (e.g. bus). Surprisingly, on the other hand, the average time spent on watching video with mobile devices is 3:46 min, which is slightly more compared with 3:39 min on Windows and 2:19 min on Macintosh. This means students tend to spend a little longer on watching video through mobile devices, and perhaps it is very likely that the students were on taking public transportation where they had less choices to be distracted by walking away from the video compared to home environment.

12.4.4 Students' Perceptions on Mobile Learning Analytics

Table 12.3 shows the selected responses of students from the questionnaire, which illustrate various perspectives concerning the influence of learning analytics. Based on the coding analysis, the three major learning dispositions were reflected: learning motivation and engagement, learning styles and attitude, which are aligned with the Shum and Crick's model suggested by Tempelaar et al. (2015). Some general perceptions of learning analytics impacted to their learning are exemplified from responses (a) to (g), which include learning effectiveness, student' activeness and engagement, learning progress tracking, increase of competitiveness among students, formative assessment for better improvement in classroom teaching, and evidence of students performance. Yet, some students may not have seen the possibilities with the learning analytics because they do not think it poses any impact on their learning with merely the data itself. As being exemplified by the

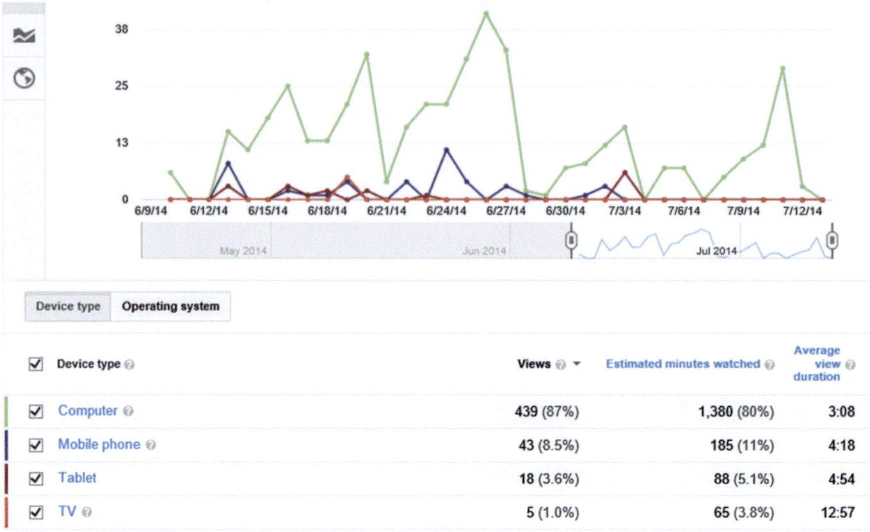

Fig. 12.10 Mobile device types versus date in YouTube

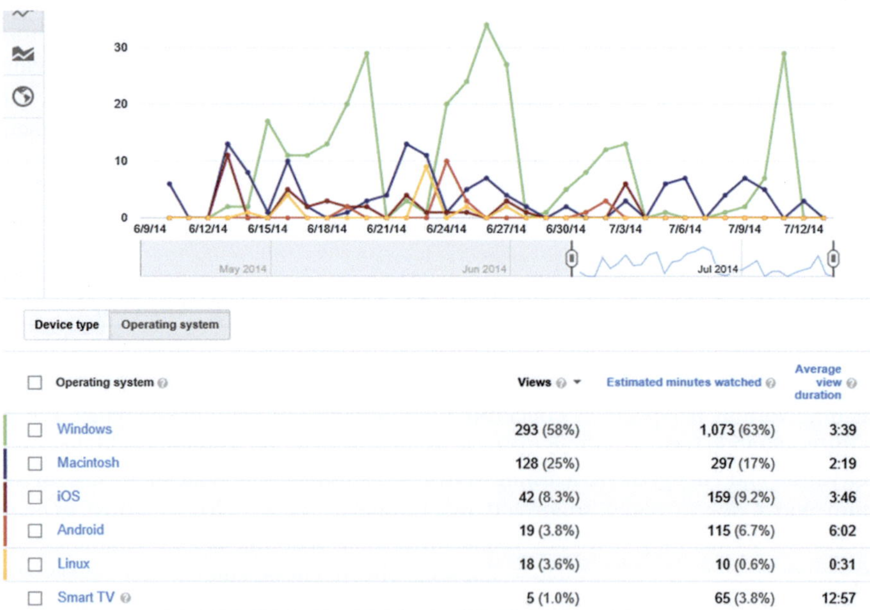

Fig. 12.11 Mobile device operating systems versus date in YouTube

Table 12.3 Selected excerpts from the questionnaire of students' perceptions

Theme	Sub-theme	Examples
General perceptions	Positive	(a) "It can help identify the level of learning activeness among the students." [#08] (b) "The teacher can know our learning progress through reading these [analytics] data so as to adjust the teaching strategies." [#12] (c) "It encourages us to place more time to complete the planned activities by the teachers and hope to do a better job." [#17]
	Negative	(d) "The teacher can track when I login, when I finish [the discussion post or assignment], it creates an intangible pressure psychologically." [#7]
	Neutral	(e) "…[it] neither creates a negative influence nor a positive influence to my learning effectiveness. Because I think learning effectiveness is really based on the teaching quality and the desire or learning motivation of students." [#02]
	Mixed	(f) "I think it will have positive influence, but it depends on whether the teacher will really analyze the data after collecting them." [#01] (g) "Because the teacher can base on our usage of different learning resources [on Schoology] to understand our needs, and then change the teaching plan and approach in order to improve our learning effectiveness. But, in the perspective of students, these data may not clearly help us realize our learning effectiveness." [#3]
Learning motivation and engagement	Active learning	(h) "We will be more active and enthusiastic to use the resources and participate in discussion [forum] on Schoology." [#03] (i) "It motivates me to complete some assignments like the video reviews far more days before the deadline…I know the teacher can know the exact time of when I complete the task, which in term provides me the incentive to complete the assignment as earlier as possible." [#14] (j) "I will spend more time to complete the [learning] activity, and constantly check if there is any updated information, so from this I know I have more motivation to learn." [#17]
	Passive learning	(k) "When I know the teacher can see the frequency of student login and time, it does increase my frequency to login to Schoology, however, it is not because I want to learn, it is just because I am afraid Schoology has no record of my login." [#01]

(continued)

Table 12.3 (continued)

Theme	Sub-theme	Examples
		(l) "No…because most of the time I will use only Schoology when I need to complete assignment or classroom activity or feel interested in the lesson. So the collected data [currently available in Schoology] does not impact my learning motivation." [#09]
	Neutral	(m) "It may help make a small increase [of my learning motivation], but it is not obvious." [#02]
Learning styles	Traditional learning	(n) "Learning methods depend on personal habit, and it is hard to be changed unless being trained from the early beginning or by requirement." [#01] (o) "I think the collected data has nothing to do with my learning methods." [#06] (p) "I believe my learning methods is to follow the teacher's particular instructions to complete the assignments to learn different knowledge." [#07]
	New learning	(q) "Yes…it has positive effects on my way of study, mostly due to the increase in my learning motivation. As I mentioned, it motivates me to finish the assignment many more days or even weeks before the deadline. I have learned to organize my time better." [#14] (r) "[Learning analytics] can help me make better use of technology and multimedia to learn, and [learning] is no longer bounded by paper and pencil." [#15] (s) "It will affect my learning methods…we are more motivated to browse the materials by the teachers and use them to learn." [#17]
Attitude	Feedback and interactions	(t) "Knowing that collecting these data only because the teacher want to get to know us and compensate for our learning concern, so the feeling is quite good." [#01] (u) "I like to let the teacher knows my effort and have appreciation through this interaction." [#02] (v) "I think it is a way that the teacher shows his or her care for students…[The teacher] can know how the students are going, and for students who never go to Schoology, the teacher can provide some reminder." [#14]
	Monitoring/surveillance	(w) "At the beginning, I feel being monitored that I must use Schoology more. But later on I find that I should just use it when I need it. Because the current collected data does not reflect how serious we learn, it is only used to reflect whether the learning resources are suitable for us." [#03]
	Threat	(x) "Honestly, it is a pressure [to learn that the teacher uses learning analytics], but I am still welcome for the teacher to collect [data for learning analytics]." [#07] (y) "A bit frightening at the beginning, it is getting better gradually." [#16]

responses (d), students can be annoyed by the "surveillance" using the learning analytics tool.

Learning effectiveness is related to the teacher's performance and quality according to the students' understanding as perceived through the selected response in (e). It may not be solely the presence of learning analytics. As indicated in response (f), teachers should recognize how to take the advantage of the learning data and interpret them appropriately before it may create a wrong impression to students similar to the response in (g). Nevertheless, students may not acknowledge that teachers can improve their teaching with the extra information about the students through the learning analytics. If teachers are willing to put extra time to analyse these data, it may be able to provide further positive influence to students.

In terms of learning motivation and engagement, students find that learning analytics can serve as motivating factor exemplified in responses (h) to (j). Students may see it as a tracking tool so that every task they perform on Schoology is recorded. In this regard, students see themselves more active and motivated to participate in learning with these tools. Although it seems to only develop their extrinsic motivation with a fear not to participate to cause penalty (e.g. response (k) and (l)), some students believe that teachers can learn more about the effort that students have paid to complete each task through it (e.g. response (i)). Thus, students can be motivated to do better performance in the online learning (either through mobile access or not). Yet, some students exemplified in response (m) demonstrate the neutral impact in motivation because they think the existing collected data cannot reflect their actual performance (e.g. login frequency) nor pose issues on their grade. In this sense, it does not matter if they should work hard for the "statistical numbers". In general, the extrinsic motivation is observed when learning analytics is implemented in the teaching.

Meanwhile, students in responses (n) to (s) illustrate their styles of learning through their developed habit in traditional learning without the cause from the learning analytics, and some indicate their new ways of learning and perspective in changing their learning methods because of the extra information collected and analysed by the teachers, exemplified in responses (q) to (s). Particularly, the response (r) shows that technology and multimedia are encouraged to use more frequently in learning with analytics because the tracking record on using traditional methods cannot be achieved. Thus, technology can be the complementary aspect that students can appreciate more in their learning experience.

Generally, various feelings and attitudes towards the use of learning analytics were identified in the theme. Some believe that it is a good companion tool to the teachers because they can improve their teaching quality through a better understanding of their students indicated in the responses (t) to (v). The use of learning analytics can also show the care about the students' learning experience. However, some see it as a monitoring purpose, which can generate extra pressure and threat to the students showing in the responses (w) to (y). They may worry that the teacher can use it wrongly to misinterpret their learning effort. In response (w), it is true that some data like *checkpoint analytics* may not reflect the learning progress without further analysis with *process analytics*. Nevertheless, some suggest that the teachers should inform the students with more details earlier about the use of it so that they will not feel surprised being monitored outside the classroom in online learning environment.

On the other hand, some students suggest that a similar learning analytics should be available to the individual level because of the usefulness although some students do not see its usefulness to make it accessible to students. Some even oppose it to be available publicly at the peer level because they may be afraid of the privacy issues or being monitored by their peers in the online environment. Yet, some students believe that sharing it to the peers can increase the learning motivation and learning attitude due to the influence among the peers. Among other responses, students seem to have a very limited idea of how to extend the existing features to become more appropriate in assessing the performance of students formatively. Several additional useful features in learning analytics were suggested (either for teachers or students) which may be helpful to their learning effectiveness, including

- Names of students in their browsing history record;
- Personal browsing time in each learning page/material;
- Records showing a list of activities each student performed after login;
- Learning progress of each student in assigned tasks;
- More descriptive/qualitative analytics information other than statistical information; and
- Highlight of questions/concerns/inquiry in each lesson based on students' responses.

12.4.5 Discussions and Implications

The above results indicate that the learning analytics can support mobile learning experience inside and outside the classroom positively in general and can facilitate the students' learning under a flipped learning environment. The findings exemplify how the learning analytics can be used to reveal the students' accessing patterns of the learning materials as well as their mobile learning behaviours. The students' perceptions on the existence of mobile learning analytics in the flipped classroom show a positive impact on the learning motivation, learning engagement, learning style and attitude towards learning, which are considered as the key learning dispositional data in learning analytics (Tempelaar et al. 2015). The results imply that learning analytic is a mandatory affordance to sustain flipped classroom teaching and learning, and it is needed in order to capture the whole learning process of each student for continual monitoring purposes both inside and outside the classroom (Gilboy et al. 2015). It helps the teachers to determine the activeness of participation before each lesson, partly addressing the issue raised by Butt (2014) in which the students tend not to prepare for the lessons. With the feedback from the analytic tools, the teachers could gain insights and adjust the planned agenda in each upcoming lesson to accommodate the needs of students, i.e. giving a mini lecture to recap some of the important skills in the video.

From the findings, the motivation and engagement in the absence of teachers can be sustained with the mobile learning analytics. The importance of this is that learners in flipped classroom are required to pay some attentions in self-learning activities online and outside the class. The students, particularly in Hong Kong, are highly mobile and it is imperative to design the learning activities with mobility support at all time (Lam and Duan 2012). Without the learning analytics as a way to track and monitor the learning progress or their participation, the in-class interaction will become minimal. Based on the perceptions, learners seem to be willing to spend more time on the learning activities and enjoy showing their participations virtually to gain an appreciation from teachers. In viewing this perspective, teachers in flipped classroom should ensure their facilitating roles with the help of learning analytics by knowing that students are making progress in their learning and showing their appreciation for the participation of students. In doing so, students can be more motivated with good attitude to engage in learning activities both individually and collaboratively.

However, the findings from the participants are coherent with the arguments by Campbell et al. (2007) and Dringus (2012) that learning analytics could be harmful and threatening if not being implemented appropriately. For example, students could feel being monitored and threaten when every movement and online learning behaviours is tracked. Suggested by Dietz-Uhler and Hurn (2013), the pedagogy should drive the development and implementation of learning analytics but not the other way round. Meaningful data, having transparency, yielding good algorithms, leading to effective use of the data, and informing process and practice, should be the guiding minimal principles and requirements in order to gain a success in the usage of learning analytics (Dringus 2012), which is also applicable and even more crucial to flipped classroom. The quality of formative assessment depends on how well the analytics is used to align with the learning design. As long as the usage and analysis of the learning data are implemented based on appropriate pedagogy and curriculum, learning analytics can bring positive impacts to learners (Snodgrass Rangel et al. 2015).

Another concern of negative impact on learning analytics may be affected by the scarcity of learning systems implemented with the features that demonstrate the innovative possibilities for learning and teaching (Drachsler and Greller 2012). In this manner, learners may not have a clear point of reference as, for example, in the case for LMSs (e.g. Schoology, Moodle, Blackboard…etc.) where an established choice of competitive platforms exists, which is also reflected from the participants' responses. For example, there are limitations on the existing analytical tools in Schoology for mobile learning (Aljohani and Davis 2012). Schoology could not distinguish if students spent more time on mobile access or not; it could not help students reflect upon their own learning behaviour with individual learning analytics; it did not provide its customized analytics to analyse the data about the video watching. In terms of evaluating the writing of students on discussion forum, Schoology could not help understand whether the words were correlated to other peers' comments, nor could it automatically send a reminder to students if they were absent from the system for too long. These limitations exist in other LMSs as well such as Moodle, which offers no deep insight unless more customized tools are developed (Retalis et al. 2006; Petropoulou et al. 2014). In addition, YouTube does not allow tracking of individual viewers. Nevertheless, further analysis could be conducted through in-class interactive activities or online quiz to find out if the students were making any progress after watching the video. To further enhance this idea, it may be appropriate to make use of the quiz to provide instant feedback to students so that students are able to learn some of the information again and review it, which might be neglected during the video watching. In addition, teachers can try to bring up more critical discussions and stimulate their thoughts during the face-to-face lesson. Students can also offer on-going suggestions to teachers through weekly forum discussion on their challenges and new findings through their own inquiry learning. If students find that their learning experience is enhanced with the affordances, they will become more accepted to the new learning approach.

To address these limitations technologically, learning analytics should be redesigned for flipped classroom purpose so that deeper insights can be gained. Some of the potential features are, for example, adding questions to videos (like TED-Ed), logging the video watching behaviour such as fast forwards, pause, rewind, text-mining functionality on the discussion posts and other textual submission. Some of these features already exist in standalone web applications. For example, TED-Ed allows the teachers to insert questions and discussions into the videos for students to actively reflect on what they learn during video watching. Yet, these tools are not well integrated with the commercial or free LMSs (Friesen 2013). However, very limited research work seems to address the needs of learning analytics in flipped classroom research (Giannakos and Chrisochoides 2014), and more research works need to be done in this area (O'Flaherty and Phillips 2015). Thus, both pedagogical and technical challenges need to be addressed.

12.5 Conclusions and Limitations

In this chapter, it describes the conceptual framework for mobile learning analytics in flipped classroom. An initiative of using learning analytics under flipped learning approach with the support of mobile technology in their learning was reported. Issues and implications for designing flipped learning with mobile technology and learning analytics were discussed based on the responses of the learners, and the findings showed mixed perceptions concerning learning analytics mainly in the learning motivation, learning engagement, learning style and attitude towards learning. In this study, it offers an insight of how teachers effectively may use mobile learning analytics to enhance the quality of formative assessment. In fact, the model of flipped classroom can become an excellent approach when the affordances are magnified through a careful learning activity design. But without learning analytics, students' learning cannot be monitored outside the classroom to better engage them in face-to-face contact hours in class. Indeed, teachers and students should have more interactions under this model to maximize the effectiveness of this learning approach. It would not be suitable to assume students to learn effectively with "extra resources" at home; rather teachers need to find a new way to obtain recursive feedbacks, e.g. learning analytics and formative assessments, so that the learning process of each student is clearly captured. Especially in this new mobile digital world, students should be encouraged to take the advantages of ubiquitous learning to maximize their learning anywhere at any time. By this way teacher can extend their teaching and reach out to students without the boundary of the classroom walls.

However, the issues such as the alignment of learning design and learning experience with the learning analytics will continue to exist, which sets forth the future research direction to study to understand how each component in the framework can complement for each other. Other limitations are the choice of LMS tools in this study (Schoology), the differences of teaching contents in the two participated courses for treatment and data collection, and the diversity of pedagogy and learning design for the courses. Students may also not be able to adapt to the new way of learning quickly when the same approach is not adopted in other courses taken simultaneously. This case study emphasizes the needs of mobile learning analytics and has identified the opportunities and challenges in flipped classroom teaching. From the experience, both students and teachers need to continue to interact and communicate with each other to refine the flipped learning model. More importantly, learning activities including the video lectures/tutorials need to be redesigned to help students find a better connection among these activities.

Acknowledgments Special thanks are given to the Research Assistant, Mr. Ho-yin Cheung, and the student participants who provide their contribution and support to this project. This project is a part of the initiative in flipped classroom research at the Hong Kong Institute of Education.

References

Abeysekera, L., & Dawson, P. (2015). Motivation and cognitive load in the flipped classroom: definition, rationale and a call for research. *Higher Education Research & Development, 34*(1), 1–14.

Aljohani, N. R., & Davis, H. C. (2012, September). Significance of learning analytics in enhancing the mobile and pervasive learning environments. In *2012 6th International Conference on Next Generation Mobile Applications, Services and Technologies (NGMAST)* (pp. 70–74). IEEE.

Aljohani, N. R., & Davis, H. C. (2013, September). Learning Analytics and Formative Assessment to Provide Immediate Detailed Feedback Using a Student Centered Mobile Dashboard. In *2013 Seventh International Conference on Next Generation Mobile Apps, Services and Technologies (NGMAST)* (pp. 262–267). IEEE.

Ali, L., Hatala, M., Gašević, D., & Jovanović, J. (2012). A qualitative evaluation of evolution of a learning analytics tool. *Computers & Education, 58*(1), 470–489.

Baker, R. S., & Inventado, P. S. (2014). Educational data mining and learning analytics. In *Learning Analytics* (pp. 61–75). New York: Springer.

Bergmann, J., & Sams, A. (2012). *Flip your classroom: Reach every student in every class every day*. International Society for Technology in Education.

Bishop, J. L., & Verleger, M. A. (2013, June). The flipped classroom: A survey of the research. In *ASEE National Conference Proceedings, Atlanta, GA*.

Bonwell, C. C., & Eison, J. A. (1991). *Active learning: creating excitement in the classroom. 1991 ASHE-ERIC higher education reports*. ERIC Clearinghouse on Higher Education, The George Washington University, One Dupont Circle, Suite 630, Washington, DC 20036–1183.

Boud, D., & Falchikov, N. (2006). Aligning assessment with long-term learning. *Assessment & Evaluation in Higher Education, 31*(4), 399–413.

Brown, A., & Green, T. (2010). Issues and trends in instructional technology: Growth and maturation of web-based tools in a challenging climate; Social networks gain educators' attention. In M. Orey, S. A. Jones, & R. M. Branch (Eds.), *Educational media & technology yearbook* (pp. 29–43). New York: Springer.

Butler, D. L., & Winne, P. H. (1995). Feedback and self-regulated learning: A theoretical synthesis. *Review of Educational Research, 65*(3), 245–281.

Butler, D. L. (2002). Individualizing instruction in self-regulated learning. *Theory Into Practice, 41*(2), 81–92.

Butt, A. (2014). Student views on the use of a flipped classroom approach: evidence from Australia. *Business Education & Accreditation, 6*(1), 33–43.

Campbell, J. P., DeBlois, P. B., & Oblinger, D. G. (2007). Academic analytics: A new tool for a new era.*EDUCAUSE review, 42*(4), 40.

Chen, H., Chiang, R. H., & Storey, V. C. (2012). Business intelligence and analytics: from big data to big impact. *MIS Quarterly, 36*(4), 1165–1188.

Chorianopoulos, K., Giannakos, M. N., Chrisochoides, N., & Reed, S. (2014, July). Open service for video learning analytics. In *2014 IEEE 14th International Conference on Advanced Learning Technologies (ICALT)* (pp. 28–30). IEEE.

Collins, J. W., & O'Brien, N. P. (2011). *The Greenwood dictionary of education*. ABC-CLIO.

Creswell, J. W. D. (2002). *Educational research: planning, conducting and evaluating quantitative and qualitative research*. London: Pearson.

Creswell, J. W. (2009). *Research design: Qualitative, quantitative, and mixed methods approaches*. Thousand Oaks, CA: Sage Publications.

Davies, R. S., Dean, D. L., & Ball, N. (2013). Flipping the classroom and instructional technology integration in a college-level information systems spreadsheet course. *Educational Technology Research and Development, 61*(4), 563–580.

Drachsler, H., & Greller, W. (2012, May). Confidence in learning analytics. In *LAK12: 2nd International Conference on Learning Analytics & Knowledge*.

De Liddo, A., Shum, S. B., Quinto, I., Bachler, M., & Cannavacciuolo, L. (2011, February). Discourse-centric learning analytics. In *Proceedings of the 1st International Conference on Learning Analytics and Knowledge* (pp. 23–33). ACM.

Deschacht, N., & Goeman, K. (2015). The effect of blended learning on course persistence and performance of adult learners: A difference-in-differences analysis. *Computers & Education, 87*, 83–89.

Dietz-Uhler, B., & Hurn, J. E. (2013). Using learning analytics to predict (and improve) student success: A faculty perspective. *Journal of Interactive Online Learning, 12*(1), 17–26.

Dringus, L. P. (2012). Learning Analytics Considered Harmful. *Journal of Asynchronous Learning Networks, 16*(3), 87–100.

Dweck, C. S. (2000). *Self-theories: Their role in motivation, personality, and development*. New York: Psychology Press.

Dyckhoff, A. L., Zielke, D., Bültmann, M., Chatti, M. A., & Schroeder, U. (2012). Design and implementation of a learning analytics toolkit for teachers. *Journal of Educational Technology & Society, 15*(3), 58–76.

Eraut, M. (1994). *Developing professional knowledge and competence*. London: Falmer.

Ferguson, R. (2012). Learning analytics: drivers, developments and challenges. *International Journal of Technology Enhanced Learning, 4*(5), 304–317.

Findlay-Thompson, S., & Mombourquette, P. (2014). Evaluation of a flipped classroom in an undergraduate business course. *Business Education & Accreditation, 6*(1), 63–71.

Friesen, N. (2013). Learning analytics: Readiness and rewards/L'analyse de l'apprentissage: état de préparation et récompenses. *Canadian Journal of Learning and Technology/La revue canadienne de l'apprentissage et de la technologie, 39*(4).

Fulantelli, G., Taibi, D., & Arrigo, M. (2013, November). A semantic approach to mobile learning analytics. In *Proceedings of the First International Conference on Technological Ecosystem for Enhancing Multiculturality* (pp. 287–292). ACM.

Fulantelli, G., Taibi, D., & Arrigo, M. (2015). A framework to support educational decision making in mobile learning. *Computers in Human Behavior, 47*, 50–59.

Fulton, K. (2012). Upside down and inside out: Flip your classroom to improve student learning. *Learning & Leading with Technology, 39*(8), 12–17.

Garcia, T. (1995). The role of motivational strategies in self-regulated learning. *New directions for teaching and learning, 1995*(63), 29–42.

Giannakos, M. N., Chorianopoulos, K., & Chrisochoides, N. (2015). Making sense of video analytics: Lessons learned from clickstream interactions, attitudes, and learning outcome in a video-assisted course. *The International Review of Research in Open and Distributed Learning*, *16*(1).

Gilboy, M. B., Heinerichs, S., & Pazzaglia, G. (2015). Enhancing student engagement using the flipped classroom. *Journal of nutrition education and behaviour, 47*(1), 109–114.

Greller, W., & Drachsler, H. (2012). Translating learning into numbers: A generic framework for learning analytics. *Journal of Educational Technology & Society, 15*(3), 42–57.

Gundecha, P., & Liu, H. (2012). *Mining social media: A brief introduction.* In P. Mirchandani (Ed.) INFORMS Tutorials in Operations Research, Vol. 9. Hanover, MD: INFORMS.

Herreid, C. F., & Schiller, N. A. (2013). Case studies and the flipped classroom. *Journal of College Science Teaching, 42*(5), 62–66.

Herro, D., Kiger, D., & Owens, C. (2013). Mobile technology: Case-based suggestions for classroom integration and teacher educators. *Journal of Digital Learning in Teacher Education, 30*(1), 30–40.

Irons, A. (2007). *Enhancing learning through formative assessment and feedback.* New York: Routledge.

Jensen, J. L., Kummer, T. A., & Godoy, P. D. D. M. (2015). Improvements from a flipped classroom may simply be the fruits of active learning. *CBE-Life Sciences Education*, *14*(1), ar5.

Jinlei, Z., Ying, W., & Baohui, Z. (2012). Introducing a new teaching model: flipped classroom. *Journal of Distance Education, 4*, 46–51.

Kong, S. C. (2014). Developing information literacy and critical thinking skills through domain knowledge learning in digital classrooms: An experience of practicing flipped classroom strategy. *Computers & Education, 78*, 160–173.

Kong, S. C., & Song, Y. (2015). An experience of personalized learning hub initiative embedding BYOD for reflective engagement in higher education. *Computers & Education, 88*, 227–240.

Kiger, D., Herro, D., & Prunty, D. (2012). Examining the influence of a mobile learning intervention on third grade math achievement. *Journal of Research on Technology in Education, 45*(1), 61–82.

Kumar, V. S., Somasundaram, T. S., Boulanger, D., Seanosky, J., & Vilela, M. F. (2015). Big data learning analytics: a new perpsective. In *Ubiquitous Learning Environments and Technologies* (pp. 139–158). Berlin Heidelberg: Springer.

Lage, M. J., Platt, G. J., & Treglia, M. (2000). Inverting the classroom: A gateway to creating an inclusive learning environment. *The Journal of Economic Education, 31*(1), 30–43.

Lam, J., & Duan, C. G. (2012). A review of mobile learning environment in higher education sector of Hong Kong: Technological and social perspectives. In *Hybrid Learning* (pp. 165–173). Berlin Heidelberg: Springer.

Lehmann, T., Hähnlein, I., & Ifenthaler, D. (2014). Cognitive, metacognitive and motivational perspectives on preflection in self-regulated online learning. *Computers in Human Behavior, 32*, 313–323.

Lockyer, L., Heathcote, E., & Dawson, S. (2013). Informing pedagogical action: Aligning learning analytics with learning design. *American Behavioral Scientist, 57*(10), 1439–1459.

Lu, E. H. C., Tseng, V. S., & Yu, P. S. (2011). Mining cluster-based temporal mobile sequential patterns in location-based service environments. *IEEE Transactions on Knowledge and Data Engineering, 23*(6), 914–927.

Ma, J., Han, X., Yang, J., & Cheng, J. (2015). Examining the necessary condition for engagement in an online learning environment based on learning analytics approach: The role of the instructor. *The Internet and Higher Education, 24*, 26–34.

Mason, G. S., Shuman, T. R., & Cook, K. E. (2013). Comparing the effectiveness of an inverted classroom to a traditional classroom in an upper-division engineering course. *IEEE Transactions on Education, 56*(4), 430–435.

Mazza, R., & Milani, C. (2004, November). Gismo: A graphical interactive student monitoring tool for course management systems. In *TEL'04 Technology Enhanced Learning'04 International Conference* (pp. 18–19).

McDonald, K., & Smith, C. M. (2013). The flipped classroom for professional development: part I. Benefits and strategies. *The Journal of Continuing Education in Nursing, 44*(10), 437.

Milman, N. B. (2012). The flipped classroom strategy: What is it and how can it best be used? *Distance Learning, 9*(3), 85.

Mirriahi, N., & Dawson, S. (2013, April). The pairing of lecture recording data with assessment scores: a method of discovering pedagogical impact. In *Proceedings of the Third International Conference on Learning Analytics and Knowledge* (pp. 180–184). ACM.

Musolesi, M. (2014). Big mobile data mining: good or evil? *IEEE Internet Computing, 18*(1), 78–81.

Nagalakshmi, S., & Sumathi, R. (2014, February). An efficient mobile commerce explorer for mobile user's behavior pattern mining and prediction. In *2014 International Conference on Information Communication and Embedded Systems (ICICES)* (pp. 1–7). IEEE.

Nicol, D. J., & Macfarlane-Dick, D. (2006). Formative assessment and self-regulated learning: A model and seven principles of good feedback practice. *Studies in Higher Education, 31*(2), 199–218.

O'Flaherty, J., & Phillips, C. (2015). The use of flipped classrooms in higher education: A scoping review. *The Internet and Higher Education, 25*, 85–95.

Pegrum, M. (2014). *Mobile learning: languages, literacies and cultures*. Palgrave Macmillan.

Persico, D., & Pozzi, F. (2015). Informing learning design with learning analytics to improve teacher inquiry. *British Journal of Educational Technology, 46*(2), 230–248.

Petropoulou, O., Kasimatis, K., Dimopoulos, I., & Retalis, S. (2014). LAe-R: A new learning analytics tool in Moodle for assessing students' performance. *Bulletin of the IEEE Technical Committee on Learning Technology, 16*(1), 1.

Retalis, S., Papasalouros, A., Psaromiligkos, Y., Siscos, S., & Kargidis, T. (2006). Towards networked learning analytics–A concept and a tool. In *Proceedings of the fifth international conference on networked learning*.

Roach, T. (2014). Student perceptions toward flipped learning: New methods to increase interaction and active learning in economics. *International Review of Economics Education, 17*, 74–84.

Roehl, A., Reddy, S. L., & Shannon, G. J. (2013). The flipped classroom: An opportunity to engage millennial students through active learning strategies. *Journal of Family & Consumer Sciences, 105*(2), 44–49.

Romero, C., Ventura, S., & García, E. (2008). Data mining in course management systems: Moodle case study and tutorial. *Computers & Education, 51*(1), 368–384.

Sadler, D. R. (1998). Formative assessment: Revisiting the territory. *Assessment in education, 5* (1), 77–84.

Sharma, N., Lau, C. S., Doherty, I., & Harbutt, D. (2015). How we flipped the medical classroom. *Medical Teacher, 37*(4), 327–330.

Shuler, C. (2009). Pockets of potential: Using mobile technologies to promote children's learning. New York: Joan ganz cooney center at sesame workshop.

Shum, S. B., & Crick, R. D. (2012, April). Learning dispositions and transferable competencies: pedagogy, modelling and learning analytics. In *Proceedings of the 2nd International Conference on Learning Analytics and Knowledge* (pp. 92–101). ACM.

Shoukry, L., Göbel, S., & Steinmetz, R. (2014, November). Learning analytics and serious games: trends and considerations. In *Proceedings of the 2014 ACM International Workshop on Serious Games* (pp. 21–26). ACM.

Smith, S., Brown, D., Purnell, E., & Martin, J. (2015). 'Flipping' the postgraduate classroom: Supporting the student experience. In *Global Innovation of Teaching and Learning in Higher Education* (pp. 295–315). Springer International Publishing.

Smith, C. M., & McDonald, K. (2013). The flipped classroom for professional development: Part II. Making podcasts and videos. *The Journal of Continuing Education in Nursing, 44*(11), 486.

Snodgrass Rangel, V., Bell, E. R., Monroy, C., & Whitaker, J. R. (2015). Toward a new approach to the evaluation of a digital curriculum using learning analytics. *Journal of Research on Technology in Education, 47*(2), 89–104.

Tempelaar, D. T., Rienties, B., & Giesbers, B. (2014). Computer assisted, formative assessment and dispositional learning analytics in learning mathematics and statistics. In *Computer Assisted Assessment. Research into E-Assessment* (pp. 67–78). Springer International Publishing.

Tempelaar, D. T., Rienties, B., & Giesbers, B. (2015). In search for the most informative data for feedback generation: Learning Analytics in a data-rich context. *Computers in Human Behavior, 47*, 157–167.

Yorke, M. (2003). Formative assessment in higher education: Moves towards theory and the enhancement of pedagogic practice. *Higher Education, 45*(4), 477–501.

West, D. M. (2012). Big data for education: Data mining, data analytics, and web dashboards. *Governance Studies at Brookings* (pp. 1–10).

Wong, K., & Chu, D. W. (2014). Is the flipped classroom model effective in the perspectives of students' perceptions and benefits?. In *Hybrid Learning. Theory and Practice* (pp. 93–104). Springer International Publishing.

Wong, G., & Cheung, H. (2015). Flipped classroom for student engagement in higher education. In J. Hawkins (Ed.), *Student engagement: Leadership practices, perspectives and impact of technology* (pp. 69–90). New York: Nova Science Publishers Inc.

Wong, G. (2014). Engaging students using their own mobile devices for learning mathematics in classroom discourse: A case study in Hong Kong. *International Journal of Mobile Learning and Organisation, 8*(2), 143–165.

Xing, W., Wadholm, R., Petakovic, E., & Goggins, S. (2015). Group learning assessment: Developing a theory-informed analytics. *Journal of Educational Technology & Society, 18*(2), 110–128.

Zheng, S., Xiong, S., Huang, Y., & Wu, S. (2008, August). Using methods of association rules mining optimizationin in web-based mobile-learning system. In *Electronic Commerce and Security, 2008 International Symposium on* (pp. 967–970). IEEE.

Author Biography

Gary K.W. Wong currently a Lecturer of the Department of Mathematics and Information Technology in the Hong Kong Institute of Education. His research interests include computer-aided education, mobile learning analytics, coding education, and flipped classroom. He received a BS in Computer Science from Brigham Young University—Hawaii and MPhil in Electronic and Computer Engineering from the Hong Kong University of Science and Technology. He holds a Ph.D. in Computer Science from City University of Hong Kong, and Ed.M. in Learning Design and Leadership from University of Illinois at Urbana Champaign.

Part IV
Mobile Learning Across Curriculum

Chapter 13
Mobile Learning in K-12 Education: Personal Meets Systemic

John Turner

Abstract This paper investigates one school's journey towards integrating mobile learning within its institutional structures. This includes a comparison of the school's objectives against mobile learning affordances. The approach takes into account the cultural contexts, dynamic nature of digital change, and school structural challenges that impact on providing worthwhile education outcomes. Several in-school case reviews on mobile learning use within the school look at mobile learning integration. As well as providing insights for other schools to consider, possible ways forward are presented for better understanding the dynamic relationship between mobile learning and school intentions, as well as challenges that go with ever-evolving digital technologies.

13.1 Introduction

All K-12 schools contain common characteristics, consideration of which can assist others to widen understanding. As well, there are particular aspects that are the product of unique historical and cultural developments. Digital meanwhile continues to evolve in depth, breadth and preference. Research into mobile learning has put forward contentions for new or enhanced learning. Bringing Digital and School together means both opportunities and challenges, as personal learning interacts with systemic education in new ways.

When considering the impact of mobile learning on any educational institution, as for any technology, it is important to align with cultural understanding. Recent history is littered with new technologies that have failed to meet advocate contentions for schools. Cuban (1986, 2001, 2014) has summarised such shortfalls. Yet within the wider community mobile technologies such as smartphones are in the

J. Turner (✉)
Canadian International School of Hong Kong, Hong Kong, SAR China
e-mail: johnturner@cdnis.edu.hk

D. Churchill et al. (eds.), *Mobile Learning Design*, Lecture Notes in Educational Technology, DOI 10.1007/978-981-10-0027-0_13

ascendancy (Columbus 2014) and bring with them demands for educational consideration as mobile learning platforms.

This paper details one school's approach towards considering and integrating mobile learning into its teaching and learning structures. It includes insights into an approach that has evolved with due consideration of mobile learning assertions, while working within the practicalities of a contemporary school system. While not a formal research study, it is provided as a starting point for more rigorous consideration of how mobile learning might be approached to progress school-based learning. At the heart remains a belief that digital can add value to learning through personal interactions with digital technologies, as put forward by Laouris and Eteokleous (2005) for mobile learning, and going as far back as Papert (1980).

13.2 Defining Mobile Learning

Differing interpretations of mobile learning, and what effect mobile devices will have on teaching and learning, have been an ongoing discussion for over a decade within the research community (Liu et al. 2014a, b; Laouris and Eteokleous 2005; Craig and Van Lom 2009). Some see mobile devices as distinct from personal computers because of their ubiquity and portability (Shuler et al. 2013), with Laouris and Eteokleous (2005, p. 2) identifying use of the term mobile "as synonymous to a mobile phone". Sharples (2009) draws a clear distinction between mobile learning and classroom use of desktops. He also provided a strong framework for defining mobile learning:

- May be mobile (but not necessarily if mobile devices are being used in designated spaces)
- May involve learning in non-formal settings
- May be extendable and interleaved across time and space
- May involve use across a variety of personal and institutional technologies
- Presents ethical challenges if shared access a requirement
- Can be evaluated by addressing "usability (will it work?), effectiveness (is it enhancing learning?) and satisfaction (it is liked?)" (Sharples 2009, p. 22).

Sharples (2013) also identified critical success factors as technology availability, institutionalised support, connectivity, (curriculum) integration and (learning) ownership. Laouris and Eteokleous (2005, p. 2) went on to differentiate between e-learning as relating to "multimedia, interactive, hyperlinked, media-rich environments", with mobile learning referring to the "spontaneous, intimate, connected, informal, lightweight, private, personal". They conclude that mobile learning leads to new relationships of time, space, learning environment, content, technologies, user attributes, and process. Liu et al. (2014b) updated this to focus on affordances available through mobile devices: flexibility, accessibility, interactivity, and motivation and engagement. Similarly, Baran (2014) lists mobility, access, immediacy, situativity, ubiquity, convenience and contextuality as overlapping characteristics of mobile learning. Kearney et al. (2012) sought to formalise this by putting forward mobile learning based on three primary affordances based on Time-Space considerations: authenticity, collaboration and personalisation; each linked to sub-considerations:

- Authenticity: context, situation
- Collaboration: conversation, data sharing
- Personalisation: agency, customised.

Related to school education, Churchill and Churchill (2008) provide a good list of mobile learning affordances: multimedia access tool, connectivity tool, capture tool, representation tool and analytical tool. But, as McFarlane (2015) points out, technology cannot do this on its own, and as Baran (2014, p. 17) concedes, "the diversity of research on mobile learning has made it difficult to generate a single definition or to determine generally added benefits". Laouris and Eteokleous (2005, p. 1) warn that the term can depend on "who is asking, and what the context is".

13.2.1 Mobile Learning in Schools

From within the school education sector, there is strong support for the potential of mobile learning in schools as reflected in recent New Media Consortium (2013, 2014) *Horizon Reports*. These identify mobile learning as within 12 months of general adoption in 2013, going on to identify such learning as a key element of BYO adoption, personalised learning, cloud computing, gamification and wearable technologies in 2014. Here no distinction is made between the levels of device mobility.

Clarke and Svanaes (2014) provided an updated review on research into the use of tablets in K-12 education. They concluded that while there is need for more research, some common themes are emerging. These include the portable nature, access to information, interaction with personalised learning content, cost advantages and ease of use. They drew on the UNESCO (2012) definition (Shuler et al. 2013) as learning arising from use of mobile technologies such as mobile phones, smartphones, eReaders and tablets. However, Clarke and Svanaes (2014) also point out that within K-12 schools context can vary depending on the student stage of development.

Within schools McFarlane (2015, p. 25) identifies personal mobile devices as having the potential to help:

- Facilitate individual, cooperative and interactive work in class
- Enable sharing of ideas, knowledge, ideas and responses
- Increase participation in whole-class settings
- Enable learners to revisit prior learning
- Provide opportunities for autonomy and independence
- Permit storage of work and resources in one place at hand.

But, an accompanying lesson is that this can be at odds with traditional pre-digital expectations, many of which schools continue to have to satisfy. Hand-written exam essays are a good case in point. Common testing can also impact on personal learning choices. The debate on the effect of digital devices on young brains continues (Greenfield 2015, p. 14). Issues of potential distraction (Duncan et al. 2012; McCoy 2013; Bjerede and Bondi 2012) have been raised as of concern; related to both pedagogical and personal identity development issues (particularly with adolescents).

Research in school environments does not to date appear to have gained significant traction. But as McFarlane (2015) notes, "could it be that the final step change in personal access to online resources and communications by young people using smartphones and tablets will be the factor that changes policy and therefore school attitudes to computer use?" (p. 141).

How mobile learning can best interact with school is at early stage, although there is an increasing focus on certain mobile learning affordances. Within schools this is likely to be affected by the school's approach to personal learning, choices provided re time and place, and associated values. Mobile learning affordances of choice, accessibility to content, learning interactions, and connections between contexts appear to have potential value. K-12 schools, though, are institutions that operate with a strong set of social obligations that impact on what is possible and what is valued. They also deal with a wide range of maturation, from 5 year olds or below, to 17–18 year olds in their final stages before high-stakes testing leading hopefully to further study. This needs to be carefully considered, and the teacher as a central authority has a critical part to play. This will be further examined in the next section.

In addition the following school ecosystem factors can also impact on what can be achieved through mobile learning:

- Values encapsulated in organisational vision and priorities
- Structures, including support
- Infrastructure choices
- How learning is evaluated
- What change choices the school system allows, including
- What affordances the school will commit to, and in what way(s).

As Watters (2014, p. 4) reminds, "while building new technologies is easy (or easy-ish), changing behaviors and culture is much, much harder."

13.2.2 The Role of the Teacher

Within K-12 schools a key determiner is the teacher. John Hattie, in the *Forward* to Bain and Weston's (2012) study of personal digital device use in schools, identified teacher mind frames as the most important enhancer and barrier to student learning. Bain and Weston agree with Hattie that within schools there exists a fundamental issue of conservative standardised-based systems up against personal digital learning devices geared to support connection, reflection and construction. Teachers have the potential to risk and build value if they see positive possibilities, or negate if they feel educational value is wanting. Teacher mindset can limit what might be sought through use of educational technologies (Turner 1999, Blackley and Walker 2015). Socio-cultural understanding is therefore a key consideration, as recognised by Siepold and Pachler (2011) in their examination of how such understanding can impact on mobile learning.

McFarlane (2015 p. 27) also highlights the importance of teacher buy-in, highlighting "professional development of teachers in the effective use of connected devices to support learning is fundamental to a successful implementation of 1:1 mobile computing", and that the "frequency of use of digital technologies overall was (still) dependent on school policy, access to technology and teacher practices". (p. 34) Tablet use in education is strongly aligned to teacher perceptions of the affordances of technology (Churchill et al. 2012). We need to constantly take note of Fullan's (2007, p. 21) observation on educational changes that "*all real change involves loss, anxiety and struggle*".

As mobile learning impacts on personal choice, systemic responses to the diversity that ensues will place pressure on standardised systems. The role of the teacher in integrating mobile learning, while critical, to date is short on practical advice. To support integration of the affordances identified from mobile learning research, an understanding of how best to bring together school, teacher and digital technologies in ways that add value to the overall educational enterprise is needed. With this in mind, an examination of one school's approach will seek to shed some light on possibilities.

13.3 A Short History of Mobile Learning in a School

The school referred to in this paper is a K-12 co-educational international school located in Hong Kong. Nearly all students' progress to tertiary education, often to universities spread across the world. It has high academic expectations. There are also many students who move in and out of the school, although a strong core percentage remains through most levels. It has a traditional timetable, teachers allocations and hierarchical curriculum through the International Baccalaureate (IB), Primary Years Programme (PYP), Middle Years Programme (MYP) and Diploma Programme (DP).

For the school mobile learning has been defined as learning accruing through access to digital devices at-hand within the school and beyond. Historically, this has been based on the school's 1:1 laptop program. Increasingly this is being widened to consider personal mobile devices such as phones and tablets.

A 2006 plan, Sustainable Human Networks led to the establishment of a group of educators tasked to help drive and support digital change, the introduction of a 1:1 laptop program from Grade 5 onwards, and a series of recommendations affecting curriculum, infrastructure and teacher training. The 1:1 laptop program was progressively embedded for all Grade 5 through 12 students, who own and manage their own laptop using a school provided image.

A 2011 review led to a Digital Learning Infusion (DLI) plan built around infusion, as defined by the Florida Centre for Instructional Technology (2011) Technology Integration Matrix (TIM). This sought to infuse beliefs and practices that would improve student achievement, teacher practice, and support for the school's curriculum objectives and mission. A vision that "*digital technologies enable opportunities for greater active student learning that is valued, visible, connected and progressive*" provided a focus.

The DLI led to the development of teacher digital learning certification and professional learning networks, more active student involvement, digital portfolios as more visible learning journeys, online learning environments, a digital literacy curriculum, global and environmental objectives, and strengthening of in-school research.

The 2011 plan has been updated to take on new or emerging technologies deemed to have teaching and learning potential, such as those provided through Google Educational Apps suite, eBook construction, social media developments and iPads as mobile personal learning devices. Digital Literacy curriculum development drew on Meyer and Land's (2003) threshold concepts approach to help progress teacher and student digital learning capacity. A reaffirmation of the 1:1 laptop program to support inquiry-led learning, digital portfolios, and the infused approach to digital supported or enhanced learning constituted a continuing strong commitment to the role of digital teaching and learning in the school.

The laptop remains the primary digital device for all students from Grade 5 onwards, although iPads are being increasingly integrated in earlier years, and Grade 4 is moving to each student having their own personal laptop. This was to enable younger students and their families to communicate and connect through blogging, build up a media-based record of learning, and connect to wider audiences. Research insights are being developed within the school's programs.

Construction of eBooks and apps within the school complements use of mobile device. Chinese eBooks with their use of audio and interactive media are a good example of this. An updated vision, "*constructing visible, connected and progressive learning journeys to support reflection, feedback, ownership and conceptual depth (for teachers and students)*", was developed.

The school's vision reflects that learning can be enhanced by appropriately focused use of mobile devices. This is important, because as Clarke and Svanaes (2014, p. 15) identify, "tablets specifically must be supported by a pedagogical vision in order to reach its potential impacts on learning". Ignoring the importance of a pedagogical vision has impeded gaining academic worthwhile research on the impact of tablets on education (Cochrane et al. 2013).

Allied to this, a comprehensive teacher learning program ensures all teachers are supported. Part of this involves enhancing teacher and team adaptability, and generating more flattened learning environments (and related pedagogical approaches) so that student expertise can likewise be developed and supported. This is also recognised by Clarke and Svanaes (2014) as crucial for effective integration. Depth of change is supported through groups that connect bottom-up and top-down drivers. School leadership by example also plays an important role.

The school's commitment to mobile learning is apparent in

- The student relationship with their laptop as a personal mobile learning device
- The use of digital devices to advance new and established learning
- Support structures and leadership commitment for progressing such learning

An examination of the school's learning ecosystem demonstrates:

- *Vision and purpose*—the school has a school-wide Digital Learning Infusion vision and plan which can be mapped against particular objectives: valued, visible, connected, progressive
- *Supportive structures*—dedicated support for curriculum and teacher personal development is provided
- *Connected infrastructure choices*—systemised while allowing some individual choice
- *Curriculum*—IB Curriculum with strong inquiry emphasis, with academic focus increasing into senior years
- *Learning evaluated*—teacher-centered, but with efforts to make learning more visible for student inclusion and wider considerations
- *Digital change management*—through a school-wide Learning Technologies Council, connecting support, curriculum and leadership.

The extent to which this has been successful against mobile learning contentions will be evaluated later in this paper.

13.4 Evaluating Mobile Learning

How best to see if a school's use of mobile technologies is leading to worthwhile educational value? The OECD (2013) case study methodology identifies analysis of primary documents, interviews of key stakeholders, discussion with focus groups of stakeholders and a discourse analysis of relevant media as an appropriate investigative approach. Sharples (2009), in the Mobile Learning Organisers Project, called on diary and interview methods. Traxler and Kukulska-Hulme (2005) defined a good evaluation as enabling quality sharing, reporting and embedding connections that are consistent, rigorous, scalable and ethical.

For any school, understanding the cultural context is a necessary prelude to evaluating what is and can be. For mobile learning this requires clarification of affordances that can be mapped against school objectives. In this paper, the following have been identified as of potential value and practical use within a school (Table 13.1):

Table 13.1 Mobile learning affordances v school digital learning objectives

Mobile learning affordances	School digital learning objectives
Increased access	Valued
Building personal relationships with learning	Visible
Personalization of choice and pathways	Connected
Increased accessibility to content	Progressive
Increased learning interactivity	
Connecting across contexts	

With this focus, the following reviews were undertaken to seek insights into ways that might progress mobile device affordances, and the school's vision for use of such devices.

13.5 School Case Reviews

Four projects within the school were analysed:

13.5.1 Study One: Grade 11 Parent Conferencing

Since 2013 all grade 11 students have developed and used a digital portfolio as part of their conversations with their parents on the progress they had achieved within the IB DP (Grades 11–12). This conversation covers the Community and Service, Theory of Knowledge (TOK) and Extended Essay aspects of the DP. This is bound by the IB's focus on developing approaches to learning (ATLs). In addition to sharing with parents evidence of achievement through personal construction, the folios also link with other subject portfolios (such as in Digital Art work) and support possible university interest in a student's school performance. Students choose their own digital publishing medium and put together their own selection of materials.

The student use of a personal school digital portfolio to help support parent conferencing supported the following mobile learning affordances:

- *Increased access*—Parents accessed and engaged in the conversation both in-school and beyond, thus widening student learning interactions and parent understanding of their child's progress. Students can also access each other's work
- *Building personal relationships with learning*—Students developed their own digital portfolio as a reflection of their learning journey

- *Personalization of choice and pathways*—Students chose what to put in their digital portfolio to best reflect their own learning achievements and the medium for publishing
- *Increased accessibility to content*—Students linked to other learning and drew on digital tools such as Google Educational Apps to provide examples
- *Increased learning interactivity*—Student developed their own links and obtained feedback from parents and teachers through the comments feature. TOK is one area that draws heavily on student discourse with others
- *Connecting across contexts*—Subject connections were also included and wider use, such as for university selection, is available

School objectives were supported by:

- *Valued*—The use continued through a change of DP Coordinator
- *Visible*—The digital portfolio provided a visible window into student learning appreciated by parents, peers, and teachers as a means to celebrate progress and identify areas for support
- *Connected*—Students drew on digital folio work created in-school in previous years as well as informal digital learning to enhance their digital portfolio
- *Progressive*—The addition of parental understanding of the non-academic subject aspects of the IB was progressed (Fig. 13.1).

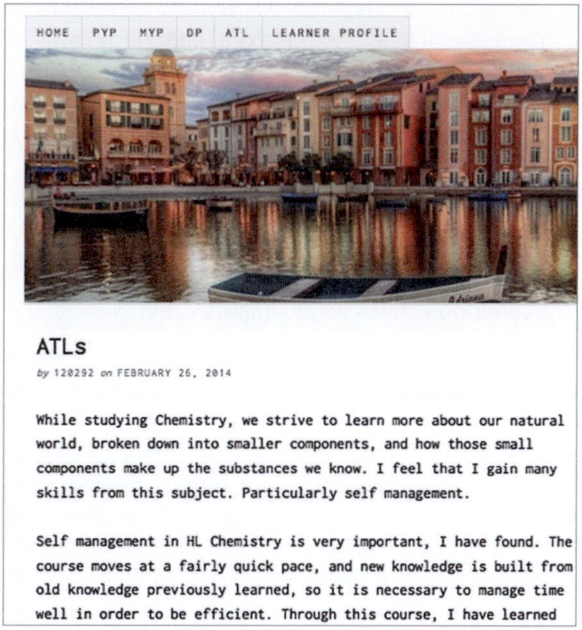

Fig. 13.1 Grade 11 digital portfolio example

13.5.2 Study Two: Grade 8 EBook and Process Journal

All Grade 8 students as part of their Science studies created an eBook on a designated authentic Science topic. This project had developed over the past 3 years, with this year's eBook on Diseases developed and evaluated with Grade 5–6 students as the intended audience. Each Grade 8 students team of three to four students completed a chapter, which was then joined into a grade-wide book. Google Docs was used to connect student group discussions and unite knowledge on both personal and group levels.

The project supported the following mobile learning affordances:

- *Increased access*—Students worked on the joint aspects even when group members are elsewhere (a critical part of group work in digital domains). Access to their work was extended through the school's Management Learning System
- *Building personal relationships with learning*—Students developed a valued relationship with software and its capabilities. The student learning of new software, iBook Author, was student led and supported by teacher understanding of student digital literacy development needs
- *Personalisation of choice and pathways*—Book design was personalised by each group within stipulated book requirements. Student choice of widgets (iBook Author internal apps) and supplementing sites such as Bookry.com were personal choices in accordance with design processes and subject standards
- *Increased accessibility to content*—Students drew on Web 2 information sources such as Bookry.com and infogr.am, as well as through their own investigations
- *Increased learning interactivity*—Students evaluated and created personalised interactive widgets available in iBook Author or Bookry. This included quizzes, galleries and interactive graphics
- *Connecting across contexts*—Students appreciation of learning, as seen through younger students, was a key part of the design process. The use of iBook Author also has been extended to other Grade 8 subjects.

School objectives were supported by:

- *Valued*—Assessed as a formal school subject project (in MYP Science and Design subjects) with learning valued extended to other students (Grade 5 and 6 students)
- *Visible*—Published within the school's Virtual Learning Environment, and available as pre-learning for future projects
- *Connected*—Group learning and problem-solving approaches supported. Collaborative publishing approaches progressed
- *Progressive*—Formed basis for learning to build deeper knowledge through publishing formal science work to different audiences (Fig. 13.2).

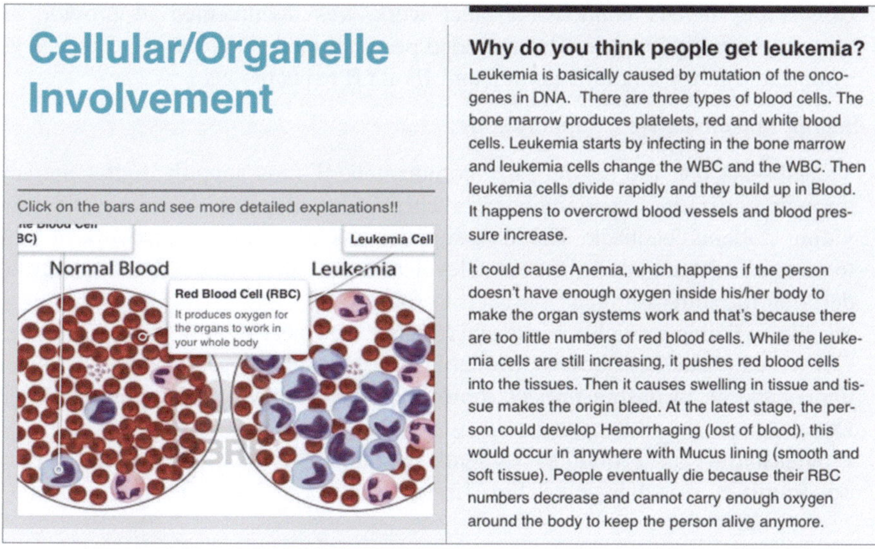

Fig. 13.2 Grade 8 science diseases eBook example

13.5.3 Study Three: Grade 6 Digital Literacy

All Grade 5 and 6 students manage their own digital portfolio (iFolio) which reports on their learning progress. In support of this a digital literacy evaluation approach was developed, where teachers provided feedback through the iFolio to each student on their digital literacy development. A digital literacy rubric applicable for teacher feedback has been developed, with a student version to support personal learning evaluation being customised by teachers.

Teacher feedback of student digital literacy through their iFolio supported the following mobile learning affordances:

- *Increased access*—Teachers, peers or parents could access student development in their own time. Teachers and students identified areas for further work as well as celebrate progress
- *Personal relationships with learning*—Students personalised within educational boundaries and developed for sharing focused areas of inquiry
- *Personalisation of choice and pathways*—Each iFolio provided avenues for personal exploration and choice, as well as a basis for further development in later years (which uses similar iFolio approaches)
- *Increased accessibility to content*—Links to new knowledge and personal inquiries were shared
- *Increased learning interactivity*—Widgets such as Flags were used to share levels of interaction Feedback provides strong learning support. Parent feedback also accessible

- *Connecting across contexts*—Project work was documented to provide an ongoing learning journey. This included personal media collections. Can also be evaluated against Digital Literacy and IB PYP expectations.

School objectives were supported by:

- Valued—As one teacher recently commented, iFolios support digital literacy through generating “possibilities for curriculum planning, teaching and providing students feedback. This also supplies teachers with a framework of how to ‘move on’ students to the next level and provides a common language to describe the differing areas we need to focus on.”
- Visible—iFolios available for teacher, parent and peer review.
- Connected—Literacy journey available for subsequent years. Forms basis for understanding of design project approaches further developed within IB MYP Design.
- Progressive—Can evolve as students encounter iFolios and digital literacy opportunities at earlier years (Fig. 13.3).

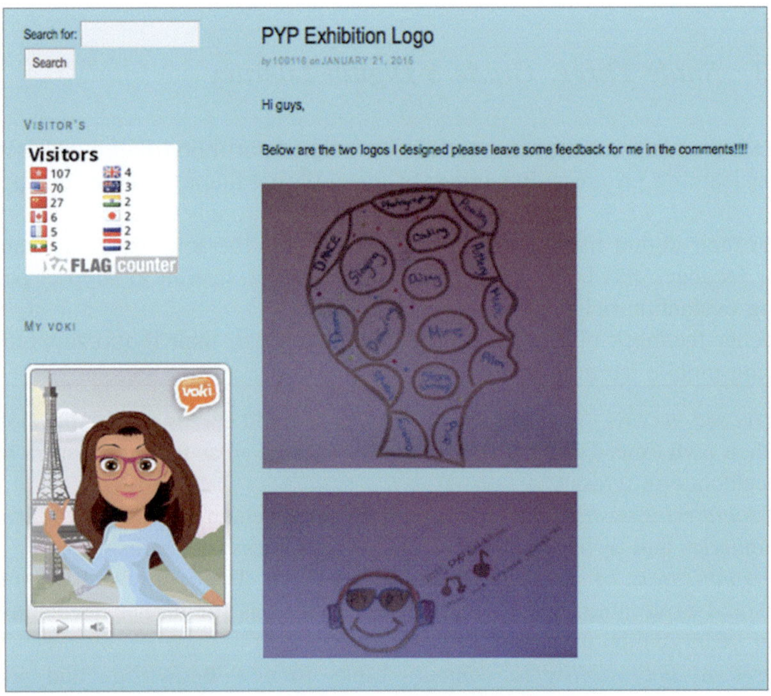

Fig. 13.3 Grade 6 iFolio example demonstrating digital literacy development

13.5.4 Study Four: Grade 1 iPad as iFolio

The recent availability of larger screen tablets has opened up new possibilities for younger students who cannot readily master keyboard technologies, enabling use of touch-screen mobile devices to engage in wider learning. While cognisant of appropriate time exposure and balance with non-digital environments critical to the young person's social and emotional well being, teachers have increasingly found that the tablet supports valued learning in new ways. While grades for younger students work with a set of six or seven school provided iPads to support learning stations, one class has been trialling the difference each student having their own personal iPad at-hand might provide. This has formed the basis for an extension to an iPad as a personal learning device for all Grade 1 students next year, building into following years as student need supported by teacher and school preparedness allows.

A core selection of apps was selected to support literacy, numeracy, communication, collaboration and media construction. For example, EasyBlog is a WordPress based app that enables young students to photograph and record through a simple click method. Other apps are selected by teachers according to student needs and learning value.

As detailed in a draft letter from the school to parents in February 2015, "by personalising the iPad and building an iFolio the student can:

- Develop confidence and competence through structured play and inquiry
- Build up a portfolio of learning through media (audio and visual) constructions
- Better engage in personalised literacy development
- Better communicate to teachers and parents
- Obtain more timely and focused feedback from a wider range of people".

The following mobile learning affordances were supported:

- *Increased access*—Teacher and parents can access student learning development anytime from multiple devices.
- *Personal relationships with learning*—Students can directly take up their iPad whenever a worthwhile learning opportunity presents.
- *Personalised of choice and pathways*—Teacher app choice is available around the core apps selected to support student creativity and personalize learning pathways. Students can build their own learning pathways through personalised use of apps.
- *Increased accessibility to content*—Students can generate and access media information.
- *Increased learning interactivity*—Students can report, reflect and communicate in more accessible media. Feedback is likewise recorded and accessible.
- *Connecting across contexts*—iFolio tags create a documented journey of curriculum value, extending classroom learning beyond the classroom.

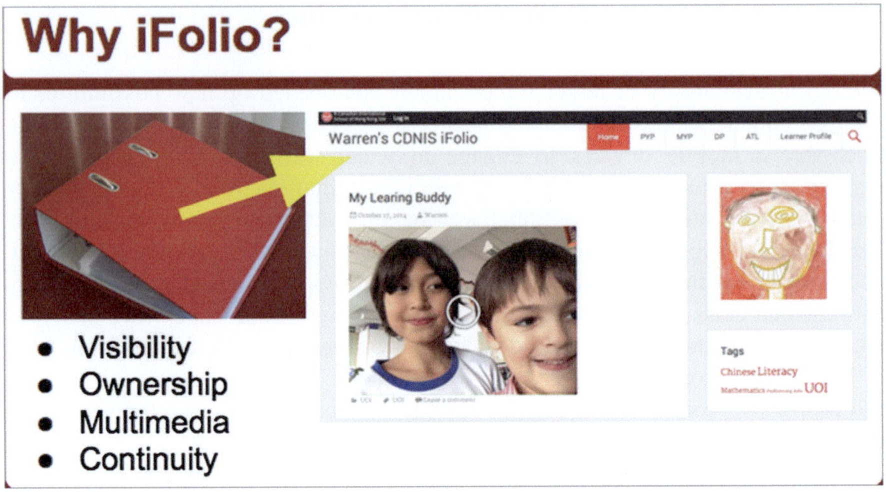

Fig. 13.4 Grade 1 iFolio example used in Grade 1 teacher presentation

School objectives were supported by:

- *Valued*—Teacher driven as worthwhile education and learning, supported by school leadership
- *Visible*—Accessible across the school community
- *Connected*—Connected to the school's iFolio approach that extends through to the DP level
- *Progressive*—iFolio moves with the student to subsequent years. New digital opportunities can be embraced and personalised learning integrated (Fig. 13.4).

13.6 Discussion

These studies support that appropriately targeted use of digital technologies can satisfy both mobile learning affordances and school intentions within a whole school framework. They provide evidence of a school coordinated approach to technology integration in line with what mobile learning research has identified as learning affordances. Teacher development and inclusion, school support and leadership, and a culture of worthwhile, evaluable risk taking are all critical aspects.

It is important continually to look deeper into any school to understand its digital ecosystems. There are many international schools spread across the world with similar surface characteristics. And as stated previously there is much that can be learned from other schools. But if one is to progress digital within a school an understanding of where the school is at, where it wants to go, and what it is willing to take on, is paramount. This includes taking into account the effect of legacy decisions. As Watters (2014, p. 3) reminds us "the future of ed-tech is shaped by the history of ed-tech—whether we realise it or not". So too the future of any school's use of ed-tech.

The mobile learning approach taken on in the school recognises several deliberations as important to progress any school's productive commitment to mobile learning. These include:

- Understanding institutional cultural and contextual foundations
- Clarifying institutional intentions
- Clearly stating what is understood by emerging concepts such as mobile learning, digital literacy and so on
- Understanding the adaptability requirements of dynamic digital systems within school structures
- Developing support structures that help facilitate a united approach to advancing personal, curriculum and school system objectives
- Importance of leadership at multiple levels, including teacher inclusion
- Looking for opportunities and challenges to be embraced and embedded as an integral part of institutional decision making
- Maintaining a willingness to integrate student personal learning facilitated by their personal devices.

The investigation in the school has positively supported the question of whether mobile learning affordances can be successfully integrated into the school learning ecosystem to provide valued learning. However, this is only a first step and somewhat limited to in-school mobile devices in early years, and personal laptop use in later years. In addition, to look deeper, more formal research is required to ascertain:

- Are there areas where more can be done, such as in level of personal choice, levels of visibility and feedback, or evaluating learning value,
- To what extent might limitations of school as a system be blocking intentions, and
- If so, what adjustments might assist and how likely to be taken up

More also needs to be done on the impact of personal smartphones and multiple devices on school-based learning. These all will relate to Sharples' (2009) identification of usability, effectiveness and satisfaction as important drivers for evaluating mobile learning (as with any digital technology change).

13.7 Conclusion

Mobile learning within school systems is at a crucial point (McFarlane 2015). Using digital to add value in education is an evolving phenomenon. How this will unfold by its very nature is unclear. But what it does tell us is that in order to progress we need to innovate and take calculated risks, particularly in fast-changing, dynamic environments. To this end schools need to continue to seek new opportunities that reflects a balanced, forward-looking focus. They also need to continually question themselves, and by doing so educate by example. All schools as Digital Age social institutions will need this trait.

This paper, by looking at one school's approach towards integrating mobile learning up against affordances identified by mobile learning research, provides pointers to consider for other schools, as well as highlights issues that all schools need to consider.

Some may prefer to stay cocoon within closed systems built around closed knowledge and minimal change controlled by individual hierarchical decision making; what some might term industrial thinking. Others may be content to focus solely on individual relationships with their personal technologies. If seeking to unite the (school) system with personal mobile learning, it is as much about the approach, the intent and the willingness to be open to the opportunities mobile device might provide. There is a long way to go, although this study provides grounds for positive expectation. It starts with a conversation; it includes risk, evaluative processes, and a recognition of the challenge and power of difference a meeting of the personal and the systemic.

References

Bain, A., & Weston, M. (2012). *The learning edge: What technology can do to educate all children*. New York: Teachers College Press.

Baran, E. (2014). A review of research on mobile learning in teacher education. *Educational Technology and Society, 17*(4), 17–32.

Blackley, S., & Walker, R. (2015). One-to-one laptop programs: Is transformation occurring in mathematics teaching? *Issues in Educational Research, 25*(2), 99–117. Retrieved from http://www.iier.org.au/iier25/blackley.pdf.

Bjerede, M., & Bondi, T. (2012). Learning is personal: Stories of android tablet use in the 5th grade. Learning Untethered. In B. Clarke & S. Svanaes (Eds.). *An updated literature review on the use of tablets in education. Tablets for Schools.* Retrieved from http://www.tabletsforschools.org.uk/wp-content/uploads/2014/04/T4S-Literature-Review-9-4-14.pdf.

Churchill, D., & Churchill, N. (2008). Educational affordances of PDAs: a study of a teacher's exploration of this technology. *Computers & Education, 50*(4), 1439–1450.

Churchill, D., Fox, B., & King, M. (2012). Study of affordances of iPads and teachers' private theories. *International Journal of Information and Education Technology., 2*(3), 4.

Clarke, B., & Svanaes, S. (2014). *Tablets for schools: An updated literature review on the use of tablets in education.* Retrieved from http://www.tabletsforschools.org.uk/wp-content/uploads/2014/04/T4S-Literature-Review-9-4-14.pdf.

Cochrane, T., Narayan, V., & Oldfield, J. (2013). iPadagogy: Appropriating the iPad within pedagogical contexts. *International Journal of Mobile Learning & Organisation, 17*(1), 18.

Columbus, L. (2014, November). Mobile is eating the world. *Forbes*. Retrieved from http://www.forbes.com/sites/louiscolumbus/2014/11/09/mobile-is-eating-the-world/.

Craig, T., & Van Lom, M. (2009) *Impact Constructivist Learning Theory and Mobile Technology Integration, Theories of Educational Technology*. EDTech, Boise State University, Boise. Retrieved from https://sites.google.com/a/boisestate.edu/edtechtheories/craig_and_vanlom.

Cuban, L. (1986). *Teachers and machines: The classroom use of technology since 1920*. New York: Teachers College Press.

Cuban, L. (2001). *Oversold and underused: Computers in the classroom*. Cambridge, MA: Harvard University Press.

Cuban, L. (2014). *Larry Cuban on school reform and classroom practice*. Retrieved from https://larrycuban.wordpress.com/.

Duncan, D., Hoekstra, A., & Wilcox, B. (2012). Digital devices, distraction, and student performance: does in-class cell phone use reduce learning? *Astronomy Education Review*, 11, 010108–1, 10.3847/AER2012011.

Florida Centre for Instructional Technology. (2011). *Technology integration matrix (TIM)*. Retrieved from http://fcit.usf.edu/matrix.

Fullan, M. (2007). *The new meaning of educational change* (4th ed.). New York: Teachers College Press.

Greenfield, S. (2015). *Mind change: How digital technologies are leaving their mark on our brains*. New York: Random House.

Kearney, M., Schuck, S., Burden, K., & Aubusson, P. (2012). Viewing mobile learning from a pedagogical perspective. *Research in Learning Technology*, 20. doi: 10.3402/rlt.v20i0.14406.

Laouris Y., & Eteokleous, N. (2005). We need an educationally relevant definition of mobile learning. In *Proceedings of the 4th World Conference on Mobile Learning, mLearn 2005*, October 25–28, Cape Town, South Africa.

Liu, M., Scordino, R., Geurtz, R., Navarrete, C., Ko, Y., & Lim, M. (2014a). A look at research on mobile learning in K-12 education From 2007 to the present. *Journal of Research on Technology in Education, 46*(4), 1–49.

Liu, M., Navarrete, C., & Wivagg, J. (2014b). Potentials of mobile technology for K-12 education: An investigation of iPod touch use for english language learners in the united states. *Journal of Educational Technology and Society, 17*(2), 115–126.

McCoy, B. (2013). Digital distractions in the classroom: Student classroom use of digital devices for non-class related purposes. *Journal of Media Education, 4*(4), 10.

McFarlane, A. (2015). *Authentic learning for the digital generation – Realising the potential of technology in the classroom*. Abingdon, UK: Rutledge.

Meyer, J., & Land, R. (2003) *Threshold concepts and troublesome knowledge: Linkages to ways of thinking and practicing within the disciplines*. Retrieved from http://www.etl.tla.ed.ac.uk/docs/ETLreport4.pdf.

New Media Consortium. (2013). *Horizon report, K-12 edition 2013*. Retrieved from http://www.nmc.org/publications/2013-horizon-report-k12.

New Media Consortium. (2014). *Horizon report, K-12 edition 2014*. Retrieved from http://www.nmc.org/publications/2014-horizon-report-k12.

OECD. (2013). *Governing complex education systems: Framework for case studies*. Retrieved from http://www.oecd.org/edu/ceri/governingcomplexeducationsystemsgces.htm OECD.

Papert, S. (1980). *Mindstorms: Children, computers and powerful ideas*. Brighton: Harvester Press.

Sharples, M. (2009). Methods for Evaluating Mobile Learning. In G. N. Vavoula, N. Pachler, & A. Kukulska-Hulme (Eds.), *Researching mobile learning: Frameworks, tools and research designs* (pp. 17–39). Oxford: Peter Lang Publishing Group.

Sharples, M. (2013). Mobile learning: Research, practice and challenges (In Chinese, J. Xiao, Trans.). *Distance Education in China*, 3:5–11, 44. Also reprinted as Sharples, M. (2013). Mobile Learning: Research, Practice and Challenges (In Chinese, Trans. J. Xiao). *Adult Education*, 9, 28–35. Retrieved from http://oro.open.ac.uk/37510/2/sharples.pdf.

Shuler, C., Winters, N., & West, N. (2013). *The future of mobile learning: Implications for policy makers and planners*. Paris: UNESCO.

Seipold, J., & Pachler, N. (2011). *Evaluating mobile learning practice - towards a framework for analysis of user-generated contexts with reference to the socio-cultural ecology of mobile learning Medienpaedagogik* (Vol. 19, pp. 1–13).

Traxler, J., & Kukulska-Hulme, A. (2005). Evaluating mobile learning: Reflections on current practice. In *Proceedings of mLearn2005: 4th World Conference on mLearning*, Cape Town, South Africa. Retrieved from http://oro.open.ac.uk/12819/1/.

Turner, J. (1999). Use of the Logo programming language as software to support constructivist learning within a post-primary school mathematics environment. *Unpublished PHD thesis.* Monash University, Victoria.

UNESCO. (2012). Turning on mobile learning: global themes. *UNESCO Working Paper Series on Mobile Learning*. (pp. 1–14).

Watters, A. (2014). *The monsters of educational technology*. Retrieved from http://hackeducation.com/2014/12/01/the-monsters-of-education-technology/.

Author Biography

John Turner is Head of Learning and Teaching Technologies at Canadian International School of Hong Kong, with nearly 30 years experience working with educational technologies across K-12. As well as providing advice on how new technologies might add value to learning, he is also actively involved in leading by example and working with teacher teams, united in support of the school's vision of digital learning infusion. John has previously led similar approaches in Australia and the Middle East. Experiences and research have been shared through local, national and international conferences and journals.

Chapter 14
Overcoming Teachers' Concerns—Where Are We in the Harnessing of Mobile Technology in K-12 Classrooms in Hong Kong?

Tianchong Wang

Abstract The emergence of "Post-PC" iPads and Galaxy tablets as global heavyweights of mobile technologies have prompted a wave of educational technology advocates and policy makers to encourage teachers in the harnessing of mobile technology into K-12 classrooms. The actual level of implementation, however, has been reported as lagging far behind these research-led initiatives and slowed down the momentum envisaged by these policies, especially in many classrooms in the public sector. Teachers as individual innovation adopters are believed to play a crucial role in this innovation change process for the adoption of mobile technologies. To better understand the reluctance of teachers to adopt these mobile technologies into their classrooms, this study assessed teachers' concerns over harnessing mobile technology in Hong Kong public sector K-12 classrooms. A total of 159 teachers participated in this study. Utilising the Stage of Concern framework, a mixed-method approach was taken. Data collection compiled self-reported Stage of Concern Questionnaires and Open-Ended Concerns Statement opinion polls. Preliminary descriptive analysis showed that teachers experienced all five categorical concerns over harnessing mobile technology in teaching practices. The Information construct underscored a more intense area of concern. From the findings, implications in terms of accessibility, time, support-related interventions, leadership issues, and further suggested interventions are discussed.

14.1 Introduction

In recent years, the proliferation of mobile computing devices, especially "Post-PC" smartphones and tablets such as Apple iPads, have had a tremendous impact on different facets of Hong Kong society. A recent market research report (TNS 2011)

T. Wang (✉)
The Chinese University of Hong Kong, Hong Kong, SAR China
e-mail: shaohua3@gmail.com

D. Churchill et al. (eds.), *Mobile Learning Design*, Lecture Notes in Educational Technology, DOI 10.1007/978-981-10-0027-0_14

established that a total of 17 percent, or one in six Hong Kong residents, owned an iPad, which is nearly six times the global average. Following suit, in order to maintain and advance the competitiveness of Hong Kong, the HKSAR government has implemented initiatives on harnessing the mobile technology in K-12 education: the government guideline (Education Bureau 2007) on the *Third Strategy on Information Technology in Education* (ITE3), named *Right Technology at the Right Time for the Right Task*, published by the Hong Kong Education Bureau, has recognised mobile learning as a trend; The most recent consultation document (Education Bureau 2014) of the forthcoming *Fourth Strategy on Information Technology in Education* (ITE4) took one step further by positioning the use of mobile technology as one of the most important strategies for students and teachers learning and teaching.

On the face of it, the Hong Kong public education system seems well-adapt at embracing the "Smart Age". However, some recent reports pointed out that this only applies to a certain number of Direct Subsidy Scheme (DSS) Schools (Yau 2015). Indeed, when it comes to those Aided Schools, the most common and grass-roots pre-tertiary schools in the Hong Kong public sector, the actual implementation of the mobile technology in classrooms has been relatively slow so far. Anecdotal reports have revealed that many teachers are unmotivated to alter their current teaching practices and to integrate mobile technology into the classroom. There appears to be discrepancy between policy and reality of classroom implementation.

The introduction of mobile technologies into the classroom requires a process of change in learning and teaching. Teachers, the front line change adopters and gatekeepers (Fullan 2007), would inevitably have concerns over adopting change (Hord et al. 2006). While some researchers have explored and underscored the potential of introducing the mobile technology as an educational tool in situations within and beyond the confines of the classroom (Wang et al. 2014), the potential cannot be fully realised due to individual teacher's concerns towards the mobile technology, which can result in resistance. Therefore, it is important to identify and understand individual concerns to reduce the possibility of resistance towards the implementation of mobile technology in Hong Kong K-12 classrooms.

14.2 Stages of Concern (SoC)

Concerns have been defined as "the composite representation of the feelings, preoccupations, thoughts and considerations given to a particular innovation-based task or issue" (Hall et al. 1977, p. 5). Stages of Concern (SoC) (Hall and Hord 1987) was a framework developed by Hall and his colleagues to describe how people acclimate to change to pave the way for successful implementation of an innovation. The original SoC was construed in seven stages, namely *Awareness*, *Informational*, *Personal*, *Management*, *Consequence*, *Collaboration*, and *Refocusing*. While Hall and Hord's (1987) SoC theory has been widely adopted in many fields, Cheung and Yip (2004) refined the SoC model to five stages to better cater for educational change. More specifically, *Informational* stage was merged with the *Personal* stage; the *Consequence* and *Collaboration* stages were combined; the *Awareness* stage was deemed irrelevant; and an extra stage called *Evaluation* was introduced. In sum, Cheung and Yip's (2004) revised stages of concern for educational change were *Evaluation*, *Information*, *Management*, *Consequence* and *Refocusing*. At stage 1 (*Evaluation*), the teacher feels uncertain about the worthiness and fairness of the innovation as well as the feasibility of putting the innovation into school practice. At stage 2 (*Information*), the teacher is concerned with some general aspects of the innovation, such as its rationale, requirements for use and moderation mechanism. At stage 3 (*Management*), the teacher raises a number of questions about the tasks and processes of implementing innovation. At stage 4 (*Consequence*), the teacher is concerned with the impact of the innovation on student learning and his/her professional development. At stage 5 (*Refocusing*), the teacher is concerned with further developments of the innovation.

Both Hord et al. (2006) and Cheung (2002) indicated that the dimensions of concerns over innovations occur in a developmental direction: in general, early concerns (1–2) are more self-oriented; when these concerns are resolved, what emerged (3) are more task-oriented; finally, when self- and task concerns are largely resolved, the participants in change can focus on impact (4–5). However, Hord et al. (2006) also emphasised that individuals do not necessarily progress through the stages step-by-step, and that they do not necessarily begin the stages at the same time or move through the stages at the same pace. Rather, Cheung and Yip (2004) pointed out that it is possible for individual teachers to experience several SoC over the innovation concurrently, but perhaps with differential degrees of intensity.

To frame the adoption of innovation according to the concerns and doubts of individual teachers requires taking an individualistic approach. Cheung and Yip's (2004) revised SoC model can provide important insight about individual teachers to understand the stages individual teachers must go through before and when they are convinced about the innovation, and therefore being adopted in this study. It is believed that, after identifying, accessing and addressing the concerns of individual teachers over an innovation, based on their intense stage(s), there would be a greater likelihood that the innovation will be effectively implemented in a sustainable manner.

14.3 Research Question

The purpose of this study was to assess teachers' concerns about the introduction of an innovation, mobile technology, into the classroom.

The study sought to answer the research question: *What are teachers' concerns as they implement mobile technology into their teaching practices*?

The results of this study will be used to assist educators and policy makers in understanding concerns involved in the implementation and integration of the mobile technology in their schools and in teaching practices for better adoption through appropriate efforts and interventions.

14.4 Participants

This research study involved 159 teachers who attended our workshop on mobile learning that was jointly organised with the Education Bureau in December 2013. The teachers came from both public sector primary and secondary schools in Hong Kong. Their subject areas varied, as did their exposure to Information and Communication Technology (ICT). Convenience sampling was used. Prior to the study, all participants were given assurances on the confidentiality and anonymity of the data and its representations.

14.5 Research Design

A mixed-method approach (Creswell 2014) was adopted by assessing teachers' Stages of Concern for harnessing the mobile technology in their classrooms, from different perspectives. The instruments used were self-reporting, including a quantitative SoC Questionnaire and a qualitative Open-Ended Concerns Statement opinion poll.

For quantitative purposes, a customised 25-item SoC Questionnaire written in the Chinese language was devised, based on Cheung (2005)'s version that had been rigorously tested for validity and reliability, was used to measure teachers' stages of concern as they adopted the mobile technology into their teaching practice Table 14.1. This SoC Questionnaire comprised five sub-scales with five items each that corresponded to the five categorical stages of concerns as refined by Cheung and Yip (2004). All 25 items appeared in the instrument in a mixed order. Each item was accompanied by a five-point Likert-scale, ranging from 1 (not concerned) to 5 (strongly concerned). The SoC Questionnaire was conducted before our workshop and participants were asked to choose the appropriate level which best expressed their concerns: high scores indicating high concern, and low scores indicating low concern.

The results were further supported by the qualitative data collected from the Open-Ended Concerns Statement opinion poll, in which the most salient issues were asked in detail. Additional insight into teachers' concerns formulated in their own words were gathered.

The findings of the questionnaire and the opinion poll would indicate what type of interventions need to be implemented to enable teachers to move forward to the next stage in the change process.

Table 14.1 The SoC questionnaire items (translated)

Stages	Item no.	Item
Evaluation	6	Whether it is worthwhile to promote mobile technology in my school
	8	Whether it is feasible to apply mobile technology in my school
	11	Whether I have the required knowledge and skills to use mobile technology
	16	Whether using mobile technology is better than other teaching tools
	21	Whether the government supports the use of mobile technology in school education
Information	2	How my role is supposed to change if I adopt mobile technology in my teaching practice
	7	How the use of mobile technology will affect my teaching workload
	9	Opportunities to learn from other teachers experience of using mobile technology in their teaching practice
	17	Knowing more instances for applying mobile technology in teaching
	22	Further provision of resource and support if I go on to apply mobile technology in my teaching practice
Management	3	Insufficiency of time to prepare and adjust my current pedagogy if adopting mobile technology
	12	How to assess my students' learning with mobile technology
	13	How to conduct teaching and learning activities with mobile technology more efficiently
	18	Extra time on dealing with non-pedagogical issues after the use of mobile technology in my teaching practice
	23	Insufficiency of time for students to reflect and summarise their learning after using mobile technology
Consequence	1	Whether my student would like to use mobile technology to learn
	4	My students attitude towards mobile technology
	14	Collaboration with other teachers to facilitate teaching with mobile technology
	19	Impact on my students after applying mobile technology in my teaching practice
	24	Reinforcing my students' understanding on their learning role in learning with mobile technology
Refocusing	5	The best use of mobile technology in my own teaching
	10	How to optimise teaching practice with mobile technology based on my own experience
	15	Revising mobile technology in education to improve its effectiveness
	20	How to modify teaching practice with mobile technology based on my students' feedback
	25	Exploring other teaching tools better than mobile technology

14.6 Results and Data Analysis

Although all of the 159 questionnaires were returned, 18 of those were partially completed. The data analysis was therefore based on 141 completed questionnaires. MATLAB, a statistics analysis computer programme, was used for quantitative data analysis. A reliability analysis was performed in the beginning. The Cronbach's alphas (Cronbach 1951) of the five constructs were 0.665, 0.691, 0.701, 0.732 and 0.705, respectively. These results indicated an adequate level of reliability of the collected data.

Descriptive statistics was used to analyse the questionnaires. The mean of each SoC construct was computed, as Table 14.2 shows. The means ranged from 3.99 to 4.23, indicating that teachers experienced all five categorical concerns over harnessing the mobile technology in their teaching practices.

A one-way within subjects ANOVA analysis was conducted. The result indicated that differences among the five constructs' means were statistically significant [$F(4, 3520) = 12.582$, $p < 0.001$]. It can be seen that the mean of the *Information* concern was the greatest (4.23). Paired-samples t-tests (Nikulin 2001) verified that the mean of the *Information* concern was statistically different from the *Evaluation* construct's (p-value = $1.2923e - 007$, <0.05), the *Management* construct's (*p*-value = $1.3020e - 009$, <0.05), the *Consequence* construct's (*p*-value ≈ 0, <0.05) and the *Refocusing* construct's (*p*-value = $1.6738e - 011$, <0.05).

In sum, the statistical analysis identified that teachers experienced all five categorical concerns over harnessing the mobile technology in teaching practices, and among those, *Information* was the peak category, which appeared to be a more intense area of concern. It is worth noting that, considering the mean value of all SoC constructs were high and the *Information*'s distinctness from other constructs was roughly 0.2, such a difference was not necessarily substantive (Carver 1978).

Qualitative data from the Open-Ended Concerns Statement opinion poll was organised into table format within Microsoft Excel, a spreadsheet software. A thematic analysis framework was adopted. Among these statements, a number of concerns were flagged by the teachers. The main themes that emerged mostly mirrored those items described in the questionnaire.

Table 14.2 Univariate descriptive statistics of the SoC constructs

Construct	Mean	Standard Deviation
Evaluation	4.06	0.77
Information	4.23	0.67
Management	4.03	0.73
Consequence	3.99	0.68
Refocusing	4.02	0.65

14.7 Preliminary Discussions and Implications for Practice

On the whole, the teachers' concurrent experience of all five categorical concerns about harnessing the mobile technology in teaching practices reflected a willingness to accept the mobile technology during the adoption and implementation process. The relatively higher intensity of the *Information* stage implied that teachers were still concerned with how mobile technology affected them individually and focused on its rationale, requirements for use and moderation mechanism. Frequently occurring concerns from the Concerns Statement opinion poll along with recommendations are put forward.

Accessibility issues could hinder teachers' decision to integrate mobile technology. Poor support networks can result in negative perceptions and ultimately resistance to mobile technology use. Concerns about the instability of the Wi-Fi network as a result of concurrent usage by students and teachers indicates to policy makers and educational leaders the need for better Wi-Fi infrastructure in all classrooms. In addition, in some schools, it was reported that the accessibility issue was still at the hardware level. This indicates that digital divide across schools in terms of accessibility still exists in Hong Kong. In these schools, despite advocacy from motivated teachers for mobile hardware, the financial cost for the school was still the fundamental barrier. As mobile devices become more ubiquitous among families with children, it is suggested that the Bring-Your-Own-Device (BYOD)[1] model (Johnson et al. 2013) is a plausible option to ensure maximum hardware accessibility. Even so, considering that many children in public schools are from low-income families that do not have the ability to shoulder the costs involved in the change with innovations, it is recommended that the government should dip deeper in order ensure that no student is left behind in the "Smart Age" just for financial reasons.

While addressing accessibility issues concerns, it is important not to use technology for technology's sake. Often in many classroom situations, mobile technology serves better as a supplement rather than replacement for traditional learning and teaching tools. The affordances (Gibson 1977; Norman 1988) of the mobile technology must be exploited in a more meaningful, contextually appropriate and efficient approach. After all, it is the combination of transformed learning design and sound pedagogical practices, rather than merely fancy technologies or technology know-how, that can effectively instigate a paradigm shift in the classroom.

Time constraint was another salient concern raised by teachers who are time deficient even without the use of technology in their existing teaching practice. Although time constraints can become an excuse for a certain group of teachers who are not technologically inclined, admittedly, with the use of technology, further time-consuming factors such as technological problems could be added.

[1]A model based on the idea that students should be encouraged to bring their personal devices, especially smartphones and tablets, to class.

Just-in-time and on-going supports from the school ICT support team should be readily available to free up teachers' time so that the integration of mobile technology can become a meaningful venture.

Even with the help from ICT support teams or educational technologists, many teachers admitted that they felt nervous incorporating the mobile technology into their teaching practices. Thus, there is a pressing need for rigorous teacher training in the mobile technology in educational practices, such as pre-service and in-service professional development courses and even one-to-one consultations, while a minority of enthusiastic teachers may develop such practices through their own resources. This is a long-term process, which involves not only the development of teachers' digital literacy but also a paradigmatic shifting of how learning and teaching with mobile technology (Churchill et al. 2012; Churchill and Wang 2014). Alongside formal training, teachers should build up informal Communities of Practice (CoPs) (Wenger 1998) where they can exchange new ideas and collect feedback with local and remote partners. For example, social networking and mobile Instant Messaging (like WhatsApp and WeChat) groups can be formed by teachers and ICT professionals to identify and share educational Apps, and apply generic Apps to creative usages. The information gained from the CoPs may serve as a starting point for many. In conjunction with the community efforts, it is hoped that individual teachers' own "mobile pedagogy" can be developed to achieve their own pedagogical purposes and student learning outcomes.

Putting technical challenges aside, some teachers maintained a critical attitude toward the mobile technology integration because of classroom disciplinary concerns. A few of them anticipated that students would be over-excited during mobile technology-supported lessons, while the others questioned if the mobile technology in classrooms would lead to off-task behaviours and distractions because of its hyperconnectivity to social media. These concerns may sound legitimate. In order to tackle them, additional provisions allowing for disciplinary measures should be given extra attention. Teachers should offer guidance students to recognise that mobile tools are more than entertainment consumption "toys" and further scaffold students to apply the mobile technology to learning tasks. Nevertheless, from the viewpoint of a teacher, we must ask whose responsibility it is when there are distractions in the classroom, irrespective of whether technologies are incorporated or not. The optimal solution to avoid off-task behaviours and distractions may be to engage the learners with interesting learning activities to begin with.

Institutional leadership plays a crucial role as several teachers were concerned about "school support", where there are still bans on student's use of mobile devices in school. Exploration and action research on the educational use of mobile technology can be hampered by restrictive institutional policies and school culture. Therefore, there must be informed institutional leadership. Particularly, school leaders must recognise that educational change associated with mobile technology is not just for the "hard" outcomes (e.g. test result improvements) but more for the "soft" outcomes such as student's acquisition of twenty-first century skills (Bellanca and Brandt 2010). Best practices for teachers may only be achieved with the openness of school leaders to change. Unfortunately, at its current stage, such an informed institutional leadership, which is needed to promote innovation with mobile technology, has not yet been widespread.

Current results are preliminary in nature. Contextual differences such as different academic disciplines may have an impact on the intensity of user concerns over the technology.

14.8 Conclusions

Change in K-12 education goes far beyond the introduction of innovations like the mobile technology, and is likely not to be a one-time "dog and pony show". Rather, change with technology must begin with innovation practitioners - the teachers, although they will raise concerns. In this study, by acknowledging and giving serious attention to the intensive areas of concerns among teachers based on the SoC framework, perhaps more meaningful interventions can be taken, as suggested in the *Preliminary Discussions and Implications for Practice* section, to enable the change process to be directly relevant to the teacher's needs. Nevertheless, harnessing the potential of mobile technology in K-12 education will require a concerted effort on the part of all stakeholders to reduce the "discomfort" aspect of the change process and eventually achieve the ideal state of the innovation implementation: as Marshall (1995) stated, *adding wings to caterpillars does not create butterflies…Butterflies are created through transformation* (p. 11).

References

Bellanca, J. A., & Brandt, R. S. (Eds.). (2010). *21st century skills: Rethinking how students learn*. Bloomington, IN: Solution Tree Press.

Carver, R. P. (1978). The case against statistical significance testing. *Harvard Education Review, 48*, 378–399.

Cheung, D. (2002). Refining a stage model for studying teacher concerns about educational innovations. *Australian Journal of Education, 46*, 305–322.

Cheung, D. (2005). Science teachers' concerns about school-based curriculum development. *Hong Kong Science Teachers' Journal, 22*, 1–7.

Cheung, D., & Yip, D. Y. (2004). How science teachers' concerns about school-based assessment of practical work vary with time: The Hong Kong experience. *Research in Science & Technological Education, 22*, 153–169.

Churchill, D., Fox, B., & King, M. (2012). Study of affordances of iPads and teachers' private theories. *International Journal of Information and Education Technology, 2*, 251–254.

Churchill, D., & Wang, T. (2014). Teacher's use of iPads in higher education. *Educational Media International, 51*, 214–225.

Creswell, J. W. (2014). *Educational research: Planning, conducting, and evaluating quantitative and qualitative research.* (5th ed.). Upper Saddle River, New Jersey: Merrill.

Cronbach, L. J. (1951). Coefficient alpha and the internal structure of tests. *Psychometrika, 16*, 297–334.

Education Bureau. (2007). *Consultation document on the third strategy on information technology in education: Right technology at the right time for the right task*. Retrieved April 26, 2015, from http://www.edb.gov.hk/index.aspx?nodeID=6140&langno=1.

Education Bureau. (2014). *Consultation document on the fourth strategy on information technology in education: Realising it potential, unleashing learning power, a holistic approach*. Retrieved from http://www.edb.gov.hk/attachment/en/edu-system/primary-secondary/applicable-to-primary-secondary/it-in-edu/it-in-edu/Policies/4th_consultation_eng.pdf.

Fullan, M. (2007). *The new meaning of educational change* (4th ed.). New York: Teachers College Press.

Gibson, J. J. (1977). The theory of affordances. In R. Shaw & J. Bransford (Eds.), *Perceiving, acting, and knowing: Toward an ecological psychology*. New Jersey: Lawrence Erlbaum.

Hall, G. E., George, A. A., & Rutherford, W. A. (1977). *Measuring stages of concern about the innovation: A manual for use of the SoC questionnaire*. Austin, TX: Research and Development Center For Teacher Education, University of Texas at Austin.

Hall, G. E., & Hord, S. M. (1987). *Change in schools: Facilitating the process*. Albany, NY: State University of New York Press.

Hord, S. M., Rutherford, W. L., Huling, L., & Hall, G. E. (2006). *Taking charge of change*. Alexandria, VA: Association for Supervision and Curriculum Development.

Johnson, L., Becker, S. A., Estrada, V., & Freeman, A. (2013). *NMC Horizon report: 2013 K-12 edition*. Retrieved 26 from http://www.nmc.org/pdf/2013-horizon-report-k12.pdf.

Marshall, S. P. (1995). The vision, meaning, and language of educational transformation: How chaos, complexity, theory, and flocking behavior can inform leadership in transition. *School Administrator, January 1, 1995*, 8–15).

Nikulin, M. S. (2001). Student test. In M. Hazewinkel (Ed.), *Encyclopaedia of Mathematics*. Heidelberg: Springer.

Norman, D. A. (1988). *The design of everyday things*. New York: Doubleday.

TNS (2011). *TNS survey puts Hong Kong at forefront of expected global tablet sales explosion*. Retrieved from http://www.tnsglobal.com/press-release/tns-survey-puts-hong-kong-forefront-expected-global-tablet-sales-explosion.

Wang, T., Towey, D., & Jong, M. S. Y. (2014). Exploring young students' learning experiences with the iPad: A comparative study in Hong Kong international primary schools. *Universal Access in the Information Society, November 2014*, 1–9.

Wenger, E. (1998). *Communities of practice: Learning, meaning, and identity*. Cambridge, U.K: Cambridge University Press.

Yau, E. (2015). Schools without Wi-Fi, kids without computers in Hong Kong's digital divide. *South China Morning Post*. Retrieved from http://www.scmp.com/lifestyle/family-education/article/1708291/schools-without-wi-fi-kids-without-computers-hong-kongs.

Author Biography

Tianchong Wang is currently a Senior Research Assistant at The Hong Kong Institute of Education as well as a Doctoral Student at The Chinese University of Hong Kong. Prior to this, he received an MSc degree from the University of Hong Kong and a BA degree from Hong Kong Baptist University, respectively. Tianchong has a background in ICT, receiving qualifications such as MCP, MCSA, MCSE, MCDBA, MCTS, MCITP, ACSP, ACTC, and THXCP. Tianchong is also a passionate educator and researcher. He is a member of IEEE, HKERA, and HAAL. His research interests include blended learning, mobile learning, and flipped classrooms.

Chapter 15
Exploring the Suitability of the Book Creator for iPad App for Early Childhood Education

Monika Tavernier

Abstract Handheld mobile devices are part of young children's everyday life as they observe others use and engage with such devices. Early childhood education does not ignore the popularity of mobile touch devices and starts to investigate how tablets, especially iPads, can improve learning and teaching. This study examines if the 'Book Creator App for iPads' is a suitable app to enhance 3–6 years old children's ability to express their ideas, creativity and illustrate their understanding of the world around them. Over a period of 12 weeks, a group of 3–5 years old children familiarized with the app, completed assignments and created sophisticated digital artefacts that included drawings, photos, voice and video recordings. These artefacts reflected their interests, cognitive abilities and level of fine motor skills.

15.1 Introduction

The implementation of computers in early childhood education (ECE) showed that technology can enhance young children's learning (Clements and Sarama 2002). Mobile devices such as phones and tablets, especially iPads, are highly accepted and popular among adults and children (Chiong and Shuler 2010; Yelland and Gilbert 2012). Hence, governments promote the use of information and communication technologies (ICT) in ECE (Curriculum Development Council 2006) and a plethora of research described how children as young as 4 years old use ICT independently, purposefully and for learning purposes (Hertzog and Klein 2005; Zevenbergen 2007). But ECE educators have remained sceptical (Lindahl and Folkesson 2012a, b). The implementation of ICT in ECE is therefore progressing slowly and related

M. Tavernier (✉)
German Swiss International School, Hong Kong SAR, People's Republic of China
e-mail: monikamenzel@hotmail.com

D. Churchill et al. (eds.), *Mobile Learning Design*, Lecture Notes in Educational Technology, DOI 10.1007/978-981-10-0027-0_15

activities remain basic (Edwards–Groves and Langley 2009). Prejudices and misconceptions about the effects that ICT has on children and a lack of positive ICT teaching experiences lead to the teachers' opposed attitude (Ertmer 2005; Lindahl and Folkesson 2011).

Recent tablet and iPad studies investigated the viability of such devices (Yelland and Gilbert 2012; Michael Cohen Group Llc 2011), the children's use behaviour (Falloon 2013; Hutchison et al. 2012) and the impact that applications (apps) and app interface have on learning (Falloon 2013). The findings of these studies and the wide acceptance and use of iPads, their intuitive operation, and the easy access to a wide range of low cost apps may change the current ICT use in ECE (Chiong and Shuler 2010). This study introduced 27 3–5 years old children to the 'Book Creator for iPads' app (hereafter: Book Creator) to examine the viability of the app for ECE.

15.2 Literature Review

Today's children and parents use mobile phones and tablets on a daily basis (O'Mara and Laidlaw 2011). Chiong and Shuler (2010) found that young children are given mobile phones to entertain them while the family is travelling. Older children use handheld mobile devices to play, look at or take photos and videos, or use so-called educational apps. Chiong and Shuler called this phenomenon 'pass-back effect' and Prenksy (2001) called this generation of children 'digital natives'. According to him, these children may think and learn differently, and new ways of teaching may be required to accommodate their ICT skills (Zevenbergen 2007). Prensky's preposition to adjust the current educational approaches to incorporate children's ICT knowledge stands in great contrast to the views of many early childhood teachers (Lindahl and Folkesson 2012a, b). Teachers' unquestioned assumption that ICT-related activities are naturally more interesting for young children than traditional play or outdoor activities (Lindahl and Folkesson 2011, 2012a) suggests that ICT may threaten children's healthy development if introduced too young. As a result, many early childhood teachers try to protect their young students and avoid using ICT (Lindahl and Folkesson 2012a, b).

Some teachers took on the challenge to implement ICT in their early childhood classrooms, but the use of technology remained basic (Edwards–Groves and Langley 2009). Hence, Yelland and Gilbert (2012) suggest that teachers should rethink their current technology use. They envisage that teachers go "beyond using new tablet technologies as playthings like blocks, puzzles or construction toys … [and] be aware of the wider range of uses of tablets to enable learners to become creators, innovators and to support them in their reflections about the things around them" (p. 1). The use of apps like the Book Creator within the context of ECE may support the realization of this mission.

15.3 Creating to Learn

For many years, researchers and teachers have wondered how technology fits in an ECE environment (Lindahl and Folkesson 2012b), because young children learn through experiences (Andresen et al. 2000), and interactions with the environment and the people around them (Vygotsky 1987). Today, we know that ICTs do not hinder the children's natural approaches to learning. ICTs are additional resources for learning (Sarama and Clements 2004; Herztog and Klein 2005). They can help children practice and reinforce specific content (e.g. Plowman and Stephen 2007; Clements and Nastasi 1993), and enhance their ability to (1) create original content, (2) express ideas and (3) present knowledge in sophisticated ways (Scardamalia and Bereiter 2006; Couse and Chen 2010).

Paintings and drawings are traditional ways for children to express their knowledge and thoughts (Lancaster 2012). Until recently, the interpretation of these lay in the hands of the teachers. They may use the child-created artefacts and teacher-created photos (Broadmann 2007), videos and written documentation to assess the children's development and learning (Couse and Chen 2010). But their analysis of the children's work may be incomplete or wrong, because the children's own explanation is missing (Einardottir 2005).

Einardottir's (2005) photo research is one among the few studies that attempted to understand the motivation behind child-made photos. Pre-school children were given cameras to take photos of things that are important to them in the school. Einardottir found that the neutral viewer could not identify which element of the photo was important to the young photographer without his or her explanation. So a visual artefact does not provide enough information to understand what children try to communicate. The Book Creator may improve this situation, because it allows the user to draw, type, take photos, create videos, create voice recordings, or a combination of all of these, and add these creations to their digital artefacts. So for example, a child may enhance his drawing by adding a voice recording where he explains he drew a train, and by adding a photo he took of his toy train. This would clarify the real meaning of his drawing to the neutral viewer.

Given that tablets provide a unique opportunity for young children to be in control of the device without lengthy pre-use training (Couse and Chen 2010), it is interesting to investigate how children use open-ended and complex creating apps that allow them to present and explain their ideas. This study consequently poses one main research questions: To what extent is the Book Creator app a viable tool for early childhood education?

15.4 Methods

This exploratory study used a qualitative methods. A qualitative approach was used to understand the children's use behaviour in depth. The data collection occurred within a 12 weeks period. The research instruments included narrative observation records, daily log book entries, weekly video recordings and the analysis of children's artefacts. The quantitative component used the video data of multiple single-subject case studies (Creswell 2002) to examine how individual children used the app. The video data from 27 children was analysed and critical incidents were tagged and summarized into quantitative data. These 27 sets of data were compared to determine if there are any differences in the ways the children used the app. The analysis of gender specific differences and the impact of children's character are in process and not included in this paper.

15.4.1 Early Childhood Setting

Twenty-seven children of the German section of an international Kindergarten in Hong Kong participated. The school values social play, outdoor activities, and art and crafts. Teachers use computers and digital cameras on a daily basis to communicate with the parents and to document the children's learning. The participating teachers were not familiar with the iPad and did not implement it as a part of the children's daily learning prior to the study. They designed their daily schedule freely and were only restricted by activities that were conducted by special subject teachers (e.g. music). Table 15.1 illustrates the distribution of age and gender across the participants.

The school charges 130,100HKD per year (school year 2014/2015). Therefore, it can be assumed that the social economic background of all children is high. The school was selected, because it was assumed that these children are familiar with the touch operated devices. This pre-condition allowed the researcher to spend less time on explaining how to use the iPad. Instead, she could focus on the implementation of the Book Creator.

15.4.2 Ethical Considerations

Prior to the study, the school, teachers and parents were asked to complete a consent form that confirmed their and their children's participation in the study and their understanding of (1) the study's purpose, aims and activities, (2) the basis of a voluntary participation and (3) associated risks and their right to withdraw their participation at any time. To avoid that some children feel excluded, all children could engage in the iPad-related activities, but only children that had the parent's consent

Table 15.1 Participating children

Age	Boys	Girls
3–4 year olds	8	6
4–5 year olds	7	6

could operate the iPad. The children's participation was voluntary and no child was forced to join. The researcher invited them and respected their choice to decline.

15.4.3 Selection and Evaluation of App

The app suitability was assessed using Diaz's (2013) matrix for educational eBooks. Since the Book Creator allowed the user to create eBooks, this tool was deemed suitable. Diaz suggested evaluating the:

- Richness: the information volume, access richness, diversity of presentation and interaction styles, kinds of exercises and interactive activities, as well as the scope of the activity.
- Completeness: the number of content and interaction mechanisms to cope with the goals of different kinds of users.
- Motivation: how students are motivated to use the system and to learn more about the subject being addressed.
- Autonomy: the degree of navigation freedom offered to the user and the degree of interaction freedom.
- Competence: the ability to navigate through the system and to reach a particular goal.
- Flexibility: The ease with which the system can be used.
- Aesthetic: How the inclusion of multimedia information is harmonized and used to enhance the comprehension of concepts.
- Consistency: the extent to which elements that are conceptually similar are treated equally by the application, while those that are different are treated differently.
- Ease of Use: how easily users can guess the meaning and purpose of things with which they are presented.

15.4.4 Procedure

The 12 week long study was conducted from mid-September to mid-December 2014. The researcher participated in all iPad-related activities. In an attempt to relate the study activities to the curriculum and class topics, the researcher co-designed the activities with the teachers and involved them in all planning and revision processes. The data collection was separated into three phases: (1) Familiarization phase, consisting of experimentation and explorative activities, (2) application phase including imposed and structured activities and (3) creative phase that allowed the children to use their new skills to create artefacts independently. Each phase consisted of three independent small projects that aimed to provide the children with opportunities to develop, reinforce and apply their Book Creator competencies (see Table 15.2). The children's participation was not regular since it was voluntary.

Table 15.2 Intervention overview

Phase and description	Theme and main function	Duration
Familiarization phase	1. iPad: App Introduction (Drawing)	2 weeks
> Researcher led	2. Weather and Weekdays (Photo Taking)	1 week
> Teacher-centred	3. Autumn Songs (Voice Recording)	1 week
Application Phase	1. Magazine (Multimedia)	1 week
> Researcher led	2. Daily News (Multimedia)	2 weeks
> Student-centred	3. Kids Talk—St. Martin's Day (Video Recording)	1 week
Creating Phase	1. Kids' Creations	1 weeks
> Researcher-assisted	2. Advent—the time to reflect	1 weeks
> Student-centred	3. Christmas	2 weeks

Consequently, their familiarity with the different functions varied and the researcher decided on a day-to-day basis when a child was ready to learn about a new function. All activities took place within the ecosystem of the class and during the free play period. One iPad was shared throughout the study. Occasionally, the group used two additional iPads. The emerging findings indicated that it is more effective to use only one to two iPads, rather than having one per child.

To ensure that all children had time to experiment and familiarize with the Book Creator the researcher joined the children's free play phase every day during the familiarization and application phase. The activities during these two phases were a combination of teacher- and student-centred activities that allowed the children to explore the app freely. The researcher assisted their learning by guiding the children through the different functions and demonstrating the effective application of each one of them. This practice was inspired by Plowman and Stephen's (2007) guided interactions approach to learning the effective use of ICT in ECE. The children could experience and practice newly acquired skills within a safe environment and according to their individual pace.

During the creating stage, the researcher joined the children three times a week, because the children needed less time to realize their ideas. Further, the children needed time to plan their iPad-related activities in order to diversify their artefacts. Each session lasted for 30 to 45 min.

15.4.5 Data Collection Procedure and Analysis

The data was collected during (video recording) and after each session. After each session, the researcher summarized the activities and experiences (narrative observation records and journal entries). The notes included her reflections and remarks. Further, once to twice a week, the researcher video recorded the activities and transcribed important periods of child behaviour and interactions. The videos and written documentation were used to complete rating scales that monitored

(1) the children's level of involvement, (2) their level of tablet use and (3) their actions. This data helped to determine the viability of the Book Creator for ECE for child-initiated use during free play. The changes of the children's use behaviour were assessed based on four elements of the unified theory of acceptance and use of technology (UTAUT; Venkatesh et al. 2003): the social influence, hedonic motivation, the child's attitude towards technology and self-efficacy.

Venkatesh et al. (2003) unified eight models that examine the acceptance of technology, one of which is the widely accepted and used technology acceptance model (TAM; Davis 1989). UTAUT was deemed more suitable for this study because McCoy et al. (2007) found that the TAM assumptions do not hold in cultures that have low uncertainty avoidance levels, a more collective cultural orientation, high power distance scores, or high masculinity scores and according to Hofstede (2014), Hong Kong has low uncertainty avoidance, has a more collective cultural orientation, scores high in power distance and a high score at the masculinity, so TAM may not be the right instrument to evaluate use behaviour in Hong Kong.

This study evaluated three areas: children's tablet use behaviour, level of involvement and level of tablet use (see Table 15.3) to examine the suitability of the Book Creator for ECE. The Leuven Involvement Scale (LIS; Laevers 1994) measures children's involvement in a given activity on a scale ranging from 1 (extremely low involvement) to 5 (extremely high involvement) to assess whether a child experienced deep learning (Laevers 1994). Marsh et al. (2005) used LIS to good effect in an ECE study. The children's level of tablet use was coded according to Couse and Chen's (2010) three stages of tablet use—exploring/experimenting, investigating and creating. 'On-looking' (see Table 15.4) was added because children who watch their peers have been observed to then apply their observational knowledge (Tavernier 2013).

Clarke and Clarke (2009) suggested using Bloom's Digital Taxonomy (BDT; Churches 2008) for technology-related student assessment. In a recent study, BDT was applied and found suitable for the context of assessing young children's learning with tablets (Tavernier 2013). Each level of tablet use could be associated

Table 15.3 Overview of the evaluation areas to be collected and research instruments (adapted from Goodwin 2012)

Evaluation Area	Time and frequency	Research instrument
Children's use behaviour	Three times a week	Rating scales, which are based on Venkatesh et al.'s (2003) UTAUT and journal entries and video data were used to assess the children's use behaviour
Children's level of involvement	Three times a week	Laever's (1994) Leuven Involvement Scale, with involvement rated during a review of videos from observations
Children's level of tablet use	Once a week for every child	Couse and Chen's (2010) classification of observed actions and interactions (review of videos from observations)

Table 15.4 Levels of tablet use (Adapted from Couse and Chen 2010, with on-looking and BDT elements added)

Level of iPad use	Descriptions
On-looking	*Signals*: Child stands or sits close by another child using the tablet and watches attentively.
BDT: Remembering	*Associated actions*: recognizing, listening, describing, identifying, retrieving, naming, locating, finding
Exploring/experimenting	*Signals*: Child tries to figure out what the app can do, touching and activating different options/functions to see what happens
BDT: Remembering and understanding	*Associated actions*: recognizing, listening, describing, identifying, retrieving, naming, locating, finding and explaining, classifying, exemplifying
Investigating (intentional use)	*Signals*: Child tries to figure out how to use the options/functions to create a desired effect (e.g. How can I change the colour to draw a yellow sun?)
BDT: Applying	*Associated actions*: Implementing, carrying out, using, editing, loading
Creating	*Signals*: Child produces desired effects even if the artefact is not a realistic representation of real-life objects described by the child. The child is content with, and clear about, what is being done.
BDT: Analysing, evaluating, creating	*Associated actions*: organizing, structuring, comparing, integrating and testing, critiquing and designing, constructing, planning, producing, making, mixing, video casting, podcasting

with at least one BDT stage. Table 15.4 illustrates how some BDT-related verbs link the students' actions as described by Course and Chen (2010). So a combination of Couse and Chen's work as well as BDT served as a means to determine the individual child's daily level of use. Observed actions were classified accordingly to the described signals and associated actions.

The qualitative data from the videos was analysed after tagging and coding critical incidents (Goodwin 2012). The relevant video data was transcribed to reveal reoccurring patterns of use behaviour (e.g. experienced challenges, peer collaboration, ease of use and the provided type of support). This qualitative data was used to define and illustrate emerging phenomena (Yin 2009).

15.5 Findings and Discussion

The Book Creator is an interesting app for ECE, because it allows the user to progress from consuming content to producing content. The app is complex, but young children can learn to operate it with the assistance of more knowledgeable others. The interface is kept simple and somewhat abstract, therefore young children require help to familiarize with it. However, after some familiarization activities young children can operate the app effectively and produce sophisticated content. The combination of the app evaluation according to Diaz' work and the practical findings from the case study led to the following conclusions.

15.5.1 Post-study Evaluation of the Book Creator for ECE

The evaluation of the Book Creator according to Diaz's (2013) matrix for the evaluation of educational eBooks indicates that the Book Creator is educationally useful, because it scores high on richness, completeness, motivation, autonomy and flexibility. The interface usability is less strong, because the app uses text and misses a text to speech function which reads text aloud, helping non-readers understand. Such function would make it more suitable for young children. The ease of use is reasonably high once the children learned the meaning of the text elements (see Table 15.5).

15.5.2 Suitability of the Book Creator for ECE

The app is complex and includes many functions and editing options that the children learned to understand and use over time. The children were interested in all functions, but experienced an overload of information at first. The choice of functions overwhelmed them, and hence each function was introduced separately and in the context of an imposed task.

The app has two main interfaces: the main starting page which allows the user to create a new book or open a previously created eBook. Within this interface, previously created eBooks can be combined, duplicated, deleted, and shared. The user can also add a title, an author and set display settings. Once a book is open, the interface changes. The user has access to three main menu bars: 1. functions, 2. editing options and 3. sharing and publication options. The children in this project only made use of the menu bar of the second interface, the functions menu and part of the editing options.

Using these two menu bars could be challenging, because it required the children to memorize the location of a function within the two menus and possess the fine motor skills to select the desired functions. For example, to create a voice recording, the child selects the function menu icon, which is one out of three icons presented closely together in the right corner of the screen. Children need to know that the musical key is the icon for the voice recording. They touch it and a textbox opens, featuring text and a red circle. Children are prone to touch the red circle immediately and start the recording before they are actually ready to speak. They need to understand that they only touch the red circle once they are ready to talk and they need to remember to stop the recording when they are done.

During the recording, the red circle changes into a black circle with a red square inside. To stop the recording they touch the red square. Once the recording stopped, a textbox opens and asks the user in text form if they would like to use the

Table 15.5 Evaluation of Book Creator app for iPad for the young children's use (Adapted from Diaz 2003)

	Criteria	Parameters and descriptive evaluation
Educational usefulness	Richness	Information volume: The information provided on each interface is limited to a bare minimum The user's design of each page has no limits. The user can express their ideas through photos, videos, voice recordings, or drawings Diversity of presentation and interaction style: The range of functions allowed the user to express themselves. They could present their ideas through drawings, writing, scribbling, photos, videos and voice recordings
		Scope • The app provides a 'white' platform that embeds various digital means to produce eBooks that suit the individual's preferences. The combination of the different functions can cater for different readers and could accommodate readers and non-readers and possibly hearing impaired
	Completeness: the number of content and interaction mechanisms to cope with the goals of different kinds of users	Learning activities: • The app is open ended and does not provide any learning activities. The teacher may design tasks
		Authoring support: • All functions can be used according to ones preferences. Therefore, the user can personalize the design of the digital artefact.
		Communication support: • The user can, at any time, share the artefact via email or cloud computing
		Collaboration support: • Users can co-author books, share them and continue working on the artefacts from different devices with the same app and then share the new version
	Motivation: how students are motivated, to use the system and to learn more about the subject being addressed	Self-evaluation mechanism: • The user can review and edit their work at any time. The content of videos, photos and drawings cannot be changed or edited within the app, but they can be resized, relocated, and deleted and replaced any time
		Adaptability: • The interface cannot be adjusted to the user's abilities. Instead the teacher needs to monitor the use behaviour of the students and gradually guide the child to master the different functions
		Out-of-school activities: • The children can include their personal experiences and knowledge from out-of-school in their digital artefacts • The artefacts created in school can be shared as a video or PDF and reviewed in the home environment. If it is shared as a working file, the user can continue editing the artefact out-of-school

(continued)

Table 15.5 (continued)

	Criteria	Parameters and descriptive evaluation
	Autonomy: the degree of navigation freedom offered to the user and the degree of interaction freedom	Interaction freedom: • The user can decide when and how to make use of the different functions
		Autonomy degrees: • The user does not need to finish their work within one session. An eBook can be started, stopped, continued and edited at any time • Drawing cannot be further edited once the user indicates that he is done with it. They can only be resized, relocated or deleted afterwards • New layers of digital artefacts can be added any time.
	Competence: the ability to navigate through the system and to reach a particular goal	Use levels: • There are no use levels within the app. Instead the teacher needs to determine how many functions a child should learn at a given time. A less experienced user will need more assistance and monitoring than an advanced user, because the app includes different functions and editing options
		Help mechanism: • There are no built-in help mechanism, instead the teachers or more experienced peers may provide help
		Adaptability: • The app interface cannot be adjusted. The user needs to learn, memorize and understand where to find the functions and how to use them (e.g. placing the video and voice recordings in the front • to keep them activated)
	Flexibility: The ease with which the system can be used	Accessibility: • The app is easy to use once the user is familiar with the icons and text • The text is kept to a minimum (e.g. cancel, done, and delete) • Users that cannot read need some guidance to memorize their look and meaning
		Modularity and structure of the architecture: • The function menu is separated from the editing menu. This is difficult for young users to comprehend • Features e.g. adding a page and going from page to page are very intuitive
User interface usability	Aesthetic: How the inclusion of multimedia information is harmonized and used to enhance the comprehension of concepts	Legibility: • The design of each page depends on the user and represents her or his capabilities
		Density: • Each page is white and the user creates designs and multimedia to fill it • There are no set limitations and the amount of information on each • page depends on the user • It is possible to embed too many individual designs on one page so that a cognitive overload occurs for the viewer

(continued)

Table 15.5 (continued)

	Criteria	Parameters and descriptive evaluation
		Appropriateness: • The touch interface and the ability to use a special touch screen pen, replicates the feeling of using traditional utilities to draw • The option to create and embed multimedia transforms the look-and-feel of the book and may engage children that are less interested in books, drawing and other fine motor skills, hand-eye coordination skill development activities • The app uses a different gesture than the iPad itself to resize digital objects, which confuses the user • The ability to use drawings, writings, verbal expressions as well as gestures (video recordings) makes the app very appropriate for the diversified needs and abilities of young children
	Consistency: the extent to which elements that are conceptually similar are treated equally by the application, while those that are different are treated differently	Interface areas: • The interface is the same for every page
		Labels and messages: • The labels are the same throughout the app and functions (e.g. pressing done when the children are done with drawing or typing).
		Buttons, icons and menu items: • The icons and words and sequence of use are the same throughout the app (e.g. starting and stopping voice and video recording, selecting a new function through the same icon and related menu) • The icons may not all be clear to young children (e.g. the cross to indicate the function menu) Interface clues: • Since this is an open-ended constructive app, the user designs each page. There are no fixed interface cues
	Ease of Use: how easily users can guess the meaning and purpose of things with which they are presented	Self-contained pages: • The pages have a clear design and the user can see if it is a single or double sided page. The function menu and all else remains the same throughout the app
		Multimedia expressiveness: • The app uses several common icons to illustrate the functions (e.g. a microphone for the voice recording function)
		Meaningful naming of functions: • The naming is kept to a bare minimum and often comprises of one word only. The words are clear and unambiguous

recording. The child needs to remember which of the selection options means 'yes' or 'no', to avoid an accidental deletion of their recording. Next, the icon of a speaker appears on the current book page. The child can move it around or resize it. If another recording is added, the new speaker icon will appear again in the middle of the page. If the first recording icon was not moved to a new location, it will now be inactive, because the video and voice recordings can only be activated if they are not covered by any other features. Considering all these steps, it is clear that 3–5 years old children require a substantial amount of help to familiarize, remember and master this complex sequence of actions. Despite the high level of sophistication the children were keen to learn using all functions.

The minimalistic layout provides minimal distraction and no inspiration. The biggest issue remained the text elements within the app and the lack of verbal support features (e.g. text to speech function). To make the app more suitable for young children, it would be better to have some built-in verbal guidance that helps them decode the icons and remember the sequence of actions. The following two extracts show how children relied on the guidance of the researcher to learn and remember the meaning of the text elements.

Extract 1: The researcher is guiding the Silas through the app's photo functions

Silas would like use a self-portrait for the book cover of his book. He has used the built-in photo taking function before, but still requires some guidance to complete the action. The researcher provides verbal guidance:
Researcher "Press the cross (in the function menu bar)"
Silas looks at the screen, says "cross" and touches it. The action menu opens.
Researcher continues her verbal guidance "And now presses the camera (icon)"
Silas follows the instructions. The camera function opens and the screen is black, because the back camera is covered by the iPad cover. Silas lifts the iPad and giggles. He looks at the screen and giggles, but seems puzzled, because the screen remains black and he cannot see himself.
Researcher monitored his action and waited to see if he can solve this issue.
Silas turns the iPad to the researcher and shows her the screen and saying "Look!"
Researcher reacts and provides him verbal guidance to troubleshoot the issue: "Oops, we need to press on this symbol (pointing to the small switching cameras icon). And then you press ... Hold on, we need to turn the iPad around (otherwise Silas' hand would have been in front of the camera lens and the photo would have shown the hand that pressed the shutter release].
Silas looks at the iPad and presses the shutter release taking a photo of him smiling. (The app shows the photo and on the lower edge of the screen the app is asking the user to press 'take another photo' (lower left corner of the screen) or 'use photo' (lower right corner of the screen). Silas cannot read these instructions yet.)
Researcher points to the lower right corner of the screen and says "use photo".
Silas presses it and the app inserts the photo on the current page of the book.

Notes: Video transcription from 3rd iPad session.

Silas, 4 years old, was familiar with individual letters, but he could not read and decode the text within the app. ECE is special, because the children are often just starting to develop a reading awareness and reading skills. Decoding icons requires some degree of literacy awareness and experience. In this case, Silas used the photo function for the second time and relied heavily on the researcher's guidance. The next time he used it, he did not need the verbal guidance, but used eye contact to confirm that his actions are correct. Switching from the back to the front camera was another area of difficulty for all children.

Extract 2: The researcher guiding the children through the written words

Carl and Max work together to compile a journal entry about their day's favourite activity. Carl would like to do a voice recording. The researcher asks Max and Carl how to start the voice recording. Both look at the screen.
Max says "cross"
Carl repeats "cross"
The researcher points to the cross and confirms the boy's suggestions.
Carl touches the cross and opens the function menu. "Then you click down there (pointing to the musical key). Then we can do the voice recording.
Carl touches it and a white box with a big red circle in the middle opens. His finger moves immediately towards the red circle.
The researcher notices this and says "Before you touch it, you need to think about what you want to say. What do we want to say?" (The children and the researcher discuss what to say)
As soon as Carl knew what he wanted to say he touched the red circle and started the voice recording. The red circle changes to a black circle with a red square inside.
Carl finishes his voice recording, moves his head away from the iPad and pauses. The researcher points to the black circle and touches it to stop the recording. Then she explains: "You should stop the recording when you have said what you wanted to say." Meanwhile, a text box appears in the middle of the screen, asking if we want to use the recording 'yes' or 'no'.
Carl's finger browses over the text box and he is ready to just touch any selection. The researcher moves his hand away from the dialogue box and says 'this is no (pointing to 'no'), meaning that we do not want to use the recording and this is yes (pointing to 'yes'), meaning that we want to use the recording"

Notes: Video transcript from the sixth iPad session
There is a high chance that 3–6 years old children just touch any of the of the selection options and accidentally delete a photo, voice or video recording or clear a drawing. If they encounter situations as the one described in extract 2, they may experience frustration, because they cannot realize their idea. This may reduce the joy they experience when they create and review their own creations. Children that are less persistent and have short attention spans may quickly loose interest in the app. So the guidance and assistance of more knowledgeable others is crucial during the familiarization and application stage.

Table 15.6 Comparison of the number of activities during each stage

Phase	Drawing	Photo	Voice recording	Video recording	Text
Familiarization	78	75	30	0	18
Application	30	20	22	14	24
Creating	48	12	22	7	37

The review of the journal entries, field notes and video footage showed that the children's need for guidance remained high during the familiarization and application stage, it decreased during the creation stage. During the familiarization stage all children were introduced to the drawing and photo taking function, to ensure that they can sign their works either with a self-portrait or with their written name. All other functions were introduced to the individual child, when he or she requested to use them or when the researcher felt that the child was ready to explore them. Table 15.6 provides an overview of the total number of activities for the main functions of the Book Creator app. At any time a child could engage in more than one activity. It was not recorded how often a child used one function each day, just which functions he or she used during one observation session. Therefore, the table only indicates the most popular functions. When this data is triangulated with the videos and analysed as a whole, from the beginning to the end of the study, it sheds some light on the learning behaviour.

For instance, during the familiarization stage, the children used mainly the drawing and photo taking function, and occasionally they explored the other functions. During the application stage they had a vague idea of the different functions and had experimented with most of the functions. These experiments were often unplanned and the result of a spontaneous urge or inspiration. The children felt comfortable with the drawing and photo taking function and were ready to shift their experiments to the voice and video recording function. Once the study entered the creating stage, the children's actions were less experimental and more focused. Instead of spontaneously using many functions, they now planned the design of their artefacts and selected the functions purposefully.

The use of the Text function is interesting, because only nine children could write their or spell their name, but nearly all children experimented with the text function. During the familiarization stage their text elements often consisted of chains of random letters. Only Mia, Sara, Sonja, Victoria, Silas and Carl typed their names. During the creating stage all children tried to type their names or asked the researcher to help them find the letters to type their names. This explains the raise in the use of the text function.

The children also enjoyed the text editing functions. Most experimented with the font size and font style. Two children changed the font colour or the background of the text box. This indicated that even though many of the children were not yet familiar with letters or able to write, this function may have enhanced their interest in writing and encouraged them to include text in their creative self-expression. This function allowed these children to familiarize with letters and include them in their activities in an age appropriate, playful and unambiguous manner.

15.5.3 Children's Artefacts

The artefacts presented in table seven to nine were selected, because these three students engaged in the iPad activities regularly, so there is more data available to evaluate their use behaviour and assess the suitability of the app. This includes their learning path, experienced difficulties and their technics to overcome issues is well documented.

Lilian, 3 years old, showed a high level of interest in the iPad from the very first session. She approached the researcher often, stood close to her or a peer that was using the app and watched what they were doing. The researcher would ask her, if she would like to have a turn and worked with her later. Lilian never explained any of her works to the researcher. She spent the complete familiarization stage taking self-portraits, scribbling over her photos, erasing her scribbles and scribbling again. She also used the voice recording function, but never spoke. This may have been due to the fact that the researcher was there to supervise the whole activity. She may have felt too intimidated to speak and record herself. The researcher tried to involve her in conversations about her work. Lilian ignored these attempts to have a conversation. She also ignored suggestions regarding the creative process and only accepted support regarding the operation of the app.

During the application stage the researcher wanted to see what she could do with her skills. She interrupted Lilian's routine to introduce her to the individual functions of the app and asked her to perform small tasks (see the image associated with the application stage of Table 15.7). The researcher was by her side during all activities and provided support when necessary. It appeared that Lilian knew how to use all functions purposefully, but she required some very clear instructions that provided her a goal towards which she could work. Without this goal she was lost and could not decide what to create and used the app and its photo function as an augmentation of traditional drawing. The screen shots of her work show how her ability to use the app to express herself improved over time. She progressed from (1) self-chosen activity (photo and scribbles), (2) realizing an imposed task, to (3) being able to come up with her own meaningful designs (photo of herself and a drawing of her friend Emily).

Table 15.7 Work samples of Lilian, 3 years old

Familiarization Stage	Application Stage	Creating Stage
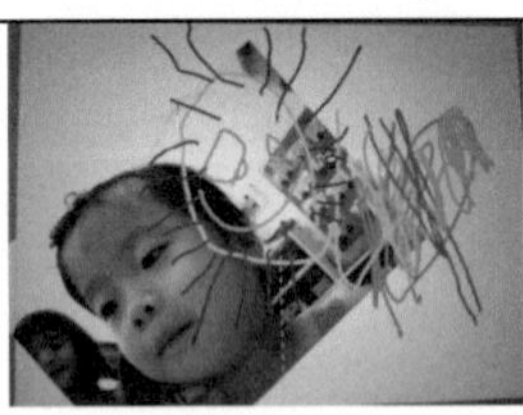	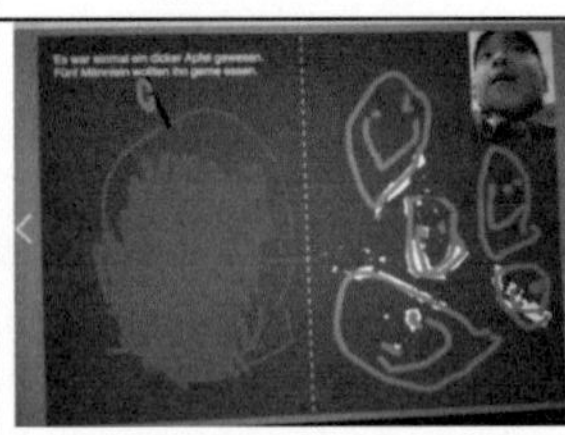	

Over the course of the study, Lilian observed her friends Sonja, Carl, Mia and Emily several times, which increased her participation. Among the group of 3 years old children, she joined the activity most often (17 times). She observed her peers during eight sessions, spent six sessions to familiarize with the app, experimented with the functions for three sessions and used three sessions to independently create works that were meaningful (outcome oriented) and could be decoded by the neutral viewer. Each of her last three works included a video sequence that explained her work, a voice recording or a caption that she dictated the researcher. Her confidence towards in her app operation skills and her self-confidence increased, and she was more ready to record herself.

The analysis and comparison of journal records, videos, and artefacts of all 3 years old children indicate that they learn to use the basic functions of the app within five intensively supervised and guided sessions. They required many opportunities to watch their peers, explore the functions and receive creative stimulations. The most remarkable impression of this age group was the persistence with which Lilian, Elena, Callestine and Sophie learned to use the app and how much the quality of the artefacts improved.

Simon, 4 years old, participated in 12 sessions. He missed some of the familiarization sessions and started to use the app during the application stage. Simon spent six sessions watching his peers, at times unintendedly. All of his works were very expressive. He spent only one session to familiarize with the app (see first photo in Table 15.8). In all other sessions, he applied his creativity and knowledge to express his ideas. He used the video and drawing function purposefully from the beginning. He completed the sentence "I am happy, if.." with a video recording that shows him say ".... if, I can play with my father". And his drawing represents his answer. He combined two different functions to fully express his view.

The analysis of journal records, videos and artefacts (see Table 15.8) indicated that 4 years old children tend to learn the operation of the Book Creator faster and may focus on the accurate representation of their ideas sooner. Simon progressed faster compared to other 4 years old children. On average, the 4 years old children participated twelve times, observed three sessions, spent four sessions each to familiarize with the app and created artefact purposefully. Their ability to link their

Table 15.8 Work samples of Simon, 4 years old

Familiarization Stage	Application Stage	Creating Stage
	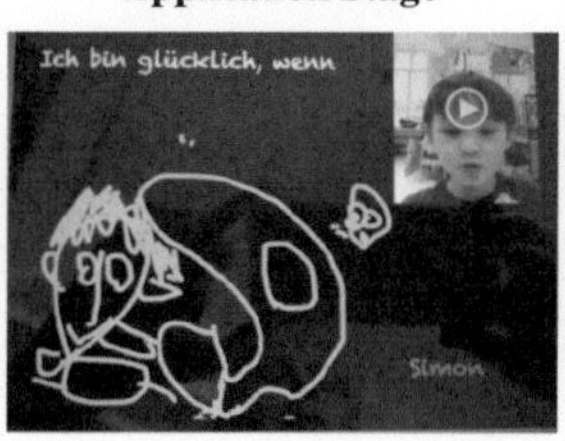	

drawings to video and voice recordings became meaningful earlier than in the case of three years old. This was to be expected, because 4 years old children are cognitively more ready to perform such task. As a result, the app helped 4 years old children to better express their understanding of the world around them and empowered them to create more meaningful multimedia artefacts.

The participation of the three 5 years old children varied a lot. At the beginning of the study, their interest in using the iPad was high, but they lost some of the initial enthusiasm as they realized that they needed to wait for their turn and were limited to the use of the Book Creator. It appears that many did not like to be limited to the use of just one app. The general interest increased again as they saw the works of their peers during group viewing sessions. These sessions were meant to demonstrate and recapitalize what we had done with the app and had the side effect to inspire the children to expand their personal repertoire, try something new, return to the app-related activity and explore the app functions further and ask more specific questions that led towards replicating what they had seen before.

Sonja's keen interest stands in contrast to that of Lilian and Simon. She participated 21 times, and her activities ranged from on-looking, to helping her peers and to creating her own artefacts (see Table 15.9). Sonja (5 years old) was not familiar with the iPad, so she needed extra time to familiarize with both device and app.

Sonja observed the iPad activities four times, her peers only twice. When she watched her peers, she reinforced her knowledge of the Book Creator by reminding her peers of the sequence of actions. She also used her peers' creations as inspirations and tried to imitate them when she had a turn. Sonja spent seven sessions familiarizing with the app and applying and practicing her new skills. Two more than her peers, before she went on to spent ten sessions creating meaningful works (see Table 15.10).

The result of her persistence was a deep understanding of the Book Creator. Compared to her peers, she took full advantage of the app. She understood that she can combine the functions to convey meaning or use them independently from each other. Her contribution to the "I am sad, if…" book includes the above displayed work of the application stage (see image 2 in Table 15.9) and a voice recording. In

Table 15.9 Work samples of Sonja, 5 years old

Familiarization Stage	Application Stage	Creating Stage
		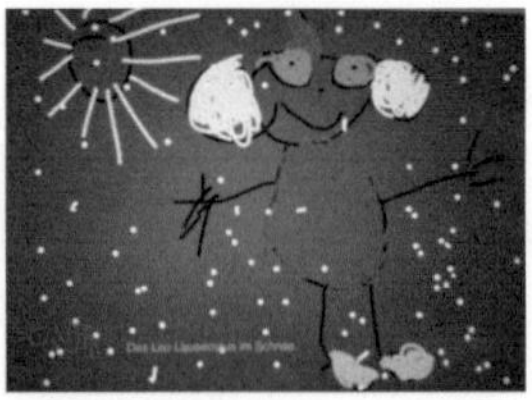

Table 15.10 Number of participations and activities of the 3 five years old participants

	Participation	Observing	Familiarizing and applying	Creating
Maximus	8	2	3	3
Sonja	21	4	7	10
Viktoria	12	2	5	5

the recording she said that "she is sad, if she loses something". When the researcher asked her why her drawing does not relate to what she said, she replied that "It does not have to. The answer to what makes me sad is provided in the recording, so now I can draw something that I like". She understood that the voice recording already conveyed the message that she wanted to send and that more means to communicate this message were not necessary.

Sara and Viktoria progressed in another direction. They created different artefacts and combined them within one page (see Fig. 15.1). For example, Viktoria first drew a picture that covers the whole page and resized the final version. Second, she took a photo of herself, resized it and added a voice recording explaining who she is and what her drawing represents. Third, she photographed the class's artworks and voice recorded which of these works was important to her. At last, she typed her name and formatted the size, font type and font colour. When she designed this complex artefact, she planned the details, having in mind that a person, who does not know her, may look at it and may need the additional information. During the whole process, the researcher looked on and was not asked to help. All of the above indicates that the Book Creator is very suitable for 5 years old children and empowers them to convey complex ideas in a way that others can understand and relate to.

The overall findings suggest that the Book Creator App has the potential to empower children to be creators of meaningful digital artefacts. The app is complex and includes many functions, which can encourage the children to fully express

Fig. 15.1 Work sample of a 5 years old (Creation stage)

their thoughts and ideas by complementing their drawings or photos with audio or video element. The range of available functions may also hinder some children's ability to express their ideas. These children may be overwhelmed by the choices they need to make and the procedures they need to follow to realize an idea. To unleash the full potential of the app and fully grasp all its functions, a substantial amount of practice is necessary. Practice will improve the level of mastery and understanding of the app over time.

15.6 Conclusion

The findings indicate that once the children overcome the barrier of decoding the text elements with the help of more knowledgeable others, they use the app more effectively as they get a feeling for the functions and understand that they can combine different function within one piece of work. Within a period of 12 weeks, the children's work became more meaningful and the use of the multimedia functions increased. They progressed from relying on the researcher's guidance through the functions and text to using the app independently to create. The support that the researcher provided changed from assisting the children with the operation of the app, to confirming that they were following the right sequence to achieve a certain outcome (e.g. editing their works), to providing resources that inspired them or helped them realize their own ideas (e.g. providing a real apple or a verbal description of how things look like). The children's questions changed from "How can I erase this?" (operational questions) to "Can you show me how to draw a mouse?" (creation related questions).

Considering the intense time and attention that the researcher provided to prepare, monitor and support all iPad activities, it is worth looking for a similar app which is less complex and easier for the children to operate.

If ECE teachers decide to use the Book Creator App, they will need to plan activities that help the children to learn to operate and understand the app thoroughly. Such a familiarization period may last from 1 to 7 sessions per child and can be very time consuming. During this period, the teacher will need to monitor and manage the activities, support the children's app operation skills, decode text, establish routines, understand technical rules (e.g. voice and video recordings only work if they are located in the front layer) and realize the creative opportunities that the app provides.

Considering the great interest shown by the children to design text and their growing confidence to create voice and video recordings, teachers may consider including the app in their literacy learning repertoire. Listening to oneself may help realize weaknesses in pronunciation and facilitate a greater awareness of speech. The children loved watching and hearing their recorded selves. Additionally, the artefacts are a great source to capture, document and evaluate the learning of children. Teachers can create very closed or open-ended tasks and cater for the individual preferences of their students.

References

Andresen, L., Boud, D., & Cohen, R. (2000). Experience-based learning. *Understanding Adult Education and Training, 2*, 225–239.

Boardman, M. (2007). "I Know How Much This Child Has Learned. I Have Proof!": Employing digital technologies for documentation processes in kindergarten. *Australian Journal of Early Childhood, 32*(3), 59–66. Retrieved from http://search.proquest.com/docview/61960953?accountid=14700.

Chiong, C., & Shuler, C. (2010). Learning: Is there an app for that. In *Investigations of young children's usage and learning with mobile devices and apps.* New York: The Joan Ganz Cooney Center at Sesame Workshop.

Churches, A. (2008). Bloom's taxonomy blooms digitally. *Tech & Learning, 1.*

Clarke, T., & Clarke, E. (2009). Born digital? Pedagogy and computer-assisted learning. *Education + Training, 51*(5/6), 395–407.

Clements, D. H., & Nastasi, B. K. (1993). Electronic media and early childhood education. *Handbook of research on the education of young children* (pp. 251–275).

Clements, D. H., & Sarama, J. (2002). The role of technology in early childhood learning. *Teaching Children Mathematics, 8*(6), 340.

Couse, L. J., & Chen, D. W. (2010). A tablet computer for young children? Exploring its viability for early childhood education. *Journal of Research on Technology in Education, 43*(1), 75–98.

Creswell, J. W. (2002). *Educational research: Planning, conducting, and evaluating quantitative and qualitative research.* Saddle River, NJ: Prentice Hall.

Curriculum Development Council. (2006). *Guide to the pre-primary curriculum.* Hong Kong.

Davis, F. D. (1989). Perceived usefulness, perceived ease of use, and user acceptance of information technology. *MIS Quarterly*, 319–340.

Diaz, P. (2013). Usability of hypermedia educational e-books. *D-Lib Magazine, 9*(1). Retrieved from http://www.dlib.org/dlib/march03/diaz/03diaz.html.

Edwards-Groves, C., & Langley, M. C. (2009). i-Kindy: Responding to home technoliteracies in the kindergarten classroom. In *National Conference for Teachers of English and Literacy*, (July), 1–17. Retrieved from http://www.englishliteracyconference.com.au/files/documents/hobart/conferencePapers/refereed/Edwards-Groves.complete.pdf.

Einarsdottir, J. (2005). Playschool in pictures: Children's photographs as a research method. *Early Child Development and Care, 175*(6), 523–541.

Ertmer, P. A. (2005). Teacher pedagogical beliefs: The final frontier in our quest for technology integration? *Educational Technology Research and Development, 53*(4), 25–39.

Falloon, G. (2013). Young students using iPads: App design and content influences on their learning pathways. *Computers & Education, 68*, 505–521.

Goodwin, K. (2012). Use of tablet technology in the classroom. *NSW Department of Education and Communities.*

Hertzog, N., & Klein, M. (2005). Beyond gaming: A technology explosion in early childhood classrooms. *Gifted Child Today, 28*(3), 24–31.

Hofstede, G. (2014). Cultural tools—Country comparison. Retrieved from http://geert-hofstede.com/hong-kong.html.

Hutchison, A., Beschorner, B., & Schmidt-Crawford, D. (2012). Exploring the use of the iPAD for literacy learning. *Reading Teacher, 66*(1), 15–23.

Laevers, F. (1994). The innovative project Experiential Education and the definition of quality in education. *Defining and assessing quality in early childhood education* (pp. 159–172).

Lancaster, L. (2012). Moving into literacy: How it all begins. In J. Larson, & J. Marsh (Eds.), *The SAGE handbook of early childhood literacy* (pp. 313–328). London: Sage Publications.

Lindahl, M. G., & Folkesson, A.-M. (2012a). ICT in preschool: Friend or Foe? The significance of norms in a changing practice. *International Journal of Early Years Education, 20*(4), 422–436.

Lindahl, M. G., & Folkesson, A. M. (2012b). Can we let computers change practice? Educators' interpretations of preschool tradition. *Computers in Human Behavior, 28*(5), 1728–1737.

Lindahl, M. G., & Folkesson, A.-M. (2012b). Can we let computers change practice? Educators' interpretations of preschool tradition. *Computers in Human Behavior, 28*(5), 1728–1737.
Mara, J. O., & Laidlaw, L. (2011). *Living in the iworld: Two literacy researchers reflect on the changing texts and literacy practices of childhood Joanne O'Mara, 10*(4), 149–159.
Marsh, J., Brooks, G., Hughes, J., Ritchie, L., Roberts, S., & Wright, K. (2005). *Digital beginnings: Young people's use of popular culture, media and new technologies*. Sheffield: University of Sheffield.
Michael Cohen Group Llc. (2011). Young children, apps & iPad. *Young*, 1–13. Retrieved from http://www.mcgrc.com.
McCoy, S., Galletta, D. F., & King, W. R. (2007). Applying TAM across cultures: the need for caution. *European Journal of Information Systems, 16*(1), 81–90.
Plowman, L., & Stephen, C. (2007). Guided interaction in pre-school settings. *Journal of Computer Assisted learning, 23*(1), 14–26.
Prensky, M. (2001). Digital natives, digital immigrants part 1. *On the Horizon, 9*(5), 1–6.
Sarama, J., & Clements, D. H. (2004). Building blocks for early childhood mathematics. *Early Childhood Research Quarterly, 19*(1), 181–189.
Scardamalia, M., & Bereiter, C. (2006). *Knowledge building*. New York: Springer.
Tavernier. (2013). Pre-school children learn to use the iPad to learn, document, assess and create content. Paper presentation accepted for the CITE Research Symposium 2013, The University of Hong Kong, Hong Kong.
Venkatesh, V., Morris, M. G., Davis, G. B., & Davis, F. D. (2003). User acceptance of information technology: Toward a unified view. *MIS Quarterly, 27*(3).
Vygotsky, L. S. (1987). *Mind in society: The development of higher psychological processes.* Cambridge: MA: Harvard University Press.
Yelland, N. & Gilbert, C. (2012). *iPlay, iLearn, iGrow*. London: IBM Paper. Retrieved fromhttp://www.ipadsforeducation.vic.edu.au/userfiles/files/IBM%20Report%20iPlay,%20iLearn%20%26%20iGrow.pdf.
Yin, R. K. (2009). *Case study research: Design and methods* (Vol. 5). London: Sage Publications.
Zevenbergen, R. (2007). Digital natives come to preschool: implications for early childhood practice. *Contemporary Issues in Early Childhood, 8*(1), 19.

Chapter 16
Text Messaging for Out-of-Class Communication: Impact on Immediacy and Affective Learning

Paul Hayes and Stephan Weibelzahl

Abstract While out-of-class communication between instructors and students can impact all types of student learning it has its greatest impact on student affective learning, including motivation and engagement. One of the primary reasons for this is that the out-of-class communication enhances student perception of instructor immediacy. Immediacy is defined as behaviour which increases psychological closeness between communicators. Research studies in instructional communication suggest that improved instructor immediacy is linked to enhance affective learning. A research study was conducted into the use of text messaging for out-of-class communication and the impact it had on student perception of instructor immediacy and student affective learning. Both quantitative measures of immediacy and qualitative feedback from students show that the instructor is perceived as closer, more approachable and responsive when text messaging services are offered. The student feedback also reveals that the use of text messaging has other positive effects on affective learning.

16.1 Introduction

Effective communication between instructor and student is very important in the quality of the learning experience of students in higher education. Hill et al. (2003) used student focus groups to answer the question of what quality education means to students. Four themes emerged from the study, the most important being the quality of the instructor in terms of delivery, feedback to students and relationship with students in the classroom. However, there are many factors that limit

P. Hayes (✉)
National College of Ireland, Dublin, Ireland
e-mail: Paul.Hayes@ncirl.ie

S. Weibelzahl
Private University of Applied Sciences, Göttingen, Germany
e-mail: weibelzahl@pfh.de

D. Churchill et al. (eds.), *Mobile Learning Design*, Lecture Notes in Educational Technology, DOI 10.1007/978-981-10-0027-0_16

communication between instructors and students in higher education including large class sizes, limited contact time and student reluctance to approach instructors.

While educational institutions generally place most emphasis on student cognitive learning it has been shown that affective learning is also crucial especially to the development of independent and lifelong learners. Learning in the affective domain includes the manner by which people deal with things emotionally, such as feelings, values, appreciation, enthusiasms, motivations and attitudes (Bloom 1956). While communication outside of normal class time (known as out-of-class communication) between instructors and students can impact all types of student learning it has its greatest impact on student affective learning (McCroskey 1994). The importance of out-of-class communication to student affective learning should not be underestimated. Research shows that out-of-class communication between instructors and students can help build more positive instructor-student relationships, and hence increase the quality of student learning (Noels et al. 1999; Vaughn and Baker 2004). One of the primary reasons for this is that the out-of-class communication enhances student perception of instructor immediacy (Jaasma and Koper 1999). Immediacy is defined as behaviour which increases psychological closeness between communicators (Mehrabian 1969, 1971). Research studies in instructional communication suggest that enhanced instructor immediacy has a positive effect on affective learning and is linked to more positive student-instructor relationships engendering positive attitudes, increased interest and motivation by students as well as improved attendance, retention, engagement and learning (Christensen and Menzel 1998; Ellis 2004).

The research presented here investigates the effect of the use of text messaging for out-of-class communication between instructor and student on student perception of the immediacy of their instructor and on student affective learning. An extensive review of research literature from both the fields of instructional communication and mobile learning was carried out prior to the commencement of a year-long research study. Both quantitative measures of immediacy and qualitative feedback from student participants in the study show that the instructor is perceived as closer, more approachable and responsive when text messaging services are offered. The student feedback also shows that the use of text messaging has other positive effects on student learning experience, including enhanced motivation and engagement. Some limitations of the research are also addressed as are some concerns with the use of text messaging in education.

16.2 Instructor Immediacy

The field of instructional communication is based on the assumption that verbal and nonverbal messages conveyed by instructors have the potential to significantly affect student learning outcomes (Witt 2000). When it comes to instructor communication behaviour one important construct is that of instructor immediacy. Immediacy is defined as behaviours, both verbal and nonverbal, that reduce physical or psychological distance between individuals (Andersen 1979; Mehrabian 1969, 1971, 1981). The results of a significant body of research conducted on instructor immediacy behaviours indicate that it can have a positive influence on student learning outcomes. For this reason instructor immediacy should be treated with great importance by any person or institution concerned with improving the quality of student learning (Witt 2000).

Research studies have shown a linear relationship between student reports of teacher immediacy behaviours and perceptions of state motivation, and of cognitive, affective and behavioural learning (Christensen and Menzel 1998; Pogue and Ahyun 2006; Witt and Wheeless 2001). This relationship has been shown to hold true for divergent classes (Kearney et al. 1985) and also in multi-cultural studies (McCroskey et al. 1996).

16.3 Text Messaging in Education

Text messaging has been exploited for supporting learning in a variety of educational settings. Studies of third-level students have shown that text messaging is used more regularly by students than email as it is perceived as being more personal and informal and is often students' preferred way for receiving information from their institution (Harley et al. 2007). However, because a text message is limited to only 160 characters it is more suitable for certain types of learning activities than others. A review of the current research literature shows that the ways in which text messaging has been employed in education fall generally into four categories. The first category is when text messaging is used to support and enhance classroom interactivity and dialogue (Clarke and Doody 2008; Markett et al. 2006). The second category is when text messaging is used for administrative purposes such as notifications of changes in the timetable and reminders of assessment submission dates (Naismith 2007; So 2009; Stone 2004). The third category is when text messaging is used as a means of supporting micro-teaching activities including the sending of short summaries for revision, the sending of links to a relevant page on a Virtual Learning Environment (VLE) and also the provision of quizzes and feedback to students (Stone 2004; Tretiakov and Kinshuk 2005). The final category is when text messaging is used not for learning purposes directly but rather to guide, motivate and support students, encourage participation and engagement, and promote collaboration and co-operation. This fourth category includes many examples where it is used to enhance student affective learning, develop a sense of community amongst students and positively affect student retention rates (Trifonova 2003; Harley et al. 2007). The fourth category may include messages from some of the other categories but the key difference is that the primary goal of the text messaging is to support students and enhance affective learning. As this research paper is concerned with the effect of text messaging on student affective learning it is this final category that is of primary interest.

There are many examples in the research literature where text messaging is not used specifically for the purpose of directly improving academic learning or for administrative purposes but is rather used to support and help students when they are outside class. Such out-of-class (OOC) text messaging may have the aim of enhancing affective learning and improving the learning environment, improving communications, supporting students' transition to third-level education, developing a sense of community among students or positively affecting student retention rates. The potential of the mobile phone as a communications medium in education prompted a research study by Brighton University to explore the use of mobile communication as a way of encouraging a supportive dialogue between students and relevant academic staff. The main motives behind the research were to support students' transition to third-level education and improve retention (Harley et al. 2007). Another very interesting and relevant research study by Griffith University in Australia relates the experience of a female instructor using out-of-class text messaging as a means of staying in touch with her students. The study demonstrates how it can be used as a means of providing connection and a sense of community for first-year students and also how it encourages them to persist with their studies (Horstmanshof 2004). Text messaging has also been used by the University of Ulster in Northern Ireland for supporting first-year chemistry students and for the reduction of student drop-outs. The university sends out messages to students of the type 'Sorry, we missed you today'. The students do not find the messages obtrusive in any way, and actually welcomed them (Keegan 2006).

16.4 Methodology

An empirical study was designed to investigate the impact of out-of-class communication between instructors and students using text messaging on student learning experience. The study was based in a real educational setting. It was hoped that analysis of the results of the study would provide evidence of an effect of the text messaging on student affective learning. In total 44 participants from four different classes took part in the study. The participants were all third-level undergraduate computing students who were taking between five and six modules each semester. The research study itself took two academic semesters to complete.

16.4.1 Text Messaging Service

The 44 students who volunteered to participate in the study had the use of a text messaging service for out-of-class communication with one of their instructors. The instructor used a freeware application called MyPhoneExplorer that was installed on the instructor's laptop together with a mobile phone that was connected to a USB port on the laptop. The software application on the laptop was very versatile and easy to use. In terms of text messaging it operated much like an email program allowing the creation, viewing, editing and deletion of text messages as well as the sending and received text messages via the connected mobile phone. The application could be synchronised with the mobile phone allowing access from the laptop to both the SIM and phone memory. During synchronisation contact details of participants and text messages sent and received could be copied automatically from the phone to the application and visa-versa. The application allowed the sending of text messages to individuals or groups and it also supported the archiving of text messages on the laptop.

16.4.2 Categories of Text Messages

While it was difficult to categorise some of the messages they generally fell into one or more of three main categories. The first category was for messages that were sent for administrative purposes. The vast majority of these messages were sent as broadcasts to all participants in a treatment group. Only very rarely was there a need to send a message of this type to an individual participant or sub-group of participants. Examples of the use of this type of message include class announcements and cancellations, and reminders of class tests and assignment submission dates. A few examples of text messages of this category that were sent to participants were as follows:

> Hi, I have put the final marks for your continuous assessment up on Moodle. Paul
> Don't forget you have a test on databases this Friday!
> Just to remind you that John from BT Ireland is coming into give us a talk next Tuesday. Paul
> Hi, DCN class is postponed tomorrow, I have to attend an important meeting, will make it up to you. Paul

The second category was for text messages that were specifically related to the topics covered in a module that were being delivered by the instructor and the contents of these messages were supplementary to the course material. These messages were sent as broadcasts to all participants and were used for the purpose of micro-learning activities. The messages included short summaries for revision purposes, short or multiple-choice questions and advice on how to prepare for forthcoming classes. Each message was restricted to 160 characters so the messages had to be short and precise. In the case of a text message containing a short question or a multiple-choice question the correct answer was sent as a broadcast text messages to all participants after a suitable period of time. A few examples of text messages in this category that were sent to participants were as follows:

> What is the name of each layer of the OSI network model?
> What does the letter 'S' stand for in the acronym ISDN? Answer (a) Signals (b) Services (c) Switching or (d) Segment?
> Do you have any questions on what we covered today in class?
> The lecture next week is on the topic of DSL. Please look over the lecture notes on this topic prior to coming to class. Thanks

The third category of message were those whose main purpose was to promote affective learning and included messages that were designed to motivate students in their studies, enhance interest in the subject and to encourage attendance, engagement and participation in class. While messages from the other two categories could have an indirect effect on affective learning this type of message was specifically aimed at enhancing it and included messages expressing pleasure at the effort students were putting into their studies and thanking students for their participation in class. These messages were always sent as a broadcast to all participants and care was exercised to make sure they were always positive in tone and never critical. A few examples of text messages in this category that were sent to participants were as follows:

> Thanks for all your work and study this week. Glad to hear the projects are getting off to a good start. Have a good one & c u nxt week, Paul
> You learn something every day if you pay attention ~~ Ray LeBlond
> Very enjoyable class today. I will try to sort out the issue with the timetable tomorrow. Paul

16.5 Data Collection

For the purposes of this investigation students who used the out-of-class text messaging service were asked to complete a questionnaire. The questions were formulated based on a review of the research literature on the use of text messaging to support students together with a review the findings of the preliminary studies and the use of the text messaging service to provide out-of-class support to students during the main study. The first section of the questionnaire consists of 30 specific questions about student perception of the use and impact of the text messaging service. Participants were asked to indicate their response to each question on a 7-point Likert scale. The second section of the questionnaire uses a series of open questions to give participants the opportunity of anonymously expressing their personal opinions in terms of communicating with their instructor using text messaging and its impact, if any, on them or their class in terms of learning and education and the relationship with their instructor. It was hoped that analysis of the student responses to the questionnaire would provide data on the effect of the text messaging on student affective learning.

16.6 Results and Discussion

Both quantitative and qualitative data is presented in this section from the responses of participants to the questionnaire. The data is analysed to reveal any evidence of the impact of the text messaging on affective learning. The responses by students to the open questions are especially revealing as they contain many references to the effect of the text messaging on their affective learning.

16.6.1 Levels of Participation

Participation in the study was purely voluntary and overall the rate of participation was 88 %. The total number of messages sent by the instructor during two 13-week semesters of the study was 202. The number of broadcast messages sent to groups of participants was 89 while 113 messages were sent to individual participants usually in response to individual queries.

A total of 155 messages were received by the instructor from participants indicating that participants not only received text messages but actively participated in the communication. Between broadcast messages and individual messages the total number of individual messages received by all participants during the study was 1,005. This means that on average 23 messages were sent to each of the participants and it equates to an average of less than two messages per participant per week.

16.6.2 Quantitative Results

The first section of the questionnaire consisted of 30 specific questions on the use and impact of the text messaging service. Participants were asked to indicate their response to each question on a 7-point Likert scale. In addition, for each question the percentage of responses that were scored with 5 points or more is also shown. As score of 5 points or more on any item by a respondent is taken to indicate agreement.

Analysis of the results shows that participants generally felt very positive about the introduction and use of the text messaging service with 91 % of participants agreeing that they thought that being in touch by text messaging with your instructor was a good idea and 86 % of participants agreed that they liked receiving text messages from their instructor.

In terms of the effect of the text messaging on their relationship with their instructor three-quarters of participants agreed that the text messaging service had been beneficial to their relationship with the instructor and over 80 % of participants agreed that it had both improved their attitude to their instructor and made their instructor more approachable. Just over half of participants agreed that the service had improved their attitude to the college, had increased their liking for the subject and had increased their motivation, engagement and participation.

When asked if they were concerned about the potential cost of replying to the text messages 34 % agreed that they were. However only a small number of messages sent had needed a reply and more and more students are now availing of free text messaging. While 84 % of participants did not agree that receiving text messages from your instructor was intrusive a small number of participants had responded to question 30 with a score of 5 or more. This was taken seriously and further emphasised the need for careful and judicious use of the service and the need to speak to participants about any concerns they might have and also the need to make sure they fully realised that they could withdraw from the service at any time of their choosing. When asked what they thought about the use of text messaging to support learning 86 % of participants agreed it was an effective approach.

In summary, the participants generally liked receiving the messages and they perceived that it improved their relationship with their instructor and his attitude towards them. It also made the instructor more approachable and made it more likely for them to talk to the instructor informally. Many participants agreed that the service had improved their attitude to the college, had increased their liking for the subject and had increased their motivation, engagement and participation. This was taken as evidence of an effect on student affective learning of the out-of-class text messaging.

16.7 Qualitative Results

The second section of the questionnaire gathered qualitative data from participants on their perceptions of communicating with their instructor using text messaging and its impact, if any, on them or their class in terms of learning and also in terms of the relationship with their instructor. A series of open questions were used to give participants the opportunity of anonymously expressing their opinions. The responses from the participants to the open question provided a great deal of valuable and insightful feedback into their perceptions of the effect of the text messaging service on their learning experience. Analysis of the responses provides further evidence of the effect of the text messaging on student affective learning.

The overwhelming majority of the feedback was very positive. The participants generally perceived that the text messaging had made them feel closer to the instructor and they felt more comfortable asking questions in class, or outside of class, about the course. One participant, who was a mature student, responded it "has motivated me more to come to class, has improved my attitude towards college and subjects". When asked in what ways (if any) they thought the text messaging service has been beneficial or detrimental to your class in general they again mostly responded very positively. They felt it improved communications and had improved the class' relationship with the instructor and as a result they felt they had a more comfortable atmosphere in class and they perceived that their learning was better. They also felt it had brought the instructor closer to the class, had become a talking point among them, and had brought the class closer together as a result. They also perceived that there were many benefits from it and that the class had a higher attendance as a result. When asked in what ways, if any, they thought the text messaging service has helped or hindered them in their learning some of the participants responded that it reminded them to study before class and was better than email for notifying them at short notice of any changes to the schedule.

The responses to the last question are particularly revealing in terms of the overall assessment by participants of the use of text messaging for out-of-class communication and their perceptions of the study. The participants generally responded that it was a good service to students and improved communications. They also felt that others should use text messaging as a means of communication and that it was easier to communicate by text than by email. One participant felt that it should be applied to all modules. They also felt the research study was innovative and should be developed further as it was a different approach in dealing with instructor-–student communication.

16.8 Reflections and Conclusions

Analysis of the quantitative and qualitative data provides evidence that the use of text messaging for instructor-–student out-of-class communication has a positive effect on instructor immediacy and student affective learning. The participants perceived that it made the instructor more approachable and made it more likely for them to ask questions in class and engage in discussions with the instructor. In addition, it made them feel more comfortable and at ease in the classroom and gave them a feeling that the instructor cared for them.

Enhanced immediacy is very important in terms of the quality of student learning experience and has many implications in terms of education, including improved attendance, motivation and engagement by students. This research is interdisciplinary in nature, intersecting the fields of both instructional communication and mobile learning. The findings of this research are a contribution to both fields as they demonstrate how the use of mobile technology in education can lead to enhanced instructor immediacy and improved learning experience.

16.8.1 Concerns with Instructor-Student Text Messaging

Some of the participants in the study who used the text messaging service had some concerns as was evidenced in their feedback. Their concerns were around the potential cost of replying to the text messages, the timing of the messages and the relevancy of the messages to their course. In terms of cost many of the participants used the same mobile provider as the instructor and so had no cost associated with sending messages to the instructor. However, the cost may have been a concern for some of the other participants. In the interests of fairness it was decided to give this concern serious consideration in terms of any future operation of the service. It was felt that if it could be demonstrated that the text messaging was beneficial to the learning experience of students then perhaps a way could be found to persuade the management of educational institutions to subsidise or make free text messaging available to students to support their learning. It should be noted that the cost was only a concern for those who wanted to send messages to their instructor. It must be pointed out that there is no charge for receiving a text message in Europe. However, this is not always the case in countries outside Europe.

A few participants also seemed to have concerns about the timing and relevancy of the text messages. This highlights the importance of judicious use as well as the need for guidelines on the sending and receiving of text messages. All those involved with the text messaging need to be aware of the guidelines and it is important that they be adhered to as much as possible. A few of the participants also felt that some of the text messages were not relevant. This may be explained by the fact that some of the messages were not directly course-related but were aimed at enhancing affective learning among students and encouraging interest, attendance and engagement.

Another potential concern may surround the often colloquial nature and ad hoc use of text messaging that might potentially lead to misuse of the service, a phenomenon that was observed in the early days of the introduction of email in organisations. Students should be made aware that text messages in this context are still part of the learning experience and that they need to bear in mind that it is their instructor they are communicating with and not one of their friends. There is a fine but significant line between high levels of perceived instructor immediacy and close personal friendship. Students might misinterpret the higher availability and closer interaction with the instructor as a kind of peer relationship. This may lead them to be surprised or disappointed when the instructor executes the necessary duties of their role such as disciplining students or allocating marks. The experience of the instructor is that the text messaging makes it more likely from students to communicate informally with them. While this is a positive effect it also highlights the importance for the instructor of always bearing in mind their role as an educator and not as a personal friend.

In their feedback participants indicated that they would like text messaging to be used by more instructors and some felt it should be used with every module. However, it stands to reason that the number of different modules taken by students would need to be taken into consideration as to how many text messages they can receive from each instructor and how often. While the experience of the instructor who provided the text messaging service was very positive not all members of academic staff might agree that it is worthwhile. The experience of the instructor was similar to that of Horstmanshof (2004). Some older more traditional colleagues had reservations to this approach. They argued that it would add to their work burden and they also felt the approach was 'mothering' the students and would lead to dependency. However, the amount of extra work required was minimal. Text messaging is asynchronous, and therefore, the instructor does not need to reply immediately. In addition, text messages are usually short and to the point due to their limited length. The use of a software application such as the one used for the main study also makes the text messaging very easy. However, for very large classes there could possibly be quite a bit of extra work involved. This could be explored in future work. In terms of the criticism of the text messaging as 'mothering' the students it can be argued that if students feel an affinity with their instructor and their course they are more likely to explore new areas of learning independently, especially, if encouraged to do so by their instructor. There is also a strong connection between enhanced affective learning and lifelong learning (McCombs 1991).

16.8.2 Guidelines for Instructor-Student Text Messaging

The guidelines for instructor-student out-of-class text messaging that were developed as part of the study are a very important output of the research work. The guidelines are necessary to avoid incorrect expectations of the text messaging service by students. They inform the student of the level of service they can expect and this may help to avoid misunderstandings. The student is required to read and familiarise themselves with the guidelines prior to consenting to participate in the service. The guidelines were drawn up on consideration of the feedback from the student focus groups and on reflection by the instructor as to the appropriateness and effectiveness of text messaging for communicating with students. These guidelines were used throughout the study and they worked very well in so far as there were no complaints from participants and no participant withdrew from the service.

The guidelines cover the need for informed consent for participants as well as the right of participants to withdraw from the service at any stage. They also specify the quality of service that participants can expect, including maximum limits on the number of messages as well as maximum response times and hours of operation. In addition the guidelines also include some stipulations about when text messages should not be sent to students, for example the day before an examination. This is intended so as to avoid what might be perceived as unfair advantage by some of their peers.

While this research concludes that guidelines are very important for the use of text messaging for instructor-student communication there is little doubt that the guidelines could vary somewhat from one institution to another. It is hoped that the guidelines developed as part of this research work may be of interest not only to researchers but also to practitioners who may be interested using text messaging for instructor-student out-of-class communication.

16.8.3 Conclusion

The main conclusion of this research is that the judicious use of text messaging for out-of-class communication can significantly enhance student perception of instructor immediacy and has many other benefits in terms of student learning experience. This finding is very important for all those involved in teaching students. While it is recommended that more instructors adopt the use of text messaging for out-of-class communication with students there are some barriers to mainstreaming this approach in higher education that need to be considered. As with any new development many instructors and educational institutions may be slow to adopt this form of communication. Their concerns may be well founded and this paper has attempted to show how these may be addressed. It is felt that if proper precautions are exercised, the benefits of using text messaging for instructor-student out-of-class communication far outweigh any potential risks.

References

Andersen, J. F. (1979). *Teacher immediacy as a predictor of teaching effectiveness. Communication Yearbook 3* (pp. 543–559). D. Nimmo, New Brunswick, NJ: Transaction Books.

Bloom, B. S. (1956). *Taxonomy of educational objectives: The classification of educational goals*. New York: D. McKay.

Christensen, L. J., & Menzel, K. E. (1998). The linear relationship between student reports of teacher Immediacy behaviors and perceptions of state motivation, and of cognitive, affective, and behavioral learning. *Communication Education, 47*(1), 82–90.

Clarke, E., & Doody, C. (2008). Using mobile phone text messaging as an alternative to traditional handheld audience response. In *Proceedings of 9th Annual Irish Educational Technology Users' Conference (EdTech) 2008*. Ireland: Dundalk Institute of Technology, Irish Learning & Teaching Association.

Ellis, K. (2004). The impact of perceived teacher confirmation on receiver apprehension, motivation, and learning. *Communication Education, 53*(1).

Harley, D., Winn, S., et al. (2007). Using texting to support students' transition to university. *Innovations in Education: Teaching International, 44*(3), 229–241.

Hill, Y., Lomas, L., et al. (2003). Students' perceptions of quality in higher education. *Quality Assurance in Education, 11*(1), 15.

Horstmanshof, L. (2004). Using SMS as a way of providing connection and community for first year students. In R. Atkinson, C. McBeath, D. Jonas-Dwyer, & R. Phillips (Eds.), *Beyond the Comfort Zone: Proceedings of the 21st Annual Conference of the Australasian Society for Computers in Learning in Tertiary Education* (pp. 423–427). Perth.

Jaasma, M. A., & Koper, R. J. (1999). The relationship of student/faculty out-of-class communication to instructor immediacy and trust and to student motivation. *Communication Education, 48*(1), 41–47.

Kearney, P., Plax, T. G., et al. (1985). Teacher immediacy for affective learning in divergent college classes. *Communication Quarterly, 33*(1), 61.

Keegan, D. (2006). The arrival of mobile learning. In *Proceedings of 7th Annual Irish Educational Technology Users' Conference (EdTech) 2006*. Sligo, Ireland: Irish Learning & Teaching Association.

Markett, C., Sanchez, I. A., et al. (2006). Using short message service to encourage interactivity in the classroom. *Computers & Education, 46*(3), 280–293.

McCombs, B. L. (1991). Motivation and lifelong learning. *Educational Psychologist, 26*(2), 117–127.

McCroskey, J. C. (1994). Assessment of affect toward communication and affect toward instruction in communication (pp. 55–65). In M. Morreale, S. Brooks (Eds.), *1994 SCA Summer Conference Proceedings and Prepare Remarks*.

McCroskey, J. C., Fayer, J. M., et al. (1996). A multi-cultural examination of the relationship between nonverbal immediacy and affective learning. *Communication Quarterly, 44*(3), 297.

Mehrabian, A. (1969). Some referents and measures of nonverbal behaviour. *Behavior Research Methods and Instrumentation, 1*(6), 205–207.

Mehrabian, A. (1971). *Silent messages*. Belmont: CA, Wadsworth.

Mehrabian, A. (1981). *Silent messages*. Belmont: CA, Wadsworth.

Naismith, L. (2007). Using text messaging to support administrative communication in higher education. *Active Learning in Higher Education, 8*(2), 155-155.

Noels, K. A., Clement, R., et al. (1999). Perceptions of teachers' communicative style and students' intrinsic and extrinsic motivation. *The Modern Language Journal, 83*(1), 23.

Pogue, L. L., & Ahyun, K. (2006). The effect of teacher nonverbal immediacy and credibility on student motivation and affective learning. *Communication Education, 55*(3), 331.

So, S. (2009). The development of a SMS-based teaching and learning system. *Journal of Educational Technology Development and Exchange, 2*(1), 113–124.

Stone, A. (2004). Mobile scaffolding: An experiment in using SMS text messaging to support first year university students. In *Proceedings of IEEE International Conference on Advanced Learning Technologies, ICALT 2004* (pp. 405–409). Joensuu, Finland.

Tretiakov, A., & Kinshuk, K. (2005). Creating a pervasive testing environment by using SMS messaging. In *Proceedings of IEEE International Workshop on Wireless and Mobile Technologies in Education 2005 (WMTE'05)* (pp. 62–66). IEEE.

Trifonova, A. (2003). Mobile learning-review of the literature. Technology Report No. DIT-03-009). University of Trento, Department of Information and Communication Technology Retrieved from http://citeseerx.ist.psu.edu/viewdoc/download?doi=10.1.1.75.4687&rep=rep1&type=pdf.

Vaughn, L. M., & Baker, R. C. (2004). Psychological size and distance: emphasising the interpersonal relationship as a pathway to optimal teaching and learning conditions. *Medical Education, 38*(10), 1053.

Witt, P. L. (2000). An experimental study of teachers' verbal and nonverbal immediacy, student motivation, and cognitive learning in video instruction, University of North Texas. Doctor of Philosophy, p. 206.

Witt, P. L., & Wheeless, L. R. (2001). An experimental study of teachers' verbal and nonverbal immediacy and students' affective and cognitive learning. *Communication Education, 50*(4), 327–342.

Authors Biography

Paul Hayes is a lecturer in the school of computing with the National College of Ireland. His research interests include mobile learning, data communications and Internet of Things. In 2010 he received a Ph.D. from the University of Limerick for his work on exploiting mobile communication in supporting education. Paul graduated in 1990 with an honours degree in Electronic Engineering from the University of Limerick. He subsequently completed a Master's degree by full-time research in the area of wireless networks. He was then employed by Queen's University Belfast as a Research Associate to work on a collaborative research project between the University and Nortel before joining the National College of Ireland.

Stephan Weibelzahl is Professor of Business Psychology at PFH Private University of Applied Sciences Göttingen, Germany. Prior to that, Stephan served as Lecturer and programme director in the School of Computing at National College of Ireland in Dublin and headed a research group in learning technologies at Fraunhofer Institute Experimental Software Engineering, Kaiserslautern, Germany. He holds a Diplom (Masters) in Psychology and received a Ph.D. from the University of

Trier, Germany. Stephan's main areas of expertise are personalisation in e-Learning and the evaluation of learning technologies. Exploiting artificial intelligence techniques, Stephan and his team designed, developed and evaluated a number of e-Learning systems that monitor and adapt to the user in order to improve the learning process.

Chapter 17
Implementing a Mobile App as a Personal Learning Environment for Workplace Learners

Nicole Gu

Abstract With the purpose of developing a strategy for supporting informal learning in the workplace, this study investigates the possibilities of launching an educational mobile application adopting various Web 2.0 tools as a personal learning environment. Functionalities including RSS (Really Simple Syndication), podcasting, Web searching, and microblogging were integrated into the app which is named "MobLearn@Work" to support workplace learners' individual learning activities which are task-oriented. The article reports on the details of designing and developing the app, from both technical and theoretical perspectives. Optimistic results from the study informed possible ways of implementing educational apps to enhance workplace-learning performance from theory to practice. Recommendations for further research are presented by end of the study.

17.1 Introduction

Information technologies, as well as mobile technologies have been continuously playing an important role in educational area since the last decade (Boulos et al. 2006; Cabada et al. 2009; Cochrane and Bateman 2008; Sharples et al. 2010). In the research area of adult learning, the use of information technologies deserves more attention. Driscoll (1998) has summarized the characteristics of adult learners that they have real-life experience, they prefer problem-centered learning, they are continuous learners, they have varied learning styles, they have responsibilities beyond the training situation, they expect learning to be meaningful, and they prefer to manage their own learning. The rapid change in nowadays workplace has led to a continuous learning situation among employees all over the world. The majority of what people learn at work belongs to flexible and self-regulated informal learning activities, which involves a combination of learning from others and personal

N. Gu (✉)
The University of Hong Kong, Hong Kong SAR, People's Republic of China
e-mail: nicole_gj@outlook.com

D. Churchill et al. (eds.), *Mobile Learning Design*, Lecture Notes in Educational Technology, DOI 10.1007/978-981-10-0027-0_17

experience (Cross 2003; Eraut 2004). This happens in the context of employees' attempt to deal with problems emerging in the workplace. In order to cope with the growing competition, employees in cooperate environments continuously engage in informal learning activities.

Literature informs that employees in cooperate environment continuously engage in informal learning, which describes learning without formally organized content and learning that happens outside of formally organized settings (Sefton-Green 2004). Technology has created a spectrum of possibilities for informal learning. Web 2.0 tools, such as RSS, blogging, Wiki, and social networks have provided quick access to information for employees to engage in informal learning. Currently, the commonly adopted Web 2.0 services in the adult educational field include but are not limited to RSS, podcasting, and microblogging (Garrison 1997; Holotescu and Grosseck 2010; Lee et al. 2008; Williamson 1998). RSS provides an effective approach to searching and managing the vast volume of information on the Internet. 'Podcasting' refers to online audio content that is delivered via an RSS feed. Microblogging is a combination of blogging and instant messaging that allows users to create a short message and post it on their profile. Emerging mobile technologies add a new dimension and considerably increase the possibilities for Web 2.0 technology usage in informal learning; however, the use of mobile technologies in the context of the workplace has not been researched widely, and relevant understanding is lacking.

This article introduces the process of design and development of a mobile App named MobLearn@Work, with the purpose of exploring how mobile Web 2.0 technologies assist individual employees in their work-related informal learning. An integration of several Web 2.0 tools including RSS, podcasting, Web searching, and microblogging was applied to serve as a personal learning environment. The study has also proposed a unique data collecting method via the log system in order to measure learners' actual time spent on the App. Limitations and recommendations for the future design of educational Apps in the context of workplace were both discussed at the end of the study.

17.2 Theoretical Framework

Along with the fast development of ICT technologies and mobile technologies, Apps have become prevalent nowadays. So far, there has been a number of Apps which are designed for educational uses, however, only a few have been developed for adult learning in the workplace to date. These include Apps mainly targeting areas, such as health and safety, operational efficiency, and soft skills, such as leadership or communication, and management skills (e.g., Workplace Safety, Lean Manufacturing, and How2Lead). Most of the learning Apps are standalone, which means that their contents are predefined by the developers. This feature greatly limits the life cycle of an App's. In contrast, net-based Apps that provide dynamic updates using feed technology (such as RSS) guarantee the continuous use by delivering automatic updates to users' devices. Especially for workplace learners, receiving the most up-to-date information is essential when the workers need to accumulate knowledge rapidly. In this manner, an App that allows users to manage their own learning materials is required. The following of this section introduces a framework for designing an App targeting workplace informal learning, based on the theoretical backgrounds of personal learning environments personal learning environments (PLE), as well as informal learning.

17.2.1 Personal Learning Environments (PLE)

The concept of PLE was first introduced by van Harmelen (2006), who describes PLEs as systems that help learners take control of and manage their own learning. A PLE is an individual e-learning system that provides learners access to a variety of learning resources. According to Chatti and his colleagues (2010), the concept of PLE "supports self-organized, informal, lifelong learning and network learning and translates the principles of constructivism and connectivism into actual practice" (p. 79). In PLE settings, the learner is expected to set their own learning goals, manage their learning contents, implementing learning strategies, and in the same time communicating with others with the purpose of achieving learning goals (Van Harmelen 2006). PLE is not simply an application, but rather a new approach of learning that emphasizes the importance of learner's self-direction in the process of learning via the use of technologies. Researchers have explored different means to bring the concept of PLE into practice, with an agreed standpoint that the use of a set of different tools may be feasible in the PLE pattern (Attwell 2007; Sclater 2008; Wilson et al. 2007). Chatti and his colleagues (2010) mentioned the RSS feeds (such as iGoogle) that allowed the integration of different services into a single personal platform, which to some extend matched the characteristics of PLE.

17.2.2 A Personal Informal Learning Framework in the Context of Mobile Web 2.0

Informal learning, a relatively underresearched term, is used to describe learning without formally organized contents and learning happens outside formally organized settings (Sefton-Green 2004). Schugurensky (2000) has generated three forms of informal learning, which are self-directed, incidental, and socialization. Intentionality and awareness have been used as the criteria to distinguish among the three forms. As one major form that is both intentional and self-aware, self-directed learning refers to learning activities undertaken by individuals or group of learners without the assistance of an instructor. Interpreted by Knowles (1975), the process of self-directed learning includes diagnosing and formulating goals, identifying

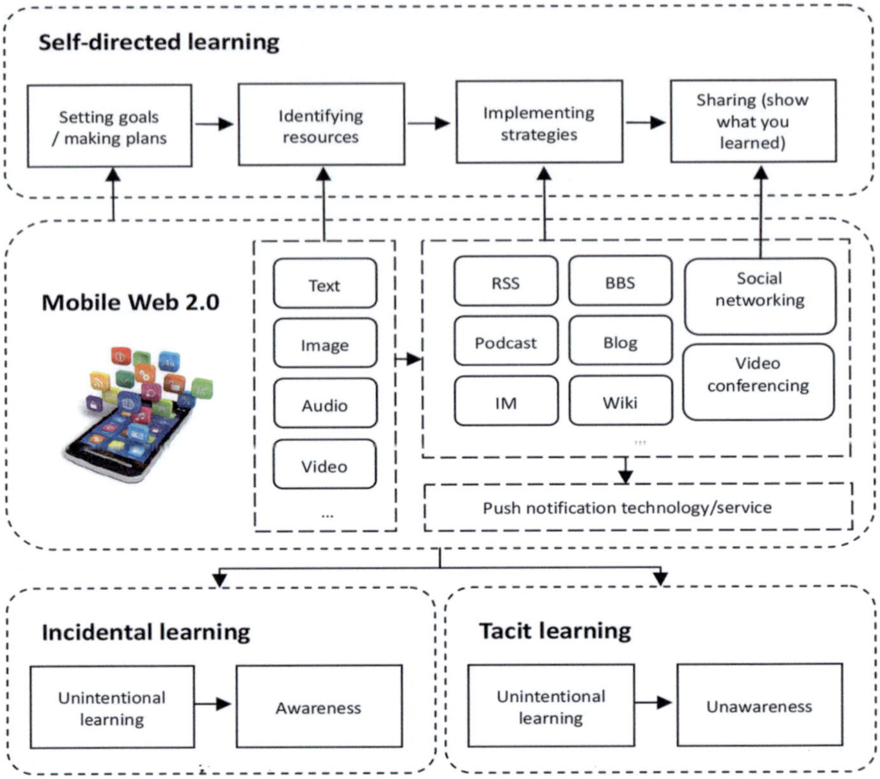

Fig. 17.1 Workplace informal learning framework in the context of mobile Web 2.0

resources, selecting and implementing suitable strategies, and evaluating outcomes. Incidental learning is not intentionally initiated by the learner, but afterwards the learner become aware that learning has occurred during the previous experience. Socialization is also known as tacit learning, which means neither does the learner intend to learn anything, nor the learner is aware of having learned anything. Based on the works of Knowles and Schugurensky, there has been summarized a design framework guiding the development of the App as shown in Fig. 17.1.

The top-dotted rectangle represents the circle of self-directed learning process, consisting of four steps:

- Setting goals and making plans. Learners decide what knowledge and skills they want to gain, when to learn, how to learn, and estimate the expected learning outcomes.
- Identifying resources. Learners make decisions about where and how to find a variety of learning resources.

- Implementing strategies. Learners find and choose the appropriate methods of learning and formulate personal learning programs according to their own situations.
- Sharing. Learners show what they have learned by sharing to other platforms. Through everyday social networking activities, new learning needs, and motivations may be triggered and a new cycle is thus formed. They may even become a subscribing source for other learners within the same circle.

Depending on the data sources from the mobile Internet, Step 2 is further divided into two categories: searching and subscribing. The action of searching is intentionally taken by the learners. Learners initiate a new search action whenever there is any learning requirement. Information is provided through rich multimedia tools (e.g., text, image, video) from various sources (e.g., Web, news, blogs, Wikis) based on the latest Web 2.0 technologies. Learners are able to select the most appropriate source according to their different search requirements. 'Subscribing' refers to the obtaining of learning materials provided by Web feeds (e.g., RSS, podcasting), which push updated contents to the learners periodically. It is a more complicated process. Although the action of subscription itself is intentional, the synchronizing and the learners' reading activities could be passive and incidental since the contents are not known before they have been downloaded for learning. Therefore, this can be considered as a 'semi-intentional learning activity.' In the context of this study, 'socialization' mainly refers to online interactions based on social networking services. Social networking services with mobile technology nowadays allow sharing of learning achievements more easily than before. From another perspective, learners can also benefit from social networking since it could be regarded as one of their subscribing sources for receiving new learning contents.

17.3 Design and Development of MobLearn@Work

Open platform technologies, such as IOS, Android, and Windows Phone have brought great opportunities to develop application software for mobile terminals; however, it currently remains difficult to satisfy the specific demands of end users and overcome the technical constraints of mobile devices. To address the constantly changing software requirements, Extreme Programming (XP) (Silva 2007) has the advantages of on-time delivery, reliance on team members' knowledge rather than documentation, short release cycles, tight customer integration, incremental design, constant communication, and coordination. Based on the XP, Abrahamsson and his colleagues (1990) developed an agile approach called 'Mobile-D' for the mobile application (Symbian platforms), which mainly emphasizes the early identification and solving of technique issues, rather than considering usability during the development process. Usability measures the quality of a user's experience when interacting with a product of system. Due to the limitations of mobile devices, user experience is of great importance during the development. In today's IOS and

Android platforms, user-centered design (UCD) is an approach for employing usability. It helps to provide the users with the real usage context, which means that users are able to touch the buttons and see how the software actually works (Rogers 1997). Hussain and his colleagues (1992) integrated the agile process and usability into his project life cycle of a mobile multimedia streaming App. The above methodologies have been successfully applied to each individual App context.

However, a complex mobile application development process is required in educational scenarios, where there usually exist three roles: the designer/tutor who may be unfamiliar with the technical implementations; the developer, who is unfamiliar with the fundamentals of education; and the end user, who has rapidly changing requirements. It is essential to separate the interactions among them during development in order to enhance efficiency regarding technical issues and usability.

17.3.1 An Agile Design Approach

As mentioned above, the modified agile development process of an educational App must consider interactions between different parties. In the context of educational Apps, a bi-loop-based software development process is proposed in order to improve the software quality and meet learners' changing requirements, which is illustrated in Fig. 17.2. Three roles in the development life cycle have been identified: the participant is the customer of the App; the designer turns the theory into practice; and the developer takes charge of the implementation. The designer and developer, however, may actually be the same person. The development process consists of two loops. The one on the left represents interaction between the participant and the designer, while the one on the right shows the relationship between the designer and the developer. The designer acts as a bridge connecting the other

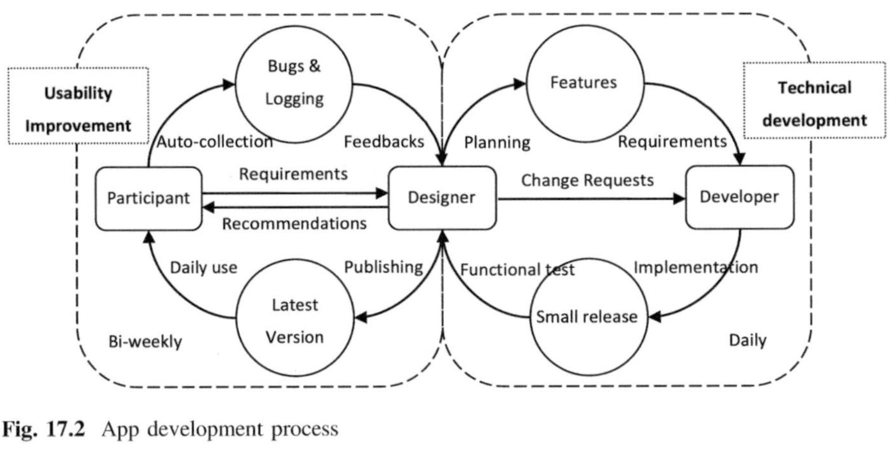

Fig. 17.2 App development process

two. The designer periodically publishes the latest version of the App for participants to use. The designer also collects feedback on users' experience through random discussions with participants. Based on this feedback, improvements, and new features are generated and transferred to the developer for an improved new version. The two loops run in a continuous manner until a satisfactory balance between the participant and designer is achieved.

The cycle on the right, referring mainly to technical development, is a traditional agile development process. The left circle is novel. Such a circle, associated with the end user and the designer, mainly focus on the improvement of usability. Unlike the traditional desktop software, the concept of UCD is crucial because the usability determines whether or not the user continuously uses the App. Therefore, the UCD design process is a separate loop at the same level as the technical development. The periodicity of the two cycles differs. The upgrading of new versions from the designer to the participant is biweekly or more frequent, while the small releases from the developer to the designer occur on a daily basis.

Regarding the emerging educational applications, the bi-loop development process for the mobile App combines usability and the agile process together, while considering different roles in the design process with the purpose of putting educational theory more effectively into practice. It has been proven to be efficient in response to the participants' learning requirements.

17.3.2 Functions of MobLearn@Work

In this section, we describe in detail how MobLearn@Work works by the use-case analysis with examples of some major activities involved in the App. Figure 17.3 describes four major functions—microblogging, RSS, podcasting, and mobile Web searching—included on the start page. With a click of the feedback button, an automatically generated email attaching the user log report is sent to the designer. The microblogging site was set-up via an open-source software tool called 'Sharetronix' (Kreber 1998).

RSS and podcasting service work similarly through Web syndication. The App features a preinstalled list of feeds, according to participants' preferences as identified during the initial interviews. Basically, when participants click the RSS/Podcasting button, they see a list of feeds. Once, they choose to read a specific item, they will find several option buttons at the bottom of the page: previous item, next item, original link, Web searching, return to homepage, share internally (to the MobLearn@Work microblog), and share externally (any external services installed on the mobile phone). Users are able to add or delete any feed items as needed. The App provides two ways of adding RSS feeds: adding by search and adding manually. Details of RSS/Podcasting can be found in Fig. 17.4 (left).

Unlike traditional mobile searching on a mobile browser, this App provides a parallel Web searching function which returns Web, image, video, news, blog, and Wiki simultaneously for a single query, thus strongly improving the searching

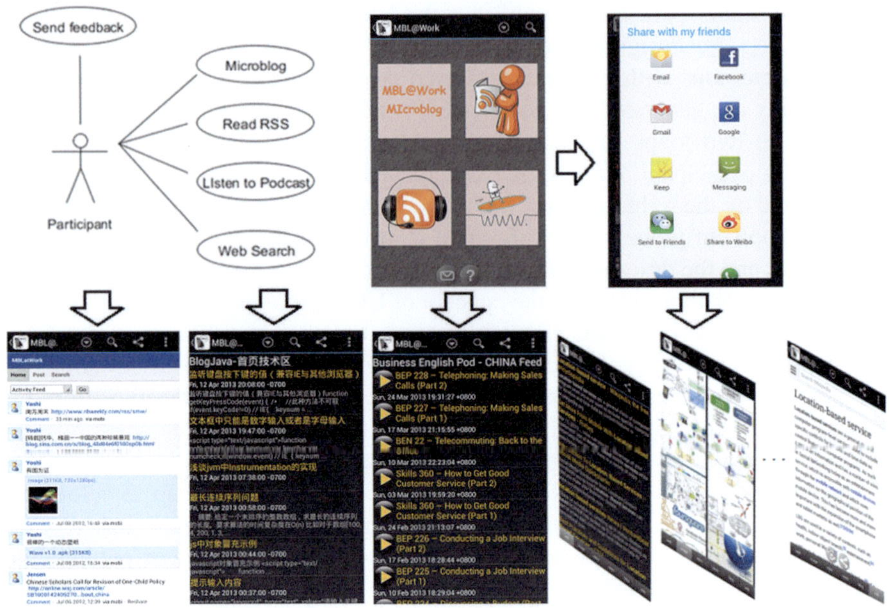

Fig. 17.3 Use case and screenshot of MobLearn@Work

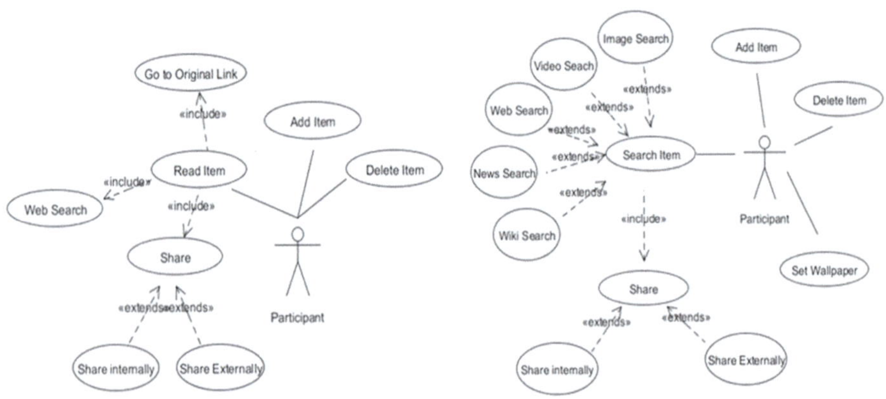

Fig. 17.4 Use case (*left* RSS/Podcasting; *right* mobile Web searching)

experience. This means that when the user searches for something, he/she may get results on all of these six aspects at the same time. Details of the searching activities can also be found in Fig. 17.4 (right). There are basically two kinds of searching activities in the study. One is initiated by clicking a keyword from the search list, and the other by adding a new search item. MobLearn@Work provides a series of online open courses for users. For example, a list of the financial open courses from MIT and Yale were provided as recommendations to help users to get started with the App.

17.3.3 UI Improvements of MobLearn@Work

With the purpose of enhancing the users' experience, we have emphasized more on the improvement of the usability rather than functionality of the App. The affordance of the App determines the activities of daily use. As a private product, the smart phone is usually installed by its user the applications that are most frequently used. Therefore, the usability of mobile Apps must be continuously improved in order to extend an App's life cycle.

Being different from desktop applications, the design of a mobile App should take into account the benefits and limitations of a mobile device. On the one hand, mobile devices obviously have some operational superiority to the PC. Recent mobile devices attract users through their excellent touch experience. The operations are usually simple and can be accomplished just with the fingers. Desktop applications often contain a complicated operational procedure, and must be manipulated by the mouse and keyboard. On the other hand, the user interfaces of mobile Apps must be well designed. Input by finger is less accurate than that by mouse and keyboard. This causes the designers of the user interface to consider the most amenable style. Therefore, the UCD is of great importance in App design.

During the development iteration, several important changes regarding the affordance of the App were made, as shown in Fig. 17.5.

- As proposed by the latest Android reference design, the action bar and swipe UI were used. Common operations are integrated into the action bar. Operations regarding switching pages were replaced by introducing a gesture (Fig. 17.5-1).
- Most content carriers were modified to be Web-based. This creates consistency in the content arrangement. Pinch open and close gestures allow users to zoom in and out the target contents, overcoming limitations of the small screen on mobile phones (Fig. 17.5-2).
- A global navigation was added to the action bar. Users can go to any of the four functions freely at any time. This saves a lot of time and makes it much easier for the users to switch between Apps (Fig. 17.5-3).
- A global search function was added to the action bar. This enables users to conduct a new Web searching activity anywhere in the App, instead of switching to the Web searching page (Fig. 17.5-4).

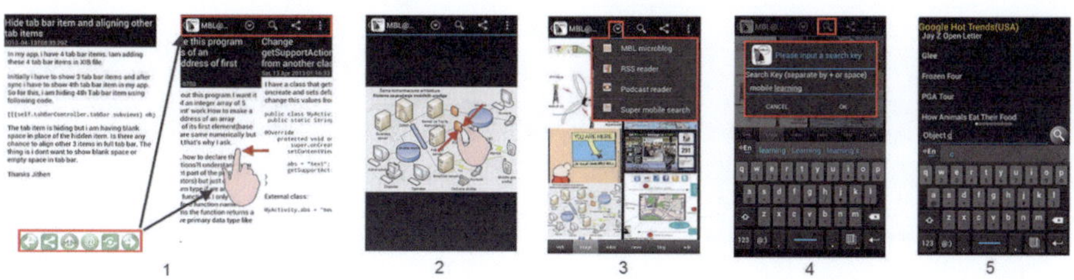

Fig. 17.5 UI Improvements of MobLear@Work

- A search widget was developed to meet users' emergent searching requirements. It also allows users to subscribe to recommendations from outside sources that are daily updated (Fig. 17.5-5).

There was a noticeable increase in the five participants' total time spent on the App after the UI improvements, especially on the Web searching function. Such a trend is clearly demonstrated by the dashed orange lines in Fig. 17.6.

According to Fig. 17.6 (top), the time of each participant spent on MobLearn@Work was around 20–38 min at the beginning of the study, which had increased to around 40-82 min by the end of the second month. The first inflection point occurred at the end of the third week when the widget of the search bar was introduced to the participants. According to Fig. 17.6 (bottom), after the search widget was published, the search count significantly increased. Participants reported

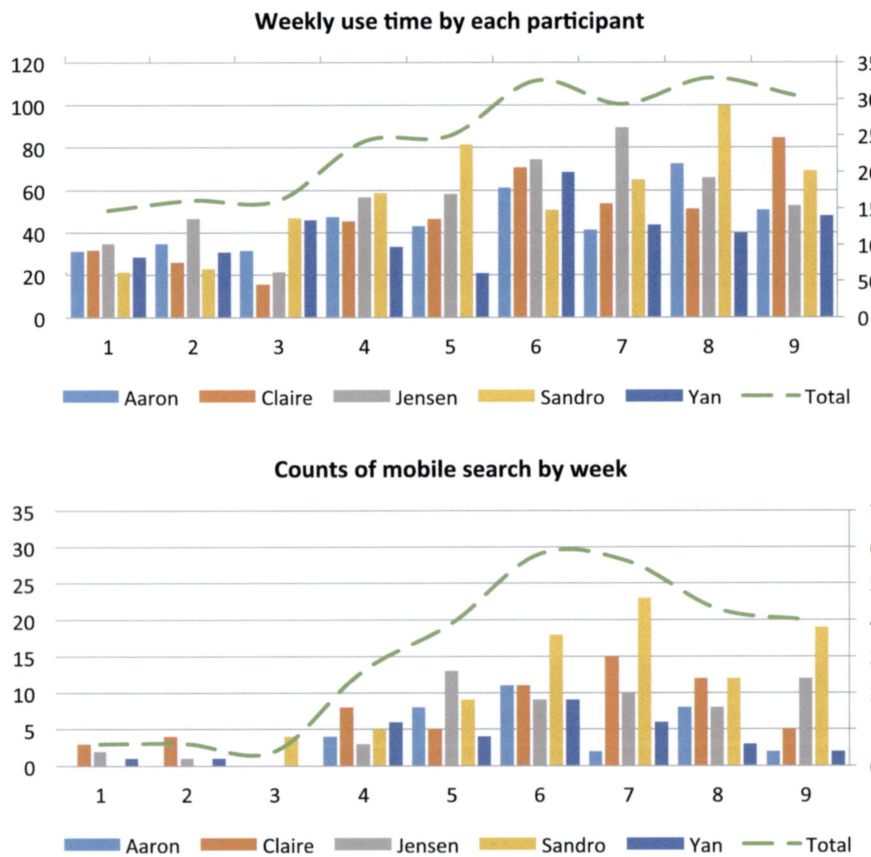

Fig. 17.6 Statistics by 9 weeks, coordinates of total are to the right axis (*top* each participant's time spent on the App; *bottom* counts of participants' mobile search activities)

that they were quite satisfied with searching for something by simply clicking the search bar widget instead of opening the App. The second inflection point occurred at the end of the fifth week when the swipe UI was published. Swipe is initiated by gestures rather than by clicks. Participants reported that the page swiping reduced the intervals of traditional page transitions that they disliked, while substantially increasing ease of use.

Another interesting phenomenon was that both curves of Fig. 17.6 fluctuate, which is especially apparent in Fig. 17.6 (top). The utility times usually rose when new updates were released and then fell after approximately 1-2 weeks. This implies that the continuous improvement of the App is a good way to increase user stickiness. Such a phenomenon is likely to occur when an App is first introduced to users. Based on Fig. 17.6 (top), the biweekly periodicity of the updates is considered to be appropriate. After several weeks, the utility time is likely to become steady, as seen in the tail of Fig. 17.6.

17.4 A Novel Method for Quantifying the Data

17.4.1 The Log System

After installing MobLearn@Work and receiving the basic training on how to use it, participants were required to start exploring the app as they want. The researcher has talked with each participant occasionally to get their feedback and suggestions with the purpose of further improving the functions and usability. Besides, a log system has been installed within the App to quantify the data by recording each learner's time spent on each function and certain activities. This complements the findings to a great extent by providing strong evidence, which leads to the improvement of the app.

The built-in logging system plays an important role in providing rich information on how each participant uses the app. Unlike the traditional bug reporting system, the logging system is used to collect the user activities towards the mobile app. Figure 17.7 is a part of a sample log report generated by learner's mobile device.

Date	Time	Function	Action	Duration
2012/7/6	10:49:57	com.cngsoftware.rss.RssActivity		15156ms
2012/7/6	10:50:12	com.cngsoftware.rss.RssItemListActivity		9258ms
2012/7/6	10:51:19	com.cngsoftware.rss.RssItemActivity	SHARE_EXTERNAL	
2012/7/6	10:52:20	com.cngsoftware.rss.RssItemActivity		27268ms
2012/7/7	11:59:52	com.cngsoftware.rss.RssItemListActivity		2734ms
2012/7/7	12:01:03	com.cngsoftware.rss.RssItemActivity		43578ms
2012/7/7	12:02:22	com.cngsoftware.rss.RssItemActivity		69929ms
2012/7/7	12:03:30	com.cngsoftware.rss.RssItemActivity		54024ms

Fig. 17.7 Screenshot of a sample log report

The life cycle of a mobile App in Android OS has been looked into in order to develop the log system. Android names every visible App window as the "activity," and every invisible part of the app which provides functions at backend as the "service." If the invisibility of windows is regarded as the moment when user releases their focuses on the app, the total time of the app use is equal to the live period of the activity. Therefore, the statistics of foreground lifetime reflects the tendency and favor of each user. Based on it, several indicators, such as the total foreground lifetime by function, the total foreground lifetime by week, and the total foreground lifetime by day, are easy to derive. Besides, users' activities in using the app are of great interests, such as adding a new RSS item, adding a new search, etc. The tendency of social sharing is also easy to obtain by counting the times and recording the sharing targets. Benefiting from Android OS, users can share their learning achievements to whatever targets that have been installed on their phones. In this study, the sharing destinies are classified into two categories, the internal microblog site built for the study, and the external platforms, e.g., Sina microblog, Facebook, Twitter, etc. The internal microblog represents for a closed community of participants while the external denotes the social networks of each individual learner.

Due to privacy issues, the detailed contents of what learners read, listen, and search are not recorded. Data collected from the log system, combined with data from interviews and other means, shall be analyzed to get a comprehensive understanding of each participant's way of using the app based on his or her own situations under certain specific contexts.

Figure shows the results from the participants' weekly log reports during the first 1 month of testing to examine the App. RSS was the mostly adopted function for gathering information by the participants, allowing participants to get what they want to learn semi-intentionally. According to conversations with all the participants during the second interview by end of the study, they all admitted that RSS had been the mostly common used application in MobLearn@Work. It thus could be concluded that reading subscribed contents anytime and anywhere contributes most to participants' learning requirements in the workplace (Fig. 17.8).

However, the usage of the other three applications are relatively lacking compared with RSS. Searching function was used to some extent but not satisfying. Microblogging was seldom used which was mainly due to the limited number of users, according to participants' reflections thereafter. Based on discussions with participants during the testing period, there existed certain difficulties while they were interacting with the app. For example, data informed that the majority of them felt not comfortable adding a new RSS feed. Aiming at this issue, a number of UI improvements had been adopted to solve the raising problems, which was discussed in the earlier part of this article.

During the testing period, it has also been found that sharing was one of the participants' major activities in the study. This informs the importance of socialization in the process of workplace informal learning. However, there has been noticed a huge difference between internal share and external share. Participants tended to repost more on their own social networks instead of sharing internally to

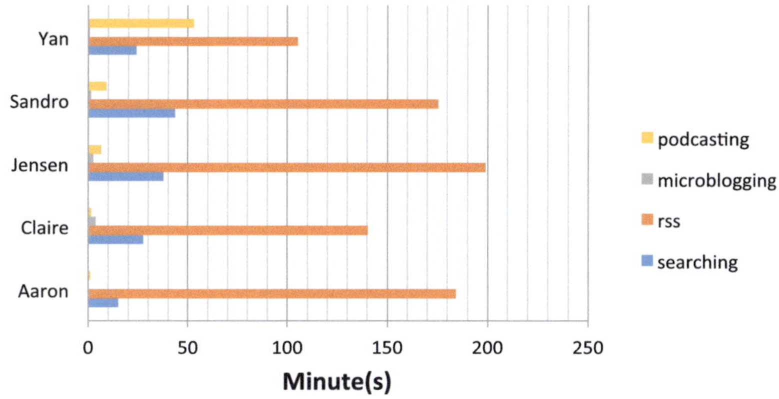

Fig. 17.8 Participants' time spent on each function separately (first month)

the MobLearn@Work's microblog. In the researcher's opinion, a newly set-up internal social networking platform with only five users is without doubt less appealing than a mature one. This situation could be improved by increasing the number of participants or establishing connection between each learner in the future. For corporations and classroom settings, such problem may be solved easily.

17.4.2 Measure of Actual Learning Time via MobLearn@Work

In fact, the logged time does not equal the actual time spent on informal learning. It is difficult to measure the accurate time of informal learning. Therefore, this study assumes a ratio η between the logged time t and the estimated time $\widetilde{t}$ for each participant, denoted by

$$\widetilde{t} = \eta \times t \tag{1}$$

The ratio was obtained by a questionnaire. The contents were classified into different categories, such as professional knowledge, communication skills, finance, business, politics, technology, entertainment, etc. Participants were required to mark the relevance of the use frequency of each category. Final result of the questionnaire was quantized into a number between $0 \sim 1$. The average learning time μ_t gathered from the log system and the estimated average learning time $\tilde{\mu}_t$ obtained from the questionnaire are summarized in Table 17.1. The estimated average learning time per week approximately equals the time of a traditional training session in most workplace environments.

Table 17.1 Weekly Use Times of MobLearn@Work (min/week)

Participants	μ_t (min)	η	$\tilde{\mu}_t$ (min)
Aaron	49.19	0.89	43.78
Claire	43.87	0.85	37.29
Jensen	52.95	0.87	46.06
Sandro	61.49	0.90	55.34
Yan	42.33	0.89	37.67

17.5 Recommendations for Future Study

MobLearn@Work has been generously adopted by participants during the period of the study in their own ways to cope with work-related learning demands. RSS was the major function used by all the participants to acquire information. Podcasting was the main source of language learning for participants with such learning requirements. The Web searching function provided participants a unique searching experience. However, the built-in microblogging site was seldom used. The reason for this situation was simply due to a poor sense of connectedness between participants. The five participants were chosen from different industries and different companies, with the purpose of providing rich and varied data in the study. The biggest problem was that they did not know each other. Although some of them shared similar learning demands, e.g., Aaron and Yan both had language learning requirements, and Sandro and Jensen was in need for massive industry news, it was still difficult to get them interact with each other via microblog. They had their own circle of friends in external platforms, such as Sina weibo by time of the study. Contrary to the situation with internal microblogging site, participants were recorded of being active in their own microblogs. However, data was mainly obtained through participants' recall of their past experience during the second interviews. It is therefore suggested that further studies be carried out within a same organization with participants who know each other, and who have already interacted or willing to interact with each other. It is in this way that more intensive and in-depth results will be achieved to explore possibilities of informal learning and team learning in online community.

Furthermore, it would be helpful to replicate the study with more experimental subjects to explore individual informal learning in the workplace. People from different industries and professional backgrounds have their own learning styles. Other factors may also affect the outcomes, e.g., gender issues, personality, and character issues, the level of self-discipline, company cultures, etc. The more we know about how different forms of informal learning can be enhanced by mobile and Web 2.0 technologies, the more we are able to broaden the opportunities of personal growth within organizations.

17.6 Conclusion

This article reports on the development of a mobile App, which is considered as a personal informal learning tool that integrates several Web 2.0 services into a single platform. A novel method for evaluating the effectiveness of the informal learning tool by analyzing learners' use time on the App has been proposed in this study. Since the logged time spent on MobLearn@Work did not equal the actual time spent on informal learning, data on the quantitative relationship between the two were collected using a simple questionnaire. Although the data collected in this way were deemed to be rough, they still provided a feasible means of measuring the time spent on informal learning. Further study needs to be conducted in this respect.

References

Attwell, G. (2007). Personal learning environments-the future of eLearning? *eLearning Papers, 2* (1), 1–8.

Boulos, M., Maramba, I., & Wheeler, S. (2006). Wikis, blogs and podcasts: a new generation of Web-based tools for virtual collaborative clinical practice and education. *BMC Medical Education, 6*(1), 41.

Cabada, R. Z., Estrada, M. L. B., Sanchez, G. S., Barrientos, E. U., Velazquez, M. O., & Garcia, C. A. R. (2009). EDUCA: A Web 2.0 collaborative, mobile and E-learning authoring system. In *Icalt: 2009 IEEE International Conference on Advanced Learning Technologies* (pp. 287–289).

Chatti, M. A., Jarke, M., & Specht, M. (2010). The 3P learning model. *Educational Technology & Society, 13*(4), 74–85.

Cochrane, T., & Bateman, R. (2008). *Engaging students with mobile web2. 0.* Paper presented at the EIT Teaching & Learning Conference, Napier, New Zealand.

Cross, J. (2003). Informal learning-the other 80 %. Retrieved from http://www.internettime.com/Learning/The%20Other%2080%25.htm.

Driscoll, M., & Alexander, L. (1998). *Web-based training: Using technology to design adult learning experiences with disk.* Jossey-Bass Inc., Publishers.

Eraut, M. (2004). Informal learning in the workplace. *Studies in Continuing Education, 26*, 247–273. doi:10.1080/158037042000225245.

Garrison, D. R. (1997). Self-directed learning: Toward a comprehensive model. *Adult Education Quarterly, 48*(1), 18–33.

Holotescu, C., & Grosseck, G. (2010). *Learning to microblog and microblogging to learn. A case study on learning scenarios in a microblogging context.*

Knowles, M. S. (1975). *Self-Directed Learning. A guide for learners and teachers.* Englewood Cliffs: Prentice Hall/Cambridge.

Kreber, C. (1998). The relationships between self-directed learning, critical thinking, and psychological type, and some implications for teaching in higher education. *Studies in Higher Education, 23*(1), 71–86.

Lee, M. J. W., McLoughlin, C., & Chan, A. (2008). Talk the talk: Learner-generated podcasts as catalysts for knowledge creation. *British Journal of Educational Technology, 39*(3), 501–521. doi:10.1111/j.1467-8535.2007.00746.x.

Rogers, A. (1997). Learning: Can we change the discourse? *Adults Learning (England), 8*(5), 116–117.

Schugurensky, D. (2000). The forms of informal learning: Towards a conceptualization of the field.
Sclater, N. (2008). Web 2.0, personal learning environments, and the future of learning management systems. *Research Bulletin, 13*(13), 1–13.
Sefton-Green, J. (2004). *Literature review in informal learning with technology outside school.* Bristol: Futurelab.
Sharples, M., Taylor, J., & Vavoula, G. (2010). A theory of learning for the mobile age. In B. Bachmair (Ed.), *Medienbildung in neuen Kulturräumen* (pp. 87–99): VS Verlag für Sozialwissenschaften.
Silva, P. M. (2007). *Epistemology of incidental learning.* Ph.D.
Van Harmelen, M. (2006). *Personal Learning Environments.* Paper presented at the ICALT.
Williamson, K. (1998). Discovered by chance: The role of incidental information acquisition in an ecological model of information use. *Library & Information Science Research, 20*(1), 23–40.
Wilson, S., Liber, O., Johnson, M. W., Beauvoir, P., Sharples, P., & Milligan, C. D. (2007). Personal learning environments: Challenging the dominant design of educational systems. *Journal of e-Learning and Knowledge Society, 3*(2), 27–38.

Author Biography

Nicole Gu is a Ph.D. candidate in the Faculty of Education, The University of Hong Kong. Her major research interests cover the areas including but not limited to mobile learning, Web 2.0 technologies and adult learning. Her current research topic is to evaluate workplace informal learning activities in the context of mobile Web 2.0 via the use of a customized App.

Part V
Mobile Learning in Subject Domains

Chapter 18
The Theory of Context-Aware Ubiquitous Learning and the Affordances of This Approach for Geometry Learners

Helen Crompton

Abstract The use of mobile learning has provided new pedagogical approaches to teaching geometry as a result of the technological affordances provided. One of the key affordances of mobile learning is the portability of the devices. This has untethered the learner from a particular environment to learn wherever and whenever the learner chooses. This chapter describes a subcategory of mobile learning called context-aware ubiquitous learning (context-aware ulearning) where learning happens in a real-world environment while using mobile devices to interact with that setting. This chapter delineates this subcategory and how this type of learning can be dichotomized into sensory and ambient context-aware ulearning. An argument is made that context-aware ulearning is a useful pedagogical approach for learning geometry.

18.1 Introduction

The use of mobile learning has provided new pedagogical approaches as a result of the new technological affordances (Crompton 2013). Crompton posited that one of the key affordances of mobile learning is the portability of the devices. This has untethered the learner from a particular environment to learn wherever and whenever the learner chooses. As educators and scholars have considered how these mobile affordances can be exploited, a new category of mobile learning has formed called context-aware ubiquitous learning (context-aware ulearning; Lonsdale et al. 2004). This term is broadly described as the use of mobile devices for learning while interacting with a real-world setting. In this chapter, context-aware ulearning is delineated and an argument is made to how a further

H. Crompton (✉)
Old Dominion University, Norfolk, VA, USA
e-mail: Crompton@odu.edu

D. Churchill et al. (eds.), *Mobile Learning Design*, Lecture Notes in Educational Technology, DOI 10.1007/978-981-10-0027-0_18

dichotomy forms sensory and ambient context-aware ulearning. This chapter also describes how context-aware ulearning is a useful pedagogical approach for extending and enhancing students understanding of geometry.

18.2 Mobile Learning

To unpack the meaning of a particular term, the definition is typically presented. However, mobile learning is a relatively new field of study and various different definitions have been published. Soloway et al. (2001) provided one of the early definitions of mobile learning as the use of a Palm as a learning device. Following Soloway et al.'s approach, many have based the definition on a particular mobile device. These ephemeral definitions exist only until new mobile technologies appear on the market.

Scholars (viz., Laouris and Eteokleous 2005; Sharples et al. 2005; Traxler 2009a) have entered into deep debates about how the definition should be based on particular attributes of mobile devices and what attributes should be included. Crompton (2013) conducted a review of the attributes described by the scholars to consolidate them into four central constructs of mobile learning: pedagogy, technological devices, context, and social interactions. This led to the construction of a recent definition of mobile learning as "learning across multiple contexts, through social and content interactions, using personal electronic devices" (Crompton 2013, p. 4). This definition does not mean that learning must be across multiple contexts and that both social and content interactions have to be included in the same learning activity, instead it provides the reader an understanding of what "can be" included in an mobile learning activity.

Another topic of great debate has been centered on which technological devices fit under the umbrella term of mobile devices (Caudill 2007; Traxler 2009b). In Crompton's (2013) definition, the term *personal electronic device* was used. This means that it is a device that is typically bought to serve the use of one person, such as a mobile phone or a tablet. However, the main consideration is if that device is able to turn on at a moment's notice to be used anywhere and at any time. Considering this, some portable technologies such as laptops would not fit under the term mobile technologies as the user typically has to wait while it turns on, which can take a few minutes. In addition, a laptop may be portable, nonetheless, it would be difficult to use a laptop while standing on a bus and with mobile phones and tablets this is much easier.

The portability of the mobile devices is the key attribute to the technologies that has led to the term *mobile learning*. Nonetheless, people conducting activities with mobile devices do not always make use of this feature. In some cases, mobile devices are used with the intention of avoiding students' mobility. For example, in schools, students can often be provided with mobile devices to enable the learner to use the device when they need it and avoid students moving around the classroom to use computers. Mobile devices are used in some locations for testing; in these cases, mobile devices can be seen as an alternative to large desktop computers that are cheaper, portable, and compact for storage. The argument could be made that this type of *testing* in not mobile as the students have to typically stay in one location to complete the test. However, the rebuttal could be that the testing can happen in rooms and other locations that testing would not be possible without a mobile device. This could be valuable in supporting students with special needs.

Learning with mobile devices in schools is a relatively new experience for mathematics teachers. In the majority of schools with mobile learning implementations, there is a lack of teacher training on how to use the technological devices to teach (viz., Cheon et al. 2012; Crow et al. 2010). Teachers often resort to traditional teaching approaches when using the devices (Hughes 2013). Drill programs can be used as teachers task the students with completing one of many mathematical games available via the Internet or application on the iPads. Students typically remain seated at their desks in the classroom as they complete these games. There are also games that encourage collaboration as two to four players can work towards a common goal on applications such as Operation Math Code Squad (2014).

Companies (e.g., Ventura Educational Systems) have developed applications to mimic traditional virtual manipulatives. For example, the Hands-On Math Geoboard (2015) application that allows the user to put virtual elastic bands on a geoboard and add color to the shape. Figure 18.1 provides a visual example of three virtual elastic bands that have been placed on the virtual board with the bottom shape filled with color. Manipulatives are particularly useful in geometry with shape and measurement as they can provide a visual representation of the concept.

Zbiek et al. (2007) postulate that when selecting manipulatives that three main points need to be considered, which are:

- Mathematical Fidelity: how the mathematical object honors the underlying mathematical properties of that object in the virtual environment.

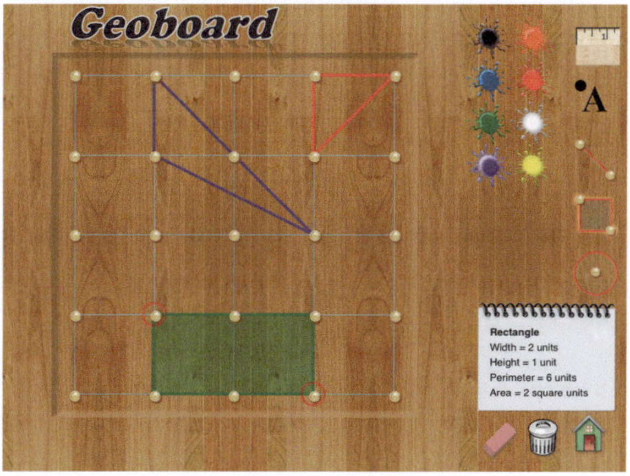

Fig. 18.1 A visual representation of the geoboard with virtual elastic bands (Hands-On Math Geoboard 2015)

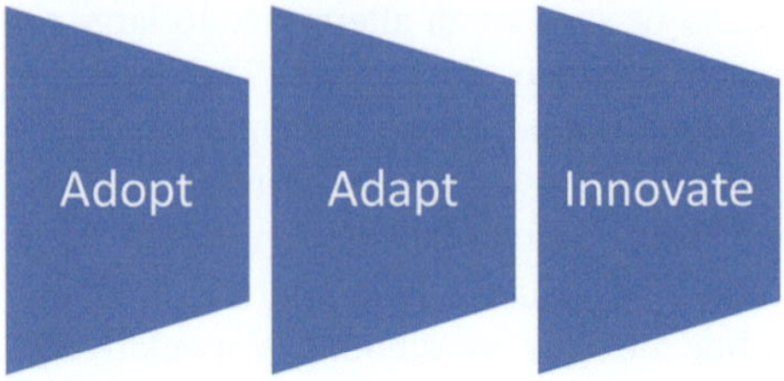

Fig. 18.2 Technology adoption framework

- Cognitive fidelity: how the virtual manipulative reflects the user's cognitive actions and choices.
- Pedagogical fidelity: how teachers and students believe that the virtual manipulative allows students to act in mathematical ways to correspond to the nature of mathematics and underlies a teacher's pedagogy.

Some mathematics applications have students use the mobility of the mobile devices to play educational games. For example, Motion Math (nd) incorporate motion sensed game applications (Motion Math nd). For these games, the student is required to make their answer choices by moving the devices from side-to-side rather than touch selecting a choice. These are different examples of mobile learning, activities that can support geometry learners.

Games and virtual manipulatives are especially helpful for mathematics teachers who are new to using mobile technologies as they can connect to traditional teaching approaches of playing board games and using concrete manipulatives. This *adoption* of familiar methods can initially be useful, but with practice and further confidence mobile learning needs to extended to *adapt* the way that technology is used and how it meets the needs of the particular learners. In this chapter, teachers are encouraged to go one step further and to *innovate* and take learning beyond what has been done in the past and fully use the affordances to push the boundaries of traditional pedagogies (Fig. 18.2).

Mobile learning's obvious innovation to teaching is that the mobile devices can also be used both inside and outside the classroom. Students can connect with the real-world environment while using the technological support of the mobile devices. This type of learning will be described in the following section on context-aware ulearning, then connections are made to how this type of learning is useful in extending and enhancing students' understanding of geometry.

18.3 Context-Aware Ubiquitous Learning

Academics and educators have advocated for centuries for students to learn educational concepts while connecting to real-world concepts (Clairaut 1741/2006; Bartolini-Bussi et al. 2010). Context-aware ulearning can be used to have students

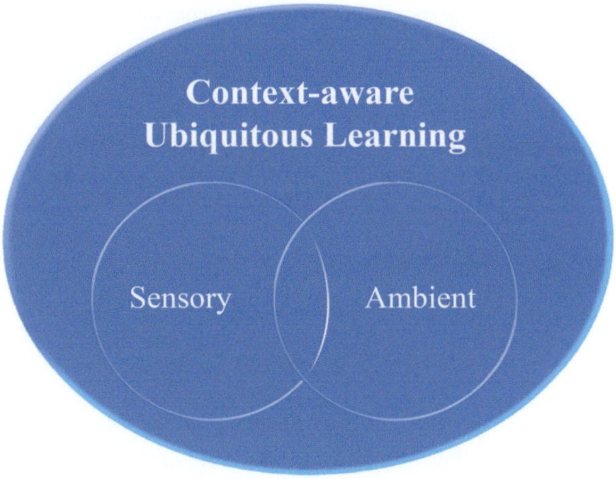

Fig. 18.3 The dichotomy of context-aware ubiquitous learning

making those connections as the students learn while interacting with the real world. Hwang et al. first defined context-aware ulearning in 2008. In this section, an argument is made for a dichotomy in context-aware ulearning to two distinct areas: ambient and sensory. See Fig. 18.3.

Hwang et al. (2008) described context-aware ulearning as:

> The learner's situation or the situation of the real-world environment in which the learner in location can be sensed, implying that the system is able to conduct the learning activities in the real world… context-aware ulearning can actively provide supports and hints to the learners in the right way, in the right place, and at the right time, based on the environmental contexts in the real world. (p. 84)

Hwang et al. went further to provide a list of examples of what could be included. Table 18.1 provides a few of those models and examples.

Many of these activities use cutting edge technologies beyond what is typically available on mobile devices. For example, in the chart above, to *Collect data in the real world* via *sensors* the application needed to *sense* water contaminants would probably require an additional peripheral technology, such as a probe, and an application designed to present that information. In the table above there are four other similar examples that require the device to "*sense*" information. This is one group of context-aware ulearning activities, called '*sensory*'. It is important to mention here that the mobile device has a full range of sensors, but again these are not accessible to all. They have to be unlocked for general use via an application. At the end of 2014, the Sense-it (2014) application was developed by the Open University boasts being the first open application to unlock the full range of sensors on mobile devices. The Sense-it application is only available for Android devices. This could be the start of easier access for educators to enable them to use the senses on mobile devices. However, this program has yet to reach general mainstream knowledge and use and the sensory descriptions of Hwang et al. (2008) require additional peripheral technologies.

Table 18.1 Models and examples of context-aware ulearning activities (The full table can be accessed at Hwang et al. 2008, p. 86.)

Model	Context- aware ubiquitous learning examples
Learning in the real world with online guidance	The students learning in the real world and are guided by the system, based on the real-world data collected by the sensors For example, for the students who takes a chemistry course, hints are provided automatically based on his or her real-world actions during the chemistry procedures
Learning in the real world with online support	The students learn in the real world, and support is automatically provided by the system based on the real-world data collected by the sensors For example, for the student who is learning to identify the types of plants on campus, relevant information concerning the features of each type of plant is provided automatically based on his or her location and the plants around him or her
Collect data in the real world via observations	The students are asked to collect data by observing objects in the real world and to transfer the data to the server via wireless communications For example, observe the plants in this area and transfer the data (including the photos you take and your own descriptions of the features of each plant) to the server
Collect data in the real world via sensors	The students are asked to collect data by sensing objects in the real world, and report what they found. For example, the students find three different samples of water and report a contaminant found by using the sensors
Identification of a real-world object	Students are asked to answer the questions concerning the identification of the real-world objects For example, what is the name of the insects shown by the teacher?
Observations of the learning environment	Students are asked to answer the questions concerning the observation of the learning environment around them For example, observe the school garden, and upload the names of all the insects you find
Cooperative data collecting	A group of students are asked to cooperatively collect data in the real world and discuss their findings with others via mobile devices For example, Cooperatively draw a map of the school by measuring each area and integrate the collected data
Cooperative problem solving	The students are asked to cooperatively solve problems in the real world by discussing through mobile devices For example, search each corner of the school and find the evidence that can be used to determine the degree of air pollution

The other group of context-aware ulearning activities refers to activities that do not need specialist additions to the mobile devices. For example, the activity from the table above titled C*ollect data in the real world* via *observations* could involve the students taking pictures of angles in the real world and uploading them to an online database (such as a Google Form to a Google spreadsheet). Activities like this are focused on the real world but do not require specialist tools and these types of activities are split into a group called '*ambient*' context-aware ulearning activities. Both the sensory and ambient categories of context-aware ulearning are elucidated below.

18.3.1 Sensory

Since 2007 (see Hwang et al. 2007), Hwang and colleagues have published on the topic of context-aware ulearning. Some of those studies, such as those in Table 18.1, connect with sensory and ambient activities, but the primary focus of their work on context-aware ulearning connects with sensory activities. Hwang et al. describe the context-aware ulearning environment as a pedagogy where individual students are guided to learn in a real-world situation with supports or instructions from a computer system or using a mobile device to access the digital content via wireless communications, where the learning system is able to detect and record the learning behaviors of the students in both the real world and the virtual world with the help of the sensor technology (Hwang et al. 2008). The following year, Liu and Hwang (2009) used this description:

> Since the early 2000s, new forms of mobile technology containing additional sensor devices have been providing new directions for technology assisted learning, and this has led to context-aware ulearning, which enables users to interact and learn with sensors and radio frequency identification (RFID) embedded objects in their surroundings. (p. 1)

These provide a clear connection to sensory technologies.

Examples of studies often involve creating systems. For example, Wu et al. (2013) developed an expert system-based context-aware ulearning approach for conducting science learning activities. This system guided students in developing and organizing knowledge for differentiating rocks in the real world. To develop this system the authors used:

> …an inference engine, a knowledge base and a web-based interface. The inference engine and web interface were implemented using Microsoft Visual Studio 2008 and Windows Mobile 6 Professional SDK. The knowledge base was developed with Microsoft SQL Server 2005. Moreover, a C# program was developed to access the RFID tag information from the reader on the Personal Digital Assistant that served as the mobile learning devices in this study (Wu et al. 2013).

This system let the students know if they were correctly identifying target rocks.

In another study using context-aware ulearning, Chen and Huang (2012) developed a context-aware ulearning environment in a museum. In this museum, the aboriginal artifacts were each labeled with an RFID tag and students were provided with mobile devices with RFID reader. As the students walked around the museum the RFID technology assisted in navigating the artifacts. The RFID tags were set up so as the student came within close proximity to the tag relevant details about the artifact were displayed on the mobile device.

The two studies are good examples of how systems were developed that enabled students to interact with the real world while being provided with additional information to support in their learning. The quote from the first study was chosen in particular to emphasis the technological competency required to develop such a system. It is essential that there are experts developing systems like these to provide unique context-aware environments that focus on sensory and RFID technologies.

18.3.2 Ambient

It is unlikely that mainstream class teachers will be able to provide these types of activities for their students. The majority will not have the technology skills, knowledge, or the time to create these types of activities for their students. However, there are other types of activities that teachers can use with their students that fit into the category of context-aware ulearning activities and these have been named ambient learning activities. For an activity to fit with this category it has to be deemed a viable option for educators to be able to use without the need for special technological skills or knowledge above the average K-12 (education for children aged ~4–18 years old) teacher.

One good example of a context-aware ulearning ambient activity is by using quick response (QR) codes. These codes are two-dimensional bar codes that can hold up to 7,089 characters of information. As these square pixelated black and white codes are scanned with a mobile device, they will take the students to website, coordinates on a map, reveal text messages, read an audio message, etc. QR codes are versatile in their use so they can fit with many different strands in mathematics. QR codes are a simple technology to use with only the basic technology skills required for teachers to scan and to make their own codes (See Crompton et al. (2012). Law and So (2010) used QR codes in a math trail activity. As the students walked along the trail, the students stopped at various locations to scan the QR codes and answer the mathematical questions. There are many different augmented reality applications available for teachers that also allow the creation of customized augmented reality activities. Augmented reality activities allow the student to view the world in real time while it is supplemented by the inclusion of a virtual addition (e.g., a video or animated cartoon character).

Crompton (2015) used another type of application to have students learn about angle and angle measurement. In this study, students aged nine to 10 years of age were introduced to angle concepts by exploring angles in the playground. In pairs, students used iPads, Sketchpad Explorer app, and an add-on program called Measure a Picture (Steketee and Crompton 2012). Within the app, the students took photographs of angles they found in the playground and were able to use the dynamic protractor to measure the angles. Elisson and Ramberg (2012) had the students involved in two context-aware ulearning activities. The first activity had students learn about the concept of volume as they played the role of architects in planning for new buildings. In the second activity, students had to relocate imaginary species of animals from a local zoo as they learned about area. Students used a measurement application for both these activities.

18.4 Geometry: Measure of the Earth

The study of geometry requires the complex understanding of properties, relationships, and transformations. Furthermore, students have to navigate geometry's interconnected network of concepts and representational systems. Mathematics is typically characterized by didactic instruction that requires rote memorization (Williams-Carlin 2009; Stigler and Hiebert 1997). This calls into question the epistemological nature of geometry as empirical findings (e.g., Prescott et al. 2002; Ubuz and Üstün 2004) indicate that this is not an effective way to learn geometry.

The Greek etymology of geometry is the measure of earth or land. Centuries ago, early scholars, (viz., Comenius 1657/1986) advocated for students to learn mathematics by connecting to real-world artifacts (environments and objects). Many since (viz., Bartolini-Bussi et al. 2010; Gainsburg 2008; Hiebert and Carpenter 1992; NCTM 2000; National Research Council 1990) have echoed that necessity to connect to the real world. Context-aware ulearning is a methodology to connect students with portable technological supports while connecting mathematical concepts to physical manifestations of those concepts in the real world.

To make a concrete connection between the study of geometry and context-aware ulearning, a set of five recommendations are presented.

- Focus on the geometry standard first and the technological tools second to that.
- Connect geometry to real-world occurrences or real-world authentic applications (e.g., architect).
- Work in local environments that students recognize.
- Connect the contextualized concepts with decontextualized versions in the classroom.
- Avoid cognitive overload by introducing new mathematical concepts and technologies at the same time.

18.5 The Future

The future of context-aware ulearning could involve wearable technologies. This type of technology would fit under the category of mobile learning as it is highly portable. Wearable technologies are on the rise, with technologies, such as the Jawbone or Fitbit, that monitor our activities, sleep, and heart rate by simply wearing a band around the wrist. While these provide numerical values that can be used to explore mathematical topics, e.g., data and algebra, the affordance of these are less obvious for geometry. Google Glass (version 1) would have been a better fit as this could provide a way for students to collect real-world data. Perhaps future versions may allow other functionalities that could use a form of augmented reality to measure angles, distance, and volume.

18.6 Conclusion

The majority of mathematics teachers have probably heard a student say "Why do I need to know this, when will I ever need it?" For centuries, scholars have advocated for the study of mathematics to connect with the real world. This will provide meaning for students as they learn to understand that mathematics is not just something that happens in the classroom, but it is all around us. The use of context-aware ulearning will connect students to that real world while providing technological supports to enhance that learning. There are many context-aware ulearning sensory activities that may require computer science experts and application developers, and there are also context-aware ulearning ambient activities that can be used by classroom teachers to extend and enhance students understanding in geometry.

References

Bartolini-Bussi, M. G., Taimina, D., & Isoda, M. (2010). Concrete models and dynamic instruments as early technology tools in classrooms at the dawn of ICMI: from Felix Klein to present application in mathematics classrooms in different parts of the world. *ZDM, 42*, 19–31.

Caudill, J. G. (2007). The growth of m-learning and the growth of mobile computing: parallel developments. *International Review of Research in Open and Distance Learning, 8*(2), 1–13.

Chen, C. C., & Huang, T. C. (2012). Learning in a u-Museum: Developing a context-aware ubiquitous learning environment. *Computers & Education, 59*, 873–883.

Cheon, J., Lee, S., Crooks, S. M., & Song, J. (2012). An investigation of mobile learning readiness in higher education based on the theory of planned behavior. *Computers & Education, 59*, 1054–1064.

Clairaut, A. C. (2006). *Elements de geometrie*. Paris: Editions Jacques Gabay (Original work published 1741).

Comenius, I. A. (1986). *Didactica Magna*. Akal Ediciones (Original work published 1657).

Crompton, H. (2013). A historical overview of mobile learning: Toward learner-centered education. In Z. L. Berge & L. Y. Muilenburg (Eds.), *Handbook of mobile learning* (pp. 3–14). Florence, KY: Routledge.

Crompton, H. (2015). Understanding angle and angle measure: A design-based research study using context-aware ubiquitous learning. *International Journal for Technology in Mathematics Education, 22*(1).

Crompton, H., LaFrance, J., & van't Hooft, M. (2012). QR Codes 101. *ISTE Learning and Leading with Technology., 39*(8), 22–25.

Crow, R., Santos, I. M., LeBaron, J., McFadden, A. T., & Osborne, C. F. (2010). Switching gears: Moving from e-learning to m-learning. *Journal of Online Learning and Teaching, 6*(1), 268–278.

Elisson, J., & Ramberg, R. (2012). Design guidelines for location-based and contextual learning supported by mobile devices. *International Journal of Handheld Computing Research, 3*(2), 26–43.

Gainsburg, J. (2008). Real-world connections in secondary mathematics teaching. *Journal of Mathematics Teacher Education, 11*(3), 199–219.

Hands-On Math Geoboard. (2015). *Ventura*. Retrieved from https://itunes.apple.com/us/app/hands-on-math-geoboard/id493388133?ls=1&mt=8.

Hiebert, J., & Carpenter, T. P. (1992). Learning and teaching mathematics with understanding. In D. A. Grouws (Ed.), *Handbook of research on mathematics teaching and learning* (pp. 65–97). New York: Macmillan.

Hughes, J. (2013). The role of teacher knowledge and learning experiences in forming technology-integrated pedagogy. *Journal of Technology and Teacher Education, 21*(4) 277–302. Retrieved December 29, 2014 from http://www.editlib.org/p/4622.

Hwang, G. J., Wu, T. T., & Chen, Y. J. (2007). Ubiquitous computing technologies in education. *Journal of Distance Education Technologies, 5*(4), 1–4.

Hwang, G., Tsai, C., & Yang, S. J. H. (2008). Criteria, strategies and research issues of context-aware ubiquitous learning. *Educational Technology & Society, 11*(2), 81–91.

Laouris, Y., & Eteokleous, N. (2005, October 25–28). *We need an educationally relevant definition of mobile learning*. Paper Presented at the 4th World Conference on mLearning, Cape Town, South Africa.

Law, C., & So, S. (2010). QR codes in education. *Journal of Educational Technology Development and Exchange, 3*(1), 85–100.

Liu, G. Z., & Hwang, G. J. (2009). A key step to understanding paradigm shifts in e-learning: Towards context-aware ubiquitous learning. *Research Express, 10*(5), 1–6.

Lonsdale, P., Baber, C., Sharples, M., & Arvanitis, T. N. (2004). A context-awareness architecture for facilitating mobile learning. In J. Attewell & C. Savill-Smith (Eds.), *Learning with mobile devices: Research and development* (pp. 79–86). London: Learning and Skills Development Agency.

Motion Math, (nd). [iOS]. Retrieved from http://motionmathgames.com/.

National Council of Teachers of Mathematics (NCTM). (2000). *Principles and standards for school mathematics*. Reston, VA: National Council of Teachers of Mathematics.

National Research Council (NRC). (1990). *Reshaping school mathematics: A philosophy and framework for curriculum*. Washington, DC: National Academy Press.

Operation Math Code Squad. (2014). *Android and iOS application*. Retrieved from https://itunes.apple.com/us/app/operation-math-code-squad/id555750694?mt=8.

Prescott, A., Mitchelmore, M., & White, P. (2002). Students' difficulties in abstracting angle concepts from physical activities with concrete material. In K. C. Irwin. B. Barton, M. Pfannkuch, & M. O. Thomas (Eds.), *Proceedings of the 25th Annual Conference of the Mathematics Education Research Group of Australasia* (pp. 583–591). Sydney, Australia: MERGA.

Sense-it. (2014). *nQuire: Young Citizen Inquiry, Open University*. Retrieved from https://play.google.com/store/apps/details?id=org.greengin.sciencetoolkit.

Sharples, M. Taylor., J., & Vavoula, G. (2005). *Towards a theory of mobile learning*. Proceedings Presented at the mLearn, Cape Town

Soloway, E., Norris, C., Curtis, M., Jansen, R., Krajcik, J., Marx, R., et al. (2001). Making palm-sized computers the PC of choice for k-12. *Learning and Leading with Technology, 28* (7), 32–57.

Steketee, S., & Crompton, H. (2012, April 13). *Measure a Picture*. An add-on program for SketchPad Explorer The Geometer's Sketchpad Sketch Exchange. Retrieved from http://sketchexchange.keypress.com/browse/topic/all-topics/by-recent/1/448/measure-a-picture.

Stigler, J. W., & Hiebert, J. (1997). Understanding and improving classroom mathematics instruction. *Phi Delta Kappan, 79*, 14–21.

Traxler, J. (2009a). The evolution of mobile learning. In Retta Guy (Ed.), *The evolution of mobile teaching and learning* (pp. 1–14). Santa Rosa, CA: Informing Science Press.

Traxler, J. (2009b). Learning in a mobile age. *International Journal of Mobile and Blended Learning, 1*(1), 1–12.

Ubuz, B., & Üstün, I. (2004). Figural and conceptual aspects in defining and identifying polygons. *Eurasian Journal of Educational Research, 16*, 15–26.

Williams-Carlin, R. (2009). A comparative study of geometry curricula (Doctoral dissertation, Louisiana State University, 2009). *Dissertation Abstracts* 1–82.

Wu, P. H., Hwang, G. J., & Tsai, W. H. (2013). An expert system-based context-aware ubiquitous learning approach for conducting science learning activities. *Educational Technology & Society, 16*(4), 217–230.
Zbiek, R. M., Heid, M. K., Blume, G. W., & Dick, T. P. (2007). Research on technology in mathematics education: A perspective of constructs. In F. K. Lester Jr (Ed.), *Second handbook of research on mathematics in teaching and learning* (pp. 1169–1208). Reston, VA: NCTM.

Author Biography

Helen Crompton (Ph.D.) is an Assistant Professor of Instructional Technology at Old Dominion University, Virginia. Professor Crompton is a keen researcher and educator in the field of instructional technology. She gained her Ph.D. in educational technology and mathematics education from the University of North Carolina at Chapel Hill. Her research is focused on mobile learning and the effective integration of technology into K-12 education. Dr. Crompton works as a consultant for two United Nations Agencies (United Nations, Educational, Scientific, and Cultural Organization: UNESCO and International Telecommunication Union: ITU) to research, author and edit publications summarizing research on mobile learning. Dr. Crompton is also a faculty member for the International Society for Technology in Education (ISTE), teaching the ISTE Standards academy, consulting, and recently designing ISTE's self –paced Mobile Learning Academy and Verizon's Mobile Learning Academy.

Chapter 19
Three Phases of Mobile Learning State of the Art and Case of Mobile Help Seeking Tool for the Health Care Sector

John Cook and Patricia Santos

Abstract The Internet mobile device enabled social networks of today stand accused of being so-called 'weapons of mass distraction' or worse. However, we point out that modern fears about the dangers of social networking are overdone. The paper goes on to present three phases of mobile learning state of the art that articulate what is possible now and in the near future for mobile learning. The Learning Layers project is used to provide a case of barriers and possibilities for mobile learning; we report on extensive initial co-design work and significant barriers with respect to the design of a mobile Help Seeking tool for the Healthcare sector (UK). We then provide an account of how the Help Seeking tool is being linked to a Social Semantic Server and report on a follow-up empirical co-design study.

19.1 Introduction

The current context is that rarely does a day go by without dire warnings and overt action to either ban mobile devices and access to social networks from the workplace or school, or for monitoring of some description to be put in place to 'police' behaviour. Put simply, social networks and mobile devices stand accused of being so-called 'weapons of mass distraction' or worse. For example, we have the following suspect claim (Infographic 2012): "Social Media Distractions Cost U.S. Economy $650 Billion". Indeed, in schools we have this recent example of 'policing' (CBSlocal 2013): "Glendale Unified School District in California is paying $40,500 to Geo Listening to collect and analyze all social media public posts of 13,000 students … even if it was done off campus". However, a McKinsey

J. Cook (✉) · P. Santos
University of the West England, Bristol, UK
e-mail: John2.Cook@uwe.ac.uk

P. Santos
e-mail: Patricia.Santosrodriguez@uwe.ac.uk

D. Churchill et al. (eds.), *Mobile Learning Design*, Lecture Notes in Educational Technology, DOI 10.1007/978-981-10-0027-0_19

Global Institute report (Chui et al. 2012) claims that social networking within companies could increase the productivity of 'knowledge workers' by 20–25 %. Modern fears about the dangers of social networking and the use of mobile devices for learning are overdone.

The first part of this chapter provides a three-phase overview of mobile learning state of the art, which incorporates a view of the emerging technologies and their pedagogical affordances, attendant barriers and how to overcome them, and which highlights emerging trends in mobile learning. The Learning Layers project is then described using the frame of design research to illustrate what is possible in social networks and mobile device-mediated learning. We report extensive initial co-design work and significant barriers with respect to a mobile Help Seeking tool for the Healthcare sector. The chapter then provides an account of how the Help Seeking tool is being linked to a Social Semantic Server and we briefly report on a follow-up empirical co-design study. We conclude by highlighting associated challenges.

19.2 Three Phases of Mobile Learning State of the Art

In this section we examine state-of-the-art mobile technologies and their pedagogical affordances, including barriers, how to overcome them, and emerging trends in mobile learning (m-learning). In order to achieve this we propose the following new three phases, which are discussed individually below but which never-the-less overlap:

- Focus on new patterns of connected social learning and work-based practices.
- Focus on designing for 'm-learning' at scale.
- Focus on the boundaries of learning that the 'm' in m-learning forces us to explore.

Each phase or focus has a key argument associated with it, which are as follows:

- A key-evolving pedagogical affordance of mobile devices is the ability to use social media and apps to enable new patterns of connected social, learning and work-based practices.
- Design research allows us to engage in inquiry surrounding the transformative possibilities for m-learning. Particularly, designing for 'm-learning' at scale is a big challenge.
- Participants in new mass communications are now actively engaged in generating their own content and contexts for learning. User/Learner-generated context for us (Cook et al. 2012) is conceived in such a way that users of mobile digital devices are being 'afforded' synergies of knowledge distributed across people, communities, locations, time (life course), social contexts, sites of practice (such as sociocultural milieus) and structures. Of particular significance for us is the way in which mobile digital devices are mediating access to

external representations of knowledge in a manner that provides (equity of) access to cultural resources.

With respect to the first state of the art (focus on new patterns of connected social learning and work-based practices), we argue that a key evolving pedagogical affordance of mobile devices is the ability to use social media and apps to enable new patterns of connected social learning and work-based practices. A review by Cochrane (2013) notes the dearth of research into social m-learning. Indeed, there is still a focus on content whereas a key theme for the future of m-learning may be augmentation. For example, FitzGerald et al. (2013) have suggested a useful taxonomy of Augmented Reality used in m-learning projects; this taxonomy has six dimensions: (i) device or technology used, (ii) mode of interaction, (iii) method of sensory feedback to the user, (iv) personal or shared experience, (v) fixed/static or portable experience, and (vi) learning activities or outcomes. Augmented Context for Development work by Cook (2010) was found to be useful both to Cochrane and FitzGerald et al. because it provides an example of a m-learning project that focussed upon augmenting the learners experience in the field by the provision of a Vygotskian environment for personal and collaborative meaning-making (we return to this below). Indeed, this and other projects point the way for Augmented Reality to be used for situated and constructivist learning, particular where collaboration and student inquiry form key aspects. However, as we point out in the introduction, social networks like Facebook and Twitter stand accused of being so-called 'weapons of mass distraction", diverting workers and students away from their 'real' work. Schools are cautious—anxious, maybe about new digital technologies and media. In Introduction, we provided key examples to counter these fears and concerns. Furthermore, recent work has examined how social media and personally owned mobile devices could be used as a means of providing a bridge from media use in everyday life to the expectations of school and higher education potentially has enormous attraction. Some research (see Cook et al. 2012 for a review) suggests that in Higher Education Facebook, for example, provides affordances that can help reduce barriers that students with lower self-esteem might experience in forming the kinds of large, heterogeneous networks that are sources of social capital. 'Trust' is a key issue in this respect. Thus, there appears to be considerable potential for mobile devices and social networks in terms of sustainability in the integration of informal and formal institutional dimensions of learning. However, although a new educational paradigm is emerging, there exists a need for more debate and further research, particularly around notions of sustainability, scalability and equity of access to opportunities to build social capital. We also need to examine how mobile devices are reconfiguring the relationships between spaces, between public spaces and private ones, and the ways in which these are penetrated by mobile virtual spaces. For example, we need to examine further the importance of the human body as the key interface in the 'interpenetration of physical and virtual spaces'.

With respect to the second state of the art (focus on designing for 'm-learning' at scale), we argue that design research (we define this below) allows us to engage in inquiry surrounding the transformative possibilities for m-learning; and particularly, this is a big challenge. Carmean et al. (2013) point out that 'features'-oriented affordances of mobile devices are not enough as a way of characterising m-learning. They propose that we need to examine the deeper affordance of mobile devices, particularly the immediacy and the connection natively built into such devices. Indeed, they propose that if we are to understand the potential for new learning experiences and support that new mobile devices afford, then we need to examine mobility + design. We fully agree with Carmean et al.'s well-argued assertion that design research allows us to bring out 'never-seen before possibilities'. However, in terms of scaling up m-learning, a key research theme, Cochrane (2013) points out that several larger m-learning projects have tended to focus on specific groups of learners, rather than developing pedagogical strategies for the integration of m-learning within tertiary education in general. It appears that the situation in m-learning research can be typified as being typically content centric, a focus on the device. We agree with Carmean et al.'s assertion that designing for m-learning can help unlock the web of individualised choices that are available by encouraging us to design for access to small chunks, and to make these customizable to individual's needs, experience and agency. Specifically, design research allows us to engage in inquiry surrounding the transformative possibilities for m-learning. Designing for 'm-learning' at scale, beyond pilots and content-centric approach, is a big challenge. To help elucidate the issues, below we describe in some detail work by the authors on an investigation of a design research approach (The Help Seeking tool). With near global ownership of mobile devices imminent, the more technical concerns surrounding 'cross-platform mobile development' becomes crucial. To summarise, the current state of cross-platform development techniques falls into two broad approaches: cross-compilation ("native" apps) and mobile web applications ("web apps") also referred to as the 'responsive web'. A trend which could significantly help scale learning is Learning Analytics (LA), this is one of the promising techniques that has been developed in recent times to effectively utilise the astonishing volume of student data available in formal education. Finally, we note that while some describe MOOCs (massively open online courses) as a fad possessing poor quality, lack of student engagement, lack of business model, and high dropout rates, others think that the MOOCs will soon become the de facto way to remediate and educate a broad swath of students in a wide variety of content areas, i.e. to assist in scaling. De Waard's (2013) work on MobiMOOC provides an example of how a mobileMOOC or mMOOC's would work. What is encouraging about de Waard's exposition of the affordances of the mMOOC is that she is clear on the reasons for the mobile additions from a reflexive pedagogical perspective.

With respect to the third state of the art (focus on the boundaries of learning that the 'm' in m-learning forces us to explore), we argue that participants in new mass communications are now actively engaged in generating their own content and contexts for learning. Indeed, for us mass media are witnessing a paradigm shift in which the 'user' can generate their own content with a mobile phone or another digital device. Thus these activities at the boundaries of learning are made up by a lot of individuals publishing user-generated content (in the form of videos that users have produced themselves or digital media that have been copied from some other source, the latter may be subject to copyright restrictions). Consequently, we suggest that what we are seeing is the emergence of 'user-generated contexts'. Not only do the twenty-first-century structures of mass communication provide a wide range of augmentations to communication but in addition, through the agency of users, the context within which communication takes place is augmented by users to suit the needs of the individual and/or the conversational community; above, we explored in detail to themes that expand on these trends:

- Bring your own device (BYOD)
- Mobile games
- Learner/User-Generated Contexts.

Examples of one-to-one programmes in US (post-compulsory) and UK (school) illustrate that there is a rising opportunity for the use of personal technologies for formal learning as well as learning in informal situations, e.g. work-related activities. Clark and Luckin (2013) conclude that the research on iPad use in the schools context report "increased motivation, enthusiasm, interest, engagement, independence and self-regulation, creativity and improved productivity". BYOD refers to the policy of permitting learners or employees to bring personally owned mobile devices (laptops, tablets, and smart phones) to their educational establishment or workplace. Although BYOD has many potential advantages, before adopting BYOD in a formal learning context there are many things to consider as these challenges or requirements will need to be met in order for you to reap the benefits, as work reported by JISC above illustrates. A case study by LaMaster and Ferries-Rowe (2013) describes the implementation of mobile technology at Brebeuf Jesuit Preparatory School using a BYOD approach; they describe a strategic plan that resulted from student, faculty and user consultation, which can be used as a template for other institutions. Also, work by Deloitte (2013) illustrates that as the buzz around BYOD dies down to some extent, Bring Your Own Application (BYOA) is likely to step to the forefront of debate. The proliferation of cloud-based applications means that BYOA does not even require installation, and so could be near-impossible to restrict.

We can expect to see widespread adoptions of games in education; this is another area that a growing number of researchers are investigating in terms of the boundaries of learning. For example, Benford et al. (2004) have evaluated systems for location-based multi-player games, seeking to understand how in situ users share location information at a distance through comparing self-reporting and GPS readings. However, very often games require that the learner maintains the high level of engagement and arousal within the game scenario. Given this, a question that arises is: can a game approach be applied to all scenarios of learning?

Finally, we draw on the phenomena (at the boundaries of learning) whereby that society is currently witnessing a significant shift away from traditional forms of mass communication and an editorial push towards user-generated content and augmented communication contexts to explore different expositions of learner-generated contexts. Of particular significance for us is the way in which mobile digital devices are mediating access to external representations of knowledge in a manner that provides access to cultural resources.

In the next section, we elaborate on design research and the application of Vygotskian theories,, in order to set up a design case study to support informal learning practices, in particular help seeking actions, in a Healthcare workplace environment. The case study is framed into a European research project called Learning Layers.

19.3 Design Research in the Learning Layers Project

19.3.1 Design Research

Design research allows us to engage in inquiry surrounding the transformative possibilities for learning technologies. In the Learning Layers Project (described below), we develop technologies that support informal learning in the workplace (Healthcare professionals in NE England and the Construction sector in North Germany). Co-design is being used with all user groups to help shape our designs and tools and to understand the context. Design research aims to have impact on real-world problems whilst providing a frame for inquiry that is rigorous and yet experimental; it has recently been characterised as follows:

> … a genre of research in which the iterative development of solutions to practical and complex educational problems also provides the context for empirical investigation, which yields theoretical understanding that can inform the work of others … [although potentially powerful] the simultaneous pursuit of theory building and practical innovation is extremely ambitious (McKenney and Reeves 2012).

19.3.2 Vygotsky

Society experienced technologically and socially driven transformations during the industrialisation of the first third of the twentieth century; it was against this background that Lev Vygotsky defined the characteristics of human development as a development which is based on the instrumental conditioning of reflexes or as the extension of the body by tools for mastering nature (Vygotsky 1930/1978, p. 19). The "higher psychological processes", as Vygotsky termed them, result from a relation "between human beings and their environment, both physical and social" (p. 19). Vygotsky considered "social interactions" to be those like 'to speak' as the transformation of practical activities such as 'to use a tool'. The leading processes are that of internalisation and that of the instrumental use of a tool; this happens where "An operation that initially represents an external activity is reconstructed and begins to occur internally" (Vygotsky 1930/1978, pp. 56–57). Vygotsky then went on to propose the Zone of Proximal Development (ZPD). This was a significant paradigm shift, because up until that point a child's mental development had been assumed to be indicated only by those things that children could achieve on their own, whereas Vygotsky took the view that "what children can do with the assistance of others ['the more capable peers'] might be in some sense even more indicative of their mental development than what they can do alone" (Vygotsky 1930/1978, p. 85). Furthermore, development in a ZPD has a forward looking, temporal and prospective dimension. Indeed, in addition to reorganising the visual-spatial field (a "centre of gravity" of current attention) Vygotsky proposed that "the child, with the help of speech, creates a time field … he can act in the present from the viewpoint of the future" (Vygotsky 1930/1978, pp. 35–36).

Recently, Cook (2010) has extended some of Vygotsky's concepts to adult learners (MA Landscape Studies, University of Sheffield, UK) to explain the way they collaborate using mediating tools (mobile phones, Augmented Reality, language). This work provides a description of the components of a 'context' that emerges at run-time (i.e. when learners engage with a task/activity using tools like mobile devices and language), whereby context is conceived as "a core construct that enables collaborative, location-based, mobile device mediated problem solving where learners generate their own 'temporal context for development' within the wider frame of Augmented Contexts for Development (ACD)" (Cook 2010). We firmly believe that tracing the links between multiple 'temporal context for development' is a key to understanding cross-contextual learning and meaning-making (this is a core notion in our proposal for the design of innovative 'recommendation services', more details described below).

The ACD appears to act as part of the substitute for what Vygotsky calls 'the more capable peer'. As Cook (2010) states, mobile devices can be used as mediators in an ACD using them as the more capable peer that is able to guide and scaffold the learners to find the solutions. The main elements to develop the ACD are: (a) the physical environment, (b) a pedagogical plan, (c) tools/devices for an augmented oriented approach, (d) learner co-constructed 'temporal context for development, and (e) collaborative learners' interpersonal interactions using tools (which overlaps with (d)).

19.3.3 Learning Layers

The context for our recent work is Learning Layers (http://learning-layers.eu/), a large European Commission co-funded project (FP7 IP) which investigates scaling the use of Technology Enhanced Learning (TEL) in workplace informal learning where users have previously been reluctant to use TEL for learning (Healthcare is the focus of this study). The consortium consists of 17 institutions from seven different countries.

19.4 Help Seeking Tool

The focus of the remainder of this chapter is from the perspective of Learning Layers work package 2 (WP2), one of six R&D work packages in the project. Figure 19.1 shows how we organise the project. All three interaction layers (i.e. WP2-4) draw on a common Social Semantic Layer (WP5) that aims to ensure that informal learning is embedded in a meaningful context.

WP2 is concerned with the 'Networked Scaffolding—Interacting with People', developing technology support for current working practices of an individual so that it is persistent over multiple work/organisational contexts and so that it extends into larger networks of people. We adopt as a basis for our work the term "scaffolding", which draws on Vygotsky's ZPD but which can be attributed to Wood et al. (1976) who described it as a "process that enables a child or a novice to solve a problem, carry out a task, or achieve a goal which would be beyond his unassisted efforts" (p. 90). Our Networked Scaffolding idea proposes on the one hand a low-barrier approach that collects and semantically analyse Q/A typically asked in practice, we create a scaffolding resource of 'solved questions' and 'similar and more capable peers' that workers have asked about a concept or problem in a particular learning context. On the other hand, from the analysis of this actions we identify learning patterns to provide support to novel informal learning challenges associated to the practice of exchanging Q/A.

This work contributes to the building of recommendations services/algorithms that are being realised with Social Semantic Server (SSS) technology from other WP5 (described below).

19.4.1 Overview of WP2 'Networked Scaffolding—Interacting with People'

In Learning Layers WP2, we have focused our design research over the last 2 years (2013 and 2014) on the study and understanding of Help Seeking in the Healthcare

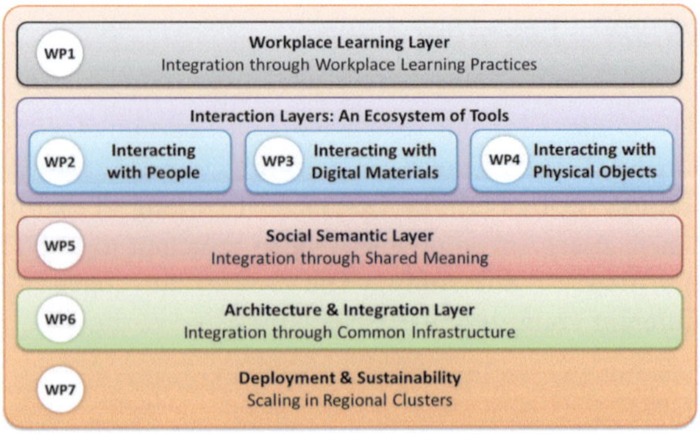

Fig. 19.1 Organisation of Learning Layers

sector (NE, England). The Help Seeking design and development team emerged from the Layers Open Design conference in February 2013 and has subsequently engaged in extensive and iterative design refinement of ideas. The co-design approach has been selected as the most suitable, because it is necessary to identify the user needs and problems, particularly because our context is one where staffs in the Healthcare sector are not confident about the use of technologies in their work practice (we elaborate on this point below).

Results derived from the analysis of the initial co-design activities in the Healthcare sector have confirmed that 'putting guidelines and pathways into practice', in particular national health guidelines, can be a problem (it represents a systemic pain point) and that as such it important to support conversations and discussions about the implementation of guidelines locally. In this context, we claim that there will be conversations over time in which these additions to the local implantation of guidelines will evolve. Our hypothesis is that these conversations will take place within Personal Learning Networks or PLN (Cook and Pachler 2012; this chapter includes a literature review of work-based practice, tagging and 'trust') and in a more organisational level through Shared Learning Networks (SLN). APLN is a group of people organised by an individual and formed by her trusted colleagues. On the other hand, a SLN is a network which contains everybody registered on the learning system.

These networks play a key role, and therefore we take the view that the development of those networks, as well as the associated help seeking of opinions in such networks, requires scaffolding. The outcomes of these conversations will feed into the local implementation of national guidelines. Therefore, the Help Seeking prototype provides peer support for workplace decision-making and problem-solving (i.e. informal learning) by scaffolding: (1) the 'building, maintaining and activating' of a trusted Personal Learning Network, and (2) the movement from using a trusted PLN to SLN. Our aim with the Help Seeking prototype is twofold: supply computer support for a range of workers in the UK's Healthcare sector to assist them in identifying (i) relevant more and trusted capable peer(s) from and with whom to learn (informal learning) and, (ii) trusted resources from their PLN. These aims are partially realised by the use of trusted recommender systems to support informal learning at work; they are typically used to build trusted networks and resolve the information overload problem, people are overwhelmed with information and have difficulties in finding the right piece of information or right person in such a space.

The Help Seeking prototype envisaged usage (i.e. a use case, see Santos et al. 2014a, for details) is as follows: a nurse uses the app to seek support in the course of her/his activities: (1) asks a question by typing a question; (2) annotates the type of problem by creating tags or selecting existing tags (from a data base of suggested problem types); (3) selects from her group of trusted colleagues (from data in her PLN) to whom the question should be circulated to. Automatically related national guidelines, peers, meeting notes and questions are recommended for her, this information is suggested by the sematic analysis of the question and corresponding tags using the Social Semantic Server or SSS (Kowald et al. 2013; Seitinger et al.

2013). The nurse checks the information and authorship of the resources and may choose to add a new person to her PLN as appropriate, adding tags to relate specific knowledge to this person. After some minutes, some colleagues provide short answer to her question.

19.4.2 Early Co-design of the Help Seeking Tool

In order to redefine our initial user stories, wireframes and various interactive prototypes, we have engaged in discuss with users in initial co-design sessions over a period of about 13 months (November 2012–December 2013).

The user stories on which our design ideas were initially based were based on empirical work done by WP1 and refined during Application Partner Days in February 2013. A specific user story was combined with findings from the Application Partner days in February 2013, to present an initial use case of a GP (General Practitioner) looking for some help with cascading national guidelines. This original user story was used to develop the first storyboards, and also fed into the designs made for the (internal) Design Conference in Helsinki in March 2013. Following the Design Conference, the iterative process of showing the designs, working through them with users, refining and reworking, then re-representing to users took place over a series of months between April and September 2013. In April 2013, having been shown the first iteration of the design idea for the Help Seeking tool, in which the example of sharing national guidelines was used, valuable feedback was noted from Healthcare staff at a specific practice. This meeting in the co-design process highlighted *uneasiness with technology, issues with trust and a reluctance to use anything like social networks*. However, it also showed that internal technology solutions, such as the intranet were being used, and although there was a reluctance to use smartphones, the issues around sharing and finding information and time constraints were clearly identified. In order to give users a clearer idea of how the technology might work in practice, a clickable, in-device wireframe was developed using Balsamiq (a rapid prototyping tool). This was tested in two selected Practices. In brief, the search for a scenario which would be useful to all the individuals across Practice A and Practice B, with its range of personnel and its different contexts, proved challenging. The final solution at the end of year 1 concentrated on a use case (given above) which describes assisting a user to develop a network of contacts which would be useful in a range of work and learning scenarios (i.e. the PLN).

As the above represents a first cut design decision, we proposed that we should use mobile devices to support the collaborative Help Seeking; this support is needed due to the lack of time and mobility issues of staff (e.g. nurses can sometimes work in different locations during the same day). A Proof of Concept (Fig. 19.2, a simulation of an Android app which is interactive and simulates certain functionalities) was constructed and demonstrated/trailed with Practices A and B.

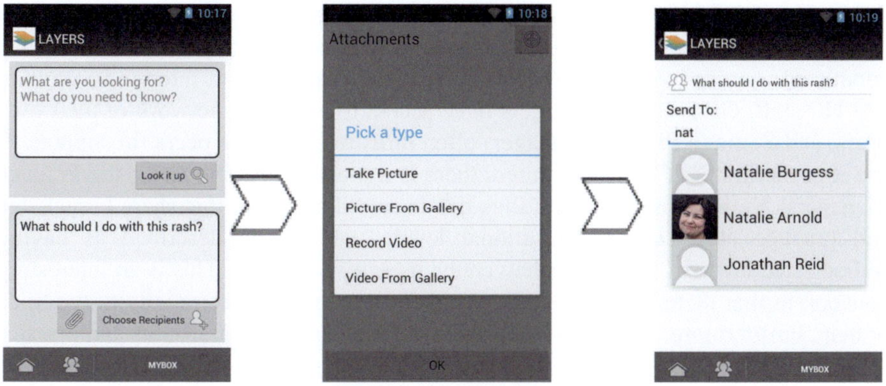

Fig. 19.2 Android proof of concept

The app allowed users to enter a question, add an attachment (image, video) and choose from a list (pre-populated for this Proof of Concept) which contacts they would send the question to. The reaction to Help Seeking using a mobile tool was mixed: some users are currently only comfortable with using desktop applications, some avoid technology, and others are very confident users. As mentioned above, since the project will evolve over a number of years, it seems essential to design not only for the current moment but also for a future in which those users who are not familiar become more confident with technology. After further co-design sessions, the design idea of the Help Seeking tool was refined. The addition of tagging of people, questions and documents in order to build a trusted network (PLN) also accords with the advances made by other technical partners on the Social Semantic Server, so that it is envisaged that the SSS will be able to analyse and recommend tags, useful contacts (similar and more capable peers), answers to questions and documents which relate to problem. Contacts will be sourced from a wider SLN or beyond via questions or key words and added to the PLN as appropriate. Tags allow the recording of other important details (e.g. specialisms, place of work). Ideally, the tool will clip data from received sources (e.g. email) and store in relation to contacts (but this is currently out of scope for technical and security reasons).

19.4.3 Help Seeking: Follow-up Empirical Co-design Study

Our initial co-design work, described above, highlights that a big problem is reticence in the Healthcare sector to use social and mobile media in workplace practice. However, as we also explained above, we believe that modern fears about the dangers of social networking are overdone. Consequently, we predicted that when we introduce a cut-down version of the Help Seeking tool to users, in a series of empirical co-design sessions, they would realise there is nothing to fear and will themselves ask for the full functionality of the Help Seeking tool to be re-added (this has been our experience on other projects). Recently in year 2014, we have conducted a cross-case analysis (Holley et al. 2014) based on interviews done to real HC staff. GP Practice Managers have access to an online network of their own peers, and Sonia (a Practice Manager) often turns to her online peers for support: “I use it in the first line”. Indeed, unless there is a practice specific issue, the Practice Managers’ Network is consulted; thus this is an online group that share knowledge and practice at a cross-organisational level, and can be described as having self-selected areas of expertise (they create their own profile). This is an interesting concept, in that there seems to be the need for mutuality or reciprocity to the trade of help. Furthermore, Sonia acknowledges that the healthcare assistants and practice nurses in the Practices lack their own peer mentoring groups of this type; and she goes on to suggest that this is in fact a barrier to learning (this points to need for Help Seeking tool) in that she has to act as a filter point for practice nurse updating. She is uneasy in this role as she is pressurised. Sonia then goes on to relay the sets of educational events, national and local guidelines that she has to deal with on a regular basis.

Two in-depth co-design workshops were conducted over the period October 2013 to October 2014, for details see Santos et al. (2014b). The research involved tool use, pre-post workshop questionnaires and workshop observations. By October 2013, we had identified three already existing healthcare networks in the north of England that were regarded as worthy of deeper inquiry. These are a well-established Practice Manager’s Network, a new network of Nurses and a group of Data Quality Leads members who want to form a network. In the first workshop, an existing Professional Networking tool (LinkedIn) was used to discuss with Healthcare staff the benefits and limitations of social tools. Concurrently, results from the initial co-design sessions (described above) were used to develop the first prototype version of the ‘Help Seeking’ tool (a WordPress based following a responsive design so it can be run on mobile devices like phone and tablets and on desktops). This version ‘beta 1’ was evaluated during the second workshop. Results from the first and second workshops were used to identify a design criterion relevant for our Help Seeking tool.

There was strong support for using tags to find relevant discussions. Our future plans include to support and facilitate the use of tags providing scaffolding mechanisms when (1) composing questions and as an alert to similar problems; and (2) when searching, finding existing groups and filtering information. For example, in order to facilitate the searching and filtering of information the use of tags seems to be generally accepted as a good solution, particularly with positive finding with respect to using tags to find a relevant discussion or groups (see above). Indeed, workshop 2 found that the Help Seeking tool should provide support to make links with similar groups: “Would also like the facility to create and link to other relevant groups (e.g. PMs) in order to open up communication channels for particular purposes”. This is in line of our idea of providing Socio-Historical tools and services where humans and the system (i.e. the Help Seeking tool and recommender system) work together connecting people with people, people with data, and data with data.

Furthermore, trust seems to be closely linked to contacts with same professional profile. An issue was raised towards the end of workshop 1 about ignoring suggestions from LinkedIn that do not relate to a person's professional identity. One of the Practice Managers commented that during the workshop they had sent an invite to connect to every Practice Manager that LinkedIn was recommending, even if they did not already know them. They commented that "it couldn't hurt to do this". However, the same person was making some choices and ignoring some recommendations—they said they were not interested in connecting to the BMA (British Medical Association) or to Practice Nurses even though LinkedIn was making these suggestions as well. Information from the users (not only the personal details provided by the individuals manually, but also the semantic analysis of their actions) will be saved as 'Key profile factors'. This information will be used by the Help seeking tool to recommend similar and more capable peers, in order to scaffold the process of building, maintaining and activating their PLN.

Overall we found, as predicted, that by 'workshop 2' participants were beginning to exhibit changes in their perception towards using social networking tools and seemed to have a clear interest in developing the Help Seeking tool to improve their current networking limitations: "This is the way forward. This is how we are going to communicate more than the once a month [that is currently achieved face-to-face] at the group Practice Manager meeting, without wading through a load of emails"; and a comment by a senior Nurse … "I didn't see the benefit of LinkedIn but I do for this [i.e. the Help Seeking tool]".

Many of the findings provided in this follow-up empirical co-design study support the direction that the Help Seeking tool is taking but also provided new requirements to be built into the next version.

19.5 Innovation in Context: Help Seeking Using the Social Semantic Server

We are currently leading on innovative work to bring the semantic approaches of WP5 (see Fig. 19.1) into the design of the Help seeking tool (Kowald et al. 2013). The SSS collective knowledge services are able to provide useful information based on human contributions and that these will get better as more people participate. Because the SSS may be unfamiliar to many readers, below we unpack some of the main ideas. This is followed by a new conceptualisation of how the Help Seeking tool and the SSS might fit together from a Vygotskian perspective; we confirm that the design research goal of 'the simultaneous pursuit of theory building and practical innovation is extremely ambitious' but attainable.

19.5.1 Social Semantic Server

The Social Semantic Information Spaces (Fig. 19.4), is claimed to be a space where "information is socially created and maintained as well as being interlinked and machine-understandable, leading to new ways to discover information on the Web" (SIOC 2009, please refer to this web page for explanation of acronyms in Fig. 19.3).

The relation between actors, activities and objects of action–activity has been a core challenge in Vygotsky-informed research and cultural-historical activity theory. Consequently, the next section presents an innovative conceptualisation of how the Vygotsky-informed research described in this chapter can be used to inform a rethink of the SSS and hence move us beyond the state of the art.

19.5.2 Layers Social Semantic Server and Help Seeking Tool in Healthcare Sector

The SSS can generate metadata to relate people and data, people and people, data and data. The goal of the following conceptualisation is to explore the integration of our Help Seeking tool's cultural-historical approach (Vygotsky) with the SSS.

In Fig. 19.4 we have three people: Patricia, Mark and Natasha. They all search for and read an article called "Registration guidelines on diabetes" which is downloaded from the Intranet onto their respective PLNs (the solid lines in Fig. 19.4). From this the SSS will begin a service known as *user event service* (or looking at what people are doing and finding patterns); in this instance, the pattern is three people have all downloaded the same document meaning they have shown an interest. From the SSS's perspective we draw a (dotted lines in Fig. 19.5)

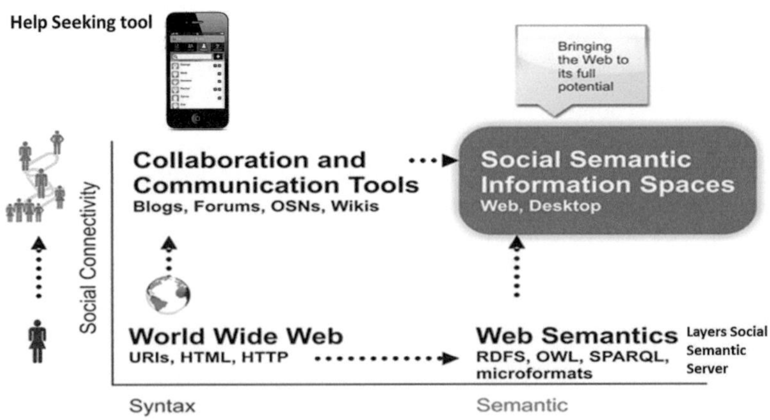

Fig. 19.3 Social semantic information spaces (with Layers tool and service included)

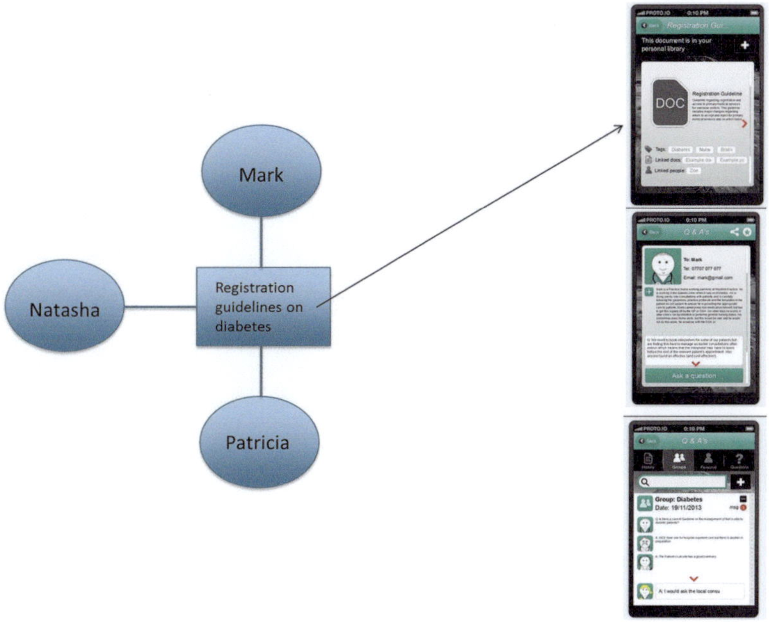

Fig. 19.4 Triggering event service

connection between the 3 people, since they all downloaded and (we assume) have read the same article.

Patricia asks Mark (who she has previously tagged in her PLN, as a 'more capable peer') a question about booking interpreters for a patient via her contacts facility in the app (the arrow in Fig. 19.5 to Patricia's PLN). For the SSS this is part of the *meaning-making system*, since they both have looked at the "Registration guidelines on diabetes" document; the SSS *user event service* draws in a *relationship between those two sets of data* (dot dash lines on the right in Fig. 19.5). Note that at the moment this relationship is detected because it is tagged by Mark and Patricia.

Now the SSS pushes a service called *recommendation service* (making links to pertinent information, Q&A or people, which is part of the *guidance service group*), because it has seen that Patricia and Mark both are in this discussion (bottom right PLN screen in Fig. 19.6). The SSS assumes that Natasha probably would like to be in the discussion too (because of the similar interests of the three persons). Consequently, the SSS suggests to Natasha that she joins the discussion (arced line across the top in Fig. 19.6); the SSS is therefore scaffolding a collaborative 'temporal context for development' or put simply creating common ground for a conversation.

In summary, Natasha discovers a discussion that she also finds useful thanks to the SSS's high-level service "recommendation". The services and connections provided/made by SSS in this example are: (1) user event service (finding a

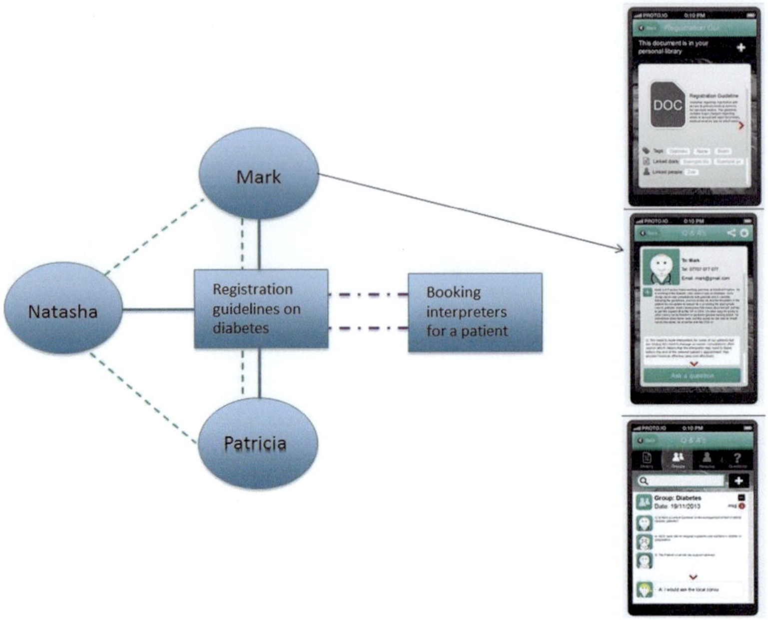

Fig. 19.5 Connection between the three people and relationship between two sets of data

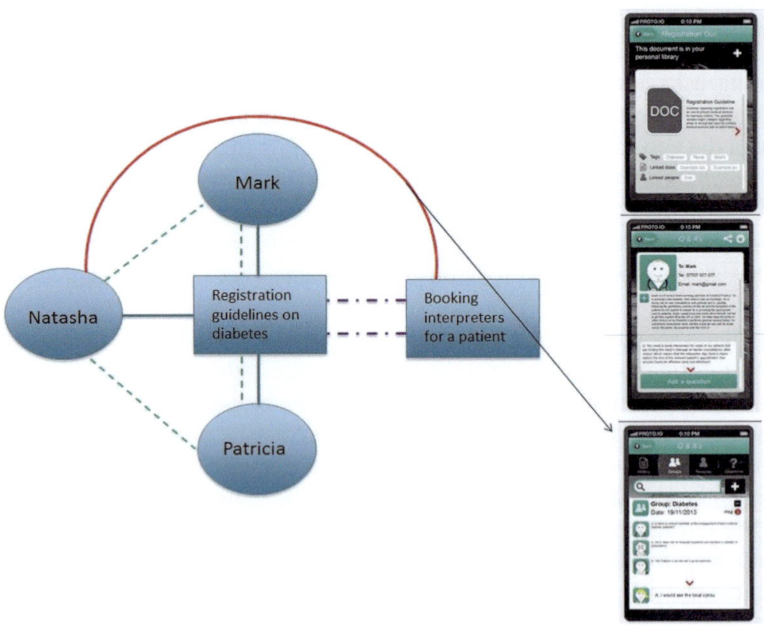

Fig. 19.6 Recommendation Service

pattern), (2) connection between the 3 people (dotted lines), (3) relationship between two sets of data (dot dash lines) and (4) recommendation service, i.e. suggesting that a person consider joining a discussion (arced line). Furthermore, in Vygotskian terms, we have in play two key concepts: More Capable Peer and Temporal Context for Development. From a conceptual point of view, we note that future work needs to hook these insights into our Vygotskyian concepts (and related notions of reciprocal collaboration) in order to refine the design of the Help Seeking tool. Key questions that arise are as follows. *Can we speculate that the centre of gravity and time field allows collaborators in the Practice Network to focus attention on future oriented and shared temporal context for development? Is some form of collaborative filtering a function that enables this context complexity to be dealt with?*

19.6 Conclusions

We conclude that from a research perspective all three of our phase argument-driven inquiries have proved a powerful lens through which to drill down into the state of the art of m-learning and also as vehicle to make connections. For example, take the second phase argument exploration (focus on designing for 'mobile learning' at scale); we are fully convinced that design research allows us to bring out 'never-seen before possibilities' of mobile learning. Designing for 'mobile learning' at scale, beyond pilots and content-centric approach, is a big challenge that is worthy of our attention. If we link across to our first phase argument (focus on new patterns of connected social learning and work-based practices) we see that designing for augmented social learning has the real potential to take us beyond content-centric views of learning, and that this has the potential to revolutionise equity of access to learning. However, although a new educational paradigm is emerging, there exists a need for more debate and further research, particularly around notions of sustainability, scalability and equity of access to opportunities to build social capital. Finally, with respect to our third phase argument exploration (focus on the boundaries of learning that the 'm' in m-learning forces us to explore) our strong belief is that not only do the twenty-first-century structures of mass communication provide a wide range of augmentations to communication but in addition, through the agency of users, the context within which communication takes place is being augmented by users to suit the needs of the individual and the conversational community and we predict that augmented social learning will give rise to a stream of innovations that will shape the modern society and culture.

Acknowledgments Acknowledgment of work used in this chapter: Tribal (designs), Owen Gray (who wrote 'SSS for dummies'), Tamsin Treasure-Jones, Micky Kerr, & various Learning Layers colleagues and Healthcare participants. Learning Layers is a 7th Framework Large-scale integrating project co-funded by the European Commission; Grant Agreement Number 318209; http://learning-layers.eu/.

References

Benford, S., Seager, W., Flintham, M., Anastasi, R., Rowland, D., Humble, J., Stanton, D., Bowers, J., Tandavanitj, N., Adams, M., Farr, J. R., Oldroyd, A, Sutton, J., et al. (2004). The error of our ways: The experience of self-reported position in a location-based game. In *Proceedings of the 6th International Conference on Ubiquitous Computing (UbiComp 2004)*, (pp. 70–87). LNCS/Springer Press.

Carmean, C., Franfort, J. L. and Salim, K. N. (2013). The power of the personal (pp. 187–195). In Berge, Z. L., & Muilenburg, L. Routledge (Eds.), *Handbook of mobile learning*, Routledge: New York.

CBSlocal. (2013). *School starts mass social media surveillance of students for their 'safety'.* Retrieved from http://tinyurl.com/ow8gdp6.

Chui, M., Manyika, J., Bughin, J., Dobbs, R., Roxburgh, C., & Sarrazin, H. (2012). *The social economy: Unlocking value and productivity through social technologies.* McKinsey Global Institute, pp. 1–170.

Clark, W., & Luckin. (2013). iPads in the Classroom, London Knowledge Lab. Retrieved from http://digitalteachingandlearning.files.wordpress.com/2013/03/ipads-in-the-classroom-report-lkl.pdf.

Cochrane, T. (2013). A summary and critique of M-Learning research and practice (pp. 24–34). In Berge, Z. L., & Muilenburg, L. Routledge (Eds.), *Handbook of mobile learning*. Routledge: New York.

Cook, J. (2010). Mobile phones as mediating tools within augmented contexts for development. *International Journal of Mobile and Blended Learning, 2*(3), 1–12. Retrieved from http://goo.gl/NFWnSZ.

Cook, J., & Pachler, N. (2012). Online people tagging: Social (mobile) network(ing) services and work-based learning. *British Journal of Education Technology, 43*(5), 711–725. Retrieved from goo.gl/S5kfgi.

Cook, J., Pachler, N., & Bachmair, B. (2012). Using social networked sites and mobile technology for Bridging social capital. In G. Trentin, & Manuela Repetto (Eds.), *Using network and mobile technology to Bridge formal and informal learning* (pp. 31–56). Chandos.

Deloitte. (2013). Understanding the bring-your-own-device landscape. Retrieved September 04, 2013, from http://www.deloitte.com/view/en_GB/uk/industries/tmt/technology/bring-your-own-device/index.htm.

de Waard, I. (2013). mMOOC Design: Ubiquitous, open leaning in the cloud. In Z. L. Berge, & Muilenburg, L. Routledge (Eds.), *Handbook of mobile learning* (pp. 356–368). Routledge: New York.

FitzGerald, E., Ferguson, R., Adams, A., Gaved, M., Mor, Y., & Thomas, R. (2013). Augmented reality and mobile learning: the state of the art. *International Journal of Mobile and Blended Learning*, *5*(4), 43–58.

Holley, D., Santos, P., & Cook, J., (2014). *'Cascades, torrents & drowning' in information: Seeking help in the contemporary GP Practice in the UK*. Designing for Digital Learners (D4DL), Research Report 1. UWE, Bristol, UK.

Infographic. (2012). *Social media distractions cost U.S. economy $650 billion [INFOGRAPHIC].* Retrieved from http://tinyurl.com/dxr3pqf.

Kowald, M., Dennerlein, S., Dieter, T., Walk, S., & Trattner, C. (2013). *The social semantic server —A framework to provide services on social semantic network data.* Graz, Austria: I-Semantics.

LaMaster, J., & Ferries-Rowe, J. D. (2013). So we had this idea: Bring your own technology at Brebeuf Jesuit. In Z. L. Berge & L. Muilenburg (Eds.), *Handbook of mobile learning* (pp. 395–404). New York: Routledge.

McKenney, S., & Reeves, T. (2012). *Conducting educational design research.* New York: Routledge.

Seitinger, P., Kowald, D., Trattner, C., & Ley, T. (2013). Recommending tags with a model of human categorization. In *Proceedings of the ACM International Conference on Information and Knowledge Management (CIKM 2013)*, ACM, New York, NY, USA.
Santos, P., Cook, J., Treasure-Jones, T., Kerr, M., and Colley, J. (2014a). Networked scaffolding: Seeking support in workplace learning contexts. In *Networked Learning Conference*, Edinburgh, UK.
Santos, P., Cook, J., Holley, D., Treasure-Jones, T., & Kerr, M. (2014b). *Going beyond your personal learning network, using recommendations and trust.* D4DL/Learning Layers Research Report No. 4, October 2014, UWE Bristol, UK. Email author for a copy.
SIOC. (2009). *Content exchange and semantic interoperability between social networks.* Retrieved from http://www.w3.org/2008/09/msnws/papers/sioc.html.
Vygotsky, L. S. (1930/1978). *Mind in society: The development of higher psychological processes.* Edited by Michael Cole, Vera John-Steiner, Sylvia Scribner and Ellen Souberman. Harvard University Press. Originally published 1930.
Wood, D., Bruner, J. S., & Ross, G. (1976). The role of tutoring in problem solving. *Journal of Child Psychology and Psychiatry, 17*(2), 89–100.

Author Biographies

John Cook is Professor of Learning Innovation at UWE, Bristol. Previous to this he was Professor of Technology Enhanced Learning (TEL) and Director at Learning Technology Research Institute, London Metropolitan University. John was part of the successful Education Research Assessment Exercise (RAE) submission; London Metropolitan University appeared in the upper quintile of the 2008 RAE Education Unit of Assessment. John has over 14 years previous experience as a full-time lecturer at various Higher Education Institutions. He has over 10 years project management experience and has been part of research and development grant proposals that have attracted over £5 million in competitive external funding. FP7 examples include MATURE IP and Learning Layers IP. In addition, he has published/presented around 280 refereed articles and invited talks in the area of TEL, having a specific interest in several related areas: hybridity in learning, informal learning, mobile learning in all sectors, augmented reality, 3D web and visualisations, social web, and work-based learning. John is a founding member of The London Mobile Learning Group. He was Chair/President of the Association for Learning Technology (2004–06) and Chair of ALT's Research Committee (2008–2012). John sits on various journal editorial boards and conducts Assessor and review work for the EU and UK Research Councils.

Patricia Santos has got a Computer Engineering degree and a PhD in Information, Communication and Media Technologies, focused on the research areas of Technology Enhanced Learning (TEL) and Human Computer Interaction (HCI). In 2011 Patricia presented her PhD thesis (focused on new representation of test and assessment using interaction contexts such as: geo-located routes of questions answered with mobile phones, use of maps for representing questions and 3D virtual worlds for assessment environments) obtaining the qualification of Summa cum Laude. Since 2007, she has been collaborating with the GTI research group (Interactive Technologies Group, UPF). Patricia has been involved in several European and national R&D projects in the field of TEL. Her role has been especially focused in the design of mobile and ubiquitous learning applications to support indoor and outdoor activities, pilots and experiments done in real educational contexts, in the evaluation of research results, in organizing R&D meetings and workshops, developing reports and deliverables, and participating in project meetings and reviews. In addition, Patricia has five years of teaching experience in different subjects (TEL, HCI, Programming…) in different Catalan universities (UPF, UAB and UOC). Since 2013 (March) Patricia is researcher in Technology enhanced Learning in UWE working in the European research project Learning Layers: http://learning-layers.eu/

Chapter 20
Mobile-Assisted Language Learning in China's College English Education: The Reality and Research

Zhuo Wang and Yang Cui

Abstract Due to the increased college enrollments in China in recent years, today's college English teachers are facing more challenges than ever. Influenced by the traditional Chinese culture, mainly Confucian, current college English classes in China are often critiqued for their teacher-centered approach, the lack of student autonomy, as well as detachment from realistic social purposes. The use of mobile technologies in language acquisition has been explored by many researchers around the world, and has the potential to spark positive changes in China's college English education, including enhanced teacher competencies, increased learner autonomy, and improved teacher–student interaction. This paper provides an overview of existing research and practices about mobile-assisted language learning in China's college English education, and proposes that certain elements should be in place to ensure its successful integration.

20.1 Introduction

With the globalization of China in recent decades, there is a growing demand for college graduates that are proficient in the use of English (Ruan and Jacob 2009), which has become "the lingua franca of the world due to its widespread use in academia, business, commerce, and technology" (Spolsky and Shohamy 1999, as cited in Lan et al. 2007, p. 130). To meet this demand, English has been made a mandatory subject for all freshmen and sophomores across the country (Xie 2013).

Z. Wang (✉)
Towson University, Baltimore, MD, USA
e-mail: jwang687@gmail.com

Y. Cui
University of Baltimore, Baltimore, MD, USA
e-mail: yang.cui@ubalt.edu

D. Churchill et al. (eds.), *Mobile Learning Design*, Lecture Notes in Educational Technology, DOI 10.1007/978-981-10-0027-0_20

Moreover, with the purpose of producing employable college graduates that are competent in various facets of English, including reading, writing, listening and speaking, college English test (CET) 4 has been integrated as an undisputable prerequisite for college graduation (Xie 2013).

Numerous research studies have revealed, however, that current college English education is far from satisfactory in producing such ideal graduates (Li 2014). On the one hand, both teachers and students are deeply influenced by the Confucius culture (Biggis and Watkins 2001), making English classes teacher-centerd and lecture-based. The passive role of students in class has led them to have little autonomy over their English learning. On the other hand, the college expansion policy has increased college enrollments dramatically and resulted in a severe shortage of competent English teachers in higher education in China (Cai 2006). Many in-service college English teachers, therefore, are often found to be lacking adequate knowledge about how English should be taught and how students learn second/foreign languages best (Chen and Goh 2011). Consequently, Chinese college English learners not only perceive China's college English education negatively (Cai 2012), but also fail to support and sustain their own learning when teachers are not present (Hurd and Xiao 2006).

With increased accessibility and awareness of using information and communications technology (ICT), Chinese policymakers have recognized the important role ICT plays in supplementing college English education. In 2002, the Chinese Ministry of Education proposed an ICT-incorporated teaching approach that aimed to not only promote students' learning autonomy but also improve teachers' efficiency and productivity (Hu and Mcgrath 2011). This proposal, despite its theoretical validity, was not implemented well in China's higher education institutions. One of the major reasons was that integrating ICT into English education required not only teachers' proficiency of utilizing technologies but also some fundamental changes regarding the roles they and their students should play, both of which can only happen with the provision of effective and continual support from their organizations (Hu and Mcgrath 2011).

Mobile technologies, while originally derived from information and communication technologies, have taken on unique characteristics with its recent developments. Aside from the benefits it brings in as a regular computing technology, it also possesses distinctive advantages, such as mobility, portability, connectivity, and ubiquity to its users (Kukulska-Hulme and Shield 2008). Simply put, mobile technologies allow its users to access resources and connect with the rest of the world from anywhere at any time with access to the Internet. The effectiveness of using mobile technologies to support language acquisition has been spotted in numerous research studies across different subjects around the world. For instance, Motallebzadeh and Ganjali (2011) investigated the effectiveness of using SMS to deliver English words to 40 Iranian university students, and it was found that those that learned with this service outperformed significantly than those who received traditional board and paper instruction, because learning content received via SMS was more convenient and accessible. In Wong and colleagues' (2010) study, 40 primary students were asked to use a camera on given smart phones to take photos of objects and/or scenes that would demonstrate their understanding of assigned English idioms. They found that mobile technology not only allowed students to create artifacts easily, but also promoted in situ learning that connected learning with their real-life context.

Mobile technologies in China, while widely accessible, have not been investigated much as a language learning tool through an academic lens. Relevant studies are not only scarce, but also problematic in certain domains, such as a lack of originality, inadequate research methodologies as well as inconsistent control of quality. This paper pinpoints current research and practices of mobile-assisted language learning (MALL) in China's college English education, with the purpose of identifying trends, gaps, and issues that may inspire future researchers and other interested parties to improve the status of MALL-related research and practical uses in related contexts. Specifically, we argue that in order to promote and integrate mobile technology as an appropriate and effective way to support college students' English learning, the capacity and culture of using mobile technology as a learning tool must be built first among all stakeholders, including college teachers, students, administrators, and policymakers through recommended ways.

20.2 Definitions of Key Terms

In order to maintain consistency throughout this paper, relevant terms are defined as below:

- MALL (Mobile-Assisted Language Learning): "Language learning enabled by the mobility of the learner and …portability of handheld devices…" (Hoven and Palalas 2011, pp. 76–77)
- Mobile Technology: Communication technologies that utilize cellular data, such as mobile phones, GPS, 4G data, etc.
- Mobile Learning or m-learning: "learning mediated via handheld devices and potentially available anytime, anywhere" (Kukulska-Hulme and Shield 2008, p. 273).
- SLA (Second Language Acquisition): SLA theories address "cognitive issues (how the brain processes information in general and language in particular), affective issues (how emotions factor into second language processing and learning), and linguistic issues (how learners interact with and internalize new language systems)" (Florez and Burt 2001, p. 1).

20.3 College English Education in China

20.3.1 Historical Context

College English language education in Mainland China has always been interweaved with China's political situations and decisions (Lam 2002; Hu 2007). For example, in 1991, after detaching from the former Soviet Union, China was facing a political situation in which a more international stance was possible (Lam 2002). This pursuit of a more international role since then has been furthered by China's constant engagement in the international arena, such as its entry into the World Trade Organization and the hosting of 2008 Olympics in Beijing. Such globalization of China demands versatile professionals that are not only experts in their fields of study, but also proficient in the use of English (Li 2014). As a result, college English curriculum in Mainland China has been reformed several times to meet this demand, namely 1980, 1986, 1999, and 2007 college English curricula (Table 20.1). From 1980 to 2007, there have been some transformative changes pertaining to teaching aims and approaches, such as a qualitative shift from

Table 20.1 College english curricula from 1980 to 2007 (from Li 2014, p. 294)

	1980 Curriculum	1986 Curriculum	1999 Curriculum	2007 Curriculum
Aim	To provide students with capability to gain some information through English	To provide students with capability to gain some information through English for their professional needs	For students to be capable of exchanging information in the target language	For students' comprehensive ability to use English to communicate effectively and to study independently, and to improve their cultural awareness in international exchanges
Objective	No specific description	Proficiency reading ability, certain listening ability and elementary speaking and writing ability	Strong reading ability and fairly good ability for listening, speaking, writing and translating	Competent in using English in a well-rounded way, especially in listening and speaking
Methodology	Teacher-centered, grammar translation	Learner centered (grammar translation and audiovisual approach in practice)	Learner centered (grammar translation and audiovisual approach in practice)	Learner centered in combination with modern technology (grammar translation and audiovisual approach in practice)
Vocabulary	From 1,500 to L800	From 1,600 to 4,000	From 4,200 to 6,500	From 4,500 to 6,500

emphasis on linguistic competence to communicative competence, and from teacher-centered to learned-centered approach (Li 2014).

20.3.2 Problems

While significant development has been achieved in college English curricula reform, "present college English language education in Mainland China is continuously criticized for failing to meet the public's demand for good English proficiency" (Li 2014, p. 292). One of the main problems is that, regardless of their theoretical soundness, the college English curriculum requirements were never executed well in practice (Li 2014). As a result, college English education in China has been rather perceived as time-consuming and ineffective (Cai 2010) by different entities.

The ineffectiveness of China's college English education can be first observed in some national studies that investigated the perceived effectiveness of current college English education. For instance, Yu and Zhong (2008) surveyed 1,615 students through random sampling in 12 universities and found out that, among all the courses they are studying, students were most unsatisfied with their improvement in English. Specifically, 11.3 % of the surveyed students considered themselves having made considerable progress, while 23.6 % reported to have made no progress and 24.6 % believed that they even digressed compared with their English proficiency in high school. Cai (2010) surveyed a total of 1,246 students from eight provinces in 16 universities about their English learning experience and the results showed that only 3.9 % of the students believed that college education improved their English capability to a great extent; 35.2 % believed that some progress was made; 25.4 % stated that not much was learned while as high as 35.1 % of the students felt that their English proficiency deteriorated from high school to college.

The reasons that have led to student dissatisfaction with college English education are multitude. To begin with, the current in-service college English teachers do not meet students' need adequately. Starting 1999, the Chinese government has implemented the expansion policy of higher education to increase the number of college graduates (Bai et al. 2012). During the 1996–2000 period, there was a total enrollment of over 11 million, while from 2001 to 2005 the number of university students was expected be up to 16 million (Meng and Tajaroensuk 2013). However, this policy has caused a severe shortage of qualified college English teachers (Cai 2006). According to a national study conducted by Dai and Zhang (2004), 32.4 % of the surveyed college English teachers had no more than five years of teaching experience. Also, Wang and Wang (2011) investigated 457 colleges in China and found out that among the surveyed 21, 065 English teachers, only 1.5 % held a doctor's degree and 60.1 % held a master's, which is below average when compared with other majors and programs.

The increased college enrollments have also resulted in heavier workloads for in-service teachers. According to Zhang (2006), the college English teacher to student ratio is nearly 1:200. Limited time and the overwhelming workload are critical factors that hinder these teachers from participating in continuous professional development (Carney 2003; Quaglia et al. 1991; Day and Gu 2010).

In terms of pedagogy, most college English teachers enter the profession without a solid understanding of SLA theories, psychology, and pedagogy (Chen and Goh 2011). The absence of such knowledge may exert an adverse influence on students' language learning experiences. For example, without being exposed to contemporary learning theories, such as Constructivism, most English instructors are still employing "a teacher-centered, textbook-reliant, grammar-translation teaching method" (Li 2014, p. 296). This traditional approach prevents students from engaging in active English learning and having ownership of their learning process. Culture, on the other hand, also has a profound influence on Chinese classroom dynamics. Chinese education is infused with Confucius beliefs and principles (Biggs and Watkins 2001; Li 2003), which hold that students should highly respect their teachers as authority figures and do as the teachers dictate (Ho 2001).

When students learn passively, however, they are less likely to be motivated to learn (Cai 2010) and may thus produce unfavorable results that harm their self-efficacy and increase their reluctance of using English in or outside of classrooms. Research indicates that many employers have complained about how college graduates often perform poorly when it comes to communication in English (Ruan and Jacob 2009), regardless of their performance in the written form of College English Tests (CET).

In order to tackle some of the above challenges, the Chinese Ministry of Education initiated a reform of college English that proposed for a "more economical and effective methodology in language teaching based on the use of information and communications technology (ICT)" was recommended in the reform (Hu and Mcgrath 2011, p. 42). The incorporation of ICT was believed to not only support and enhance language teaching and learning, but also provides students more access to resources that they can learn independently. Ideally, it would lessen teachers' workload and alleviate the tension caused by the shortage of college English teachers (Hu and Mcgrath 2011). However, the proposal was not implemented well and created even more challenges for these teachers. Hu and Mcgrath (2011) stated that

> …The reasons are manifold: inefficient CPD (college professional development), insufficient access to ICT facilities, unfavorable ICT policies, lack of technical support, improper appraisal systems related to ICT use, difficulty in changing deep-rooted roles of teachers as well as roles of schools and students, inappropriate beliefs and attitudes toward ICT use, and as noted above, lack of ICT knowledge and skills among teachers and students, and poor ICT pedagogy (O'Mahony 2003). All these issues hinder the use of ICT in schools. (p. 43)

In short, college English education in China is now facing multifaceted challenges. On a social level, deeply rooted Chinese culture (e.g., Confucian) is still influencing the roles that teachers and students, respectively, play (Tang 2009). On the institutional level, national policies and propaganda that aim to improve CE education fail to be implemented wholeheartedly due to the complexity of incorporating ICT, the lack of effective training and just-in-time support from school administrations. On an individual level, college teachers and students are both confronted with issues that prevent them from achieving desirable goals. Notably, college English teachers are expected to obtain more advanced qualifications in their profession, and enhance pertinent knowledge and skills on not only subject matters but also popular instructional technologies, while striving to maintain a balance between such expectations and their overwhelming workload. Students, on the other hand, need to transform their existing beliefs about how they are expected to learn, take a more active role in learning English, and learn to locate and utilize available resources on their own. Having a clear and comprehensive understanding of these challenges and relevant policies can help us demarcate what needs there are to be met, and if they can be met appropriately and efficiently by developing potential solutions or strategies.

Mobile technologies, which are introduced in the following section, are believed to have a tremendous potential to alleviate, if not entirely resolve, the problems and needs identified above.

20.4 Mobile Technologies

In recent years, mobile visitors have become the fastest growing web community that access web pages or locate web information (Chen 2008). Cell phones, most of which are well equipped with functionalities including Internet access, media player, digital camera and video recorder, have become the most widely used and accessible devices for almost every university student (Chirimbu and Tafazoli 2013). In China, so far 85 % of the younger urban residents (age from 18 to 30) own smartphones (NetEase News 2013). With regard to college students, around 80.8 % has at least one smartphone with Internet-connected service, which means virtually all higher education students carry some form of mobile devices (People's Daily Online 2013). The widespread ownership of mobile devices among Chinese college is an active index of its accessibility and makes its integration as a learning tool possible.

Mobile devices, such as smart phones, PDAs and tablets, provide its users with many advantages that surpass the affordances of other ICT tools. According to Klopfer and Squire (2008), such advantages include but do not limit to (1) portability—they are lightweight handheld devices that can be easily carried everywhere; (2) mobility—which indicates the accessibility of resources even while both the users and the devices are on the move; (3) connectivity—the availability of cellular data on those devices empowers its users to connect with the rest of the world from almost anywhere at any time; (4) individuality—not only can users customize the device in a way that best suits their preferences, but also seek information that is tailored to their particular needs or requests.

These characteristics of mobile technologies have an enormous potential in improving college students' language learning experience and solving many of the aforementioned problems. For instance, one of the major problems recognized above is Chinese students' low level of learner autonomy—they learn about what is being told to learn. In contrast, with Internet-connected mobile devices, students may search for and actively learn about English topics that they are sincerely interested in, rather than required by the curriculum. Driven by their innate passion, students are more likely to learn deeply and take responsibility for their own learning, and thus increase learner's autonomy (Benson 2007). At the same time, the abundance and diversity of English learning resources on the Internet allow students to acquire knowledge that may not be taught well, or at all, by their less qualified English instructors. In that sense, not only can they learn more, but also rely less on their instructors as a major learning source. Mutual benefits can also be achieved when mobile devices are used as an assessment tool. English instructors can easily create quizzes or polls on a mobile device to collect data about student learning quality, while every student can participate in a quiz or poll on their own mobile device to make their opinions count. Frequent assessment allows instructor to be accurately aware of where students are, so that corresponding adjustments in teaching methods or progress are made.

While positive findings of using MALL have been reported in numerous studies in countries like US and Japan (e.g., Hegelheimer and O'Bryan 2009; Miyakoda et al. 2011), China is a developing country that possesses its unique characteristics, including historical context, economic status, political structure, and education system. It is thus paramount to examine MALL studies that resonate with the local culture and situation of CE education in China, which may shed the most light on its future development. While China consists of provinces and districts that often vary dramatically in economic and political status, it is the author's intention to review only Mainland China where both statuses are more consistent and analogous.

20.5 MALL Research in China's Higher Education

In China's college English education, mobile-assisted language learning, while being used consciously or unconsciously, is still a new concept. For instance, the search for MALL studies in the target context yielded very few results compared with the high volume of MALL studies conducted in countries like the United States or Japan. In addition to the lack of research, the awareness of this concept among public is low as well: Most of the participants in related studies admitted to have heard of mobile learning for the first time at the time of study (e.g., Wang et al. 2009), regardless of their ownership of, and experience with, mobile devices. As a technology, which is defined by Rogers (2003) as "a design for instrumental action that reduces the uncertainty in the cause–effect relationships involved in achieving a desired outcome" (pp. 139–140), MALL in China's higher education obviously still resides in the initial stage of technology transfer—research and development (Rogers 2003). During this stage, scientific and applied researches are conducted about a problem and initial prototyping solutions are proposed by lead users (Rogers 2003). Specifically in China, problems related to current college English education have been recognized in numerous studies (e.g., Cai 2012; Wang and Wang 2011), and MALL has been advocated as a potentially viable solution to address many of the identified problems.

Current research, in terms of the purpose of studies, has primarily focused on three broad categories: theoretical rationality, user perceptions, and empirical effectiveness. Studies related to theoretical rationality are concerned with where MALL derived from and what theoretical frameworks it is built upon. Such studies provide Chinese readers with the research foundations of MALL, helping them to understand the rationale behind MALL design and a promising integration with pedagogical practices. For example, Liu and colleagues (2013) provided an overview of three relevant theories, including situated cognition theory and Construction of Cognitive Learning Theory and collaborative learning. However, their report was merely a reinstatement of important concepts often found on relevant international journals, and thus lack originality and connection with China's context.

Most of the reviewed studies have focused on the affective domain of learning, namely how students perceive MALL with regard to its usefulness and viability, and whether there is need or market for MALL. For example, Li (2014) conducted a survey on 89 undergraduates at Guangxi University, aiming to investigate their current perceptions and uses of mobile devices to support language learning. Zou (2014) used a mixed method approach to research Chinese undergraduates' perceptions of MALL, and found that 78 % of the examined students hold a positive attitude of trying MALL, but many are not aware of how to use mobile technologies to learn.

The third type of studies was the least researched among all. One of the possible causes could be that awareness of MALL is not yet high among college teachers or students, let alone using it intentionally. The search only yielded three empirical studies, which interestingly focused on different aspects of MALL. Xue (2014) explored how effective mobile technology (including MMS, mobile apps) could help increase students' test scores; Ruan and Ma (2014) reported the use of an intentionally designed mobile app to improve students' grammar learning; Yin (2013) investigated the effectiveness of using a social media tool—WeChat—to push learning information (vocabulary, grammar, etc.) as a way to prepare students for CET 4 test.

While the three categories of studies collectively provide a preliminary framework for MALL research in China's CE education, there yet has much to be done if MALL is to be integrated as a legitimate component of China's higher education system. In this section, problems associated with, or derived from, these studies are identified and some preliminary considerations are given to potential strategies for future directions.

Overall, studies about MALL in China's higher education are increasing in recent years, but are still in their infancy. First of all, the quality of reviewed studies is concerning. Most of the current studies were published in local Chinese journals that were not internationally peer-reviewed and often had a low threshold for publication. These same studies are also questionable regarding their validity and reliability, because they often do not adhere to consistent academic writing standards. For instance, it is common among these studies to cite less than ten references, or be composed to writings of less than three pages, or fail to articulate certain critical research writing components, such as limitations, instrument description, or theoretical frameworks. Also, the abstracts of those studies are unsatisfactory. According to Pyrczak (2003), an abstract should be a summary of a research that consists of a purpose of study, methodology, results and implications, or future directions. The reviewed abstracts, however, fail to include those essential components that synthesize the gist of the study; rather, they often come from the first few sentences of a study's introductory paragraphs that provide little for readers to understand the research at hand. Moreover, relevant appendices are usually missing in the reviewed studies, especially survey instruments. What questions are asked in a survey and how valid those questions prelude to a certain degree if a survey is reliable, and thus should be described and explained.

Research studies in journal articles are perceived as authoritative and reliable sources of knowledge for Chinese educators, researchers, and even the entire public. The quality of these studies, such as accuracy and validity, has an undeniable influence on readers' understanding of MALL, such as its legitimacy, prevalence, usability, etc. Therefore, editors of relevant journals should establish consistent criteria for acceptance and publication, especially about data collection, analysis, and content originality, since these are often most convincing information among all. Incentives can also be considered as a strategy to encourage related research, such as allocating grants for innovative use of MALL in higher education.

Second, some studies are found to be mere reiterations of MALL findings or trends reported in foreign language journals, with little or no originality or applicability in China's context. Such knowledge, while providing readers with an overview of what is happening worldwide, does not contribute much to the growth of MALL in China, which has its unique set of characteristics. For instance, the educational hierarchy is much different in China from that of the U.S. due to their difference in political structure. It is thus recommended that researchers synthesize research from countries that share as many similarities with China as possible, so that Chinese reformers can draw upon successful experiences from those areas when planning or initiating changes for MALL.

Third, current research studies are often limited in their scope of study. While mobile technology has the potential to benefit both students and teachers (Aubusson et al. 2009), most published studies pertaining to mobile learning have focused almost exclusively on students as the learners or consumers of mobile technology. However, for any educational change to happen, it is indispensable to involve the collective effort among all stakeholders (Fullan 2007), which in this case include not only students, but also teachers, administrators, and policymakers. For instance, to incorporate and promote MALL in regular instruction, teachers must be equipped with knowledge of MALL themselves, while administrators will have to design corresponding training and provide continual professional development for such knowledge, and policymakers have to at least not prohibit, if not promoting officially, the use of MALL in higher education. At the same time, incorporating any new educational technology may unavoidably demand additional effort and time from teachers, who already have a heavy workload to maintain. To get teachers' buy-in, the right conditions for change must be present, including clear and practice guidance for the change, support from administrative leaders, and readily accessible resources (Fullan 2007). Research on solely any of the stakeholders without making connections with others would result in a partial and even inaccurate understanding of the big picture that hinders a successful integration of MALL. Future research may turn to stakeholders other than students to collect data about their perceptions of, attitudes toward, and current uses (if any) of MALL in the context of higher education, so as to build organizational capacity, which is defined as "policy, strategy and actions taken that increases the collective efficacy of a group to improve student learning through new knowledge, enhanced resources, and greater motivation on the part of people working individually and together" (Fullan 2007, p. 58).

Methodology wise, in addition to quantitative approaches, such as survey or questionnaire, researchers are suggested to also utilize qualitative methods more, so that they can gain more in-depth and rich understanding of target research topics or populations.

Last but not least, pedagogical knowledge, which is the foundation of most educational innovations, is often missing in the reviewed MALL studies. A pedagogically competent English instructor should demonstrate mastery of diverse learning theories, sensitivity to student needs and proficiency in student assessment. However, due to the severe lack of empirical research in current MALL studies, it cannot be concluded if pedagogical components were present in MALL integration.

20.6 Recommendation

Although mobile devices have enormous potential to improve college students' English learning in numerous ways, it should be cautioned that the provision of access to technology does not ineluctably promise its successful integration into teaching and learning, especially when the learners are not motivated to use the technology (Selwyn 2007). Language instructors must carefully examine and evaluate any MALL tools they plan to integrate, and ensure that all conditions conducive to successful MALL integration are present before adopting one officially. Two key principles are recommended below:

- Identify pedagogical theories that scaffold the design of learning activities. Good learning design should characterize sound theoretical support, such as Behaviorism, Cognitive Information Processing, and Constructivism. For instance, Behaviorism is widely used to guide the design of assessments; Cognitive Information Processing dictates how learners process incoming information, and helps instructors choose mobile technologies that provide information in a manageable size, facilitate long-term memory, and increase autonomy; Constructivism denotes a series of principles that epitomize individual meaning-making, collaborative learning as well as social meaning construction. Each of these theories has a branch of summarized practical principles that can be easily applied by instructors (e.g., Driscoll 2005).

- Seek mobile technologies that afford features that are most relevant with and conducive to student learning. In their case study about PDA's educational affordances, Churchill and Churchill (2007) inductively summarized five roles of PDAs for learning: Multimedia access tool (tool for multimedia delivery and access); connectivity tool (tool for interpersonal connection and interaction); capture tool (tool for information capturing); representational tool (tool for mind mapping or knowledge conceptualization); analytical tool (tool for calculation). This categorization is also applicable to MALL with some minor changes. For instance, students with mobile devices can access videos, audios, or images that help them improve English listening, speaking, or vocabulary in a more engaging way; some mobile apps, such as WeChat and Edmodo, characterize group chat or information sharing, allow students to discuss interesting topics in English or share useful resources with just a click. As a capture tool, mobile devices can be used to take photos and record videos of human activities, objects, or scenes to help students learn about vocabularies more deeply and meaningfully. Compared with PDAs, mobile phones are more prevalent, and often have smaller screens, which makes mind mapping on them less convenient. The analytical function makes more sense to language learners when it provides information about learning progress and outcome instead of ordinary calculation for math-related subjects. For example, certain language learning apps provide statistics on how much scores a learner gains by narrating English conversations or what his or her rank is compared with other app users (e.g., Liulishuo).

Aside from teachers, school administrators are also recommended to take corresponding actions. For instance, they may consider providing free Wi-Fi access to students and teachers on campus, so that they can browse relevant knowledge on the Internet without worrying about their data plans or extra charges resulted from English learning. Incentives can also be used to encourage MALL initiatives. For example, school administrators may establish innovative teaching awards to promote MALL awareness as well as related activities. It is equally important to form offices that are specifically designated for MALL training and professional learning, so that instructors feel supported and know where to seek assistance at times of need.

20.7 Conclusion

The current college English education in China is far from satisfactory. The various affordances of mobile technologies and their wide accessibility among Chinese college students have made MALL a favorable and potential solution for some of the prominent issues identified in our earlier review. However, whether mobile technologies can become an integrated component of, or a positive catalyst for improving, China's college English education needs further and more comprehensive exploration and investigation. Current related research is not only insufficient, but also deficient in terms of quality of writing, design of methodologies, as well as scope of study. Future researchers may strive to improve upon these problematic areas, so that interested users or adopters of MALL can gain a more thorough and clear understanding of its viability in their specific contexts and compatibility with their existing practices. Instructors that are interested in adopting MALL tools should also be aware of the various affordances those tools provide and use pedagogical knowledge to make informed decisions regarding what activities can be best conducted on mobile devices and how to achieve desirable learning outcomes. Finally, university or college administrators should strive to create a positive environment for MALL integration, so as to gain sustainable development in college English education in the long run.

Acknowledgments Thanks to Dr. Liyan Song and Ms. La Tonya Dyer from Towson University, both of whom provided invaluable support during our research writing.

References

Aubusson, P., Schuck, S., & Burden, K. (2009). Mobile learning for teacher professional learning: Benefits, obstacles and issues. *Alt-J, 17*(3), 233–247. doi:10.1080/09687760903247641.

Bai, L., Millwater, J., & Hudson, P. (2012). Factors that influence Chinese TEFL academics' research capacity building: An institutional case study. *The Asia-Pacific Education Researcher, 22*(2), 119–126. doi:10.1007/s40299-012-0004-6.

Benson, P. (2007). State-of-the-art article: Autonomy in language teaching and learning. *Language Teaching, 40*(1), 21–40.

Biggs, J. B., & Watkins, D. A. (2001) Insights into teaching the Chinese learners. In D. A. Watkins & J. B. Biggs (Eds.), *Teaching the Chinese learner: Psychological and pedagogical perspectives* (pp. 277–300). Hong Kong & Melbourne: CERC & ACER.

Cai, J. (2006). College English teaching: Review, reflection and research. Shanghai: Fudan University Press.

Cai, J. (2010). The feasibility study of EMI: A case study of Public Relation Course of Fudan. *Foreign Languages in China, 6*, 61–67.

Cai, J. (2012). Reanalysis of the goal of college English teaching in the perspective of globalisation. *Foreign Languages and Their Teaching, 3*, 5–8.

Carney, S. (2003). Learning from school-based teaching training: Possibilities and constraint for experienced teachers. *Scandinavian Journal of Education Research, 47*(4), 413–429.

Chen, C. (2008). Personalized intelligent Mobile Learning System for supporting effective English learning. *Educational Technology & Society, 11*(3), 153–180.

Chen, Z., & Goh, C. (2011). Teaching oral English in higher education: Challenges to EFL teachers. *Teaching in Higher Education, 16*(3), 333–345. doi:10.1080/13562517.2010.546527.

Chirimbu, S., & Tafazoli, D. (2013). Technology & media: Applications in language classrooms (TEFL, TESL & TESOL). *Professional Communication and Translation Studies, 6*(1–2), 187–195.

Churchill, D., & Churchill, N. (2007). Educational affordances of PDAs: A study of a teacher's exploration of this technology. *Computers & Education, 50*(2008), 1439–1450.

Dai, M., & Zhang, X. (2004). An investigation of English teacher qualities in colleges and universities. *Journal of PLA University of Foreign Languages, 27*(2), 42–46.

Day, C., & Gu, Q. (2010). *The new lives of teachers*. London: Routledge.

Driscoll, M. P. (2005). *Psychology of learning and instruction* (3rd ed.). Boston, MA: Pearson Education Inc.
Florez, M. C., & Burt, M. (2001). Beginning to work with adult English language learners: Some considerations, (October) (pp. 1–5).
Fullan, M. (2007). *The new meaning of educational change* (4th ed.). London: Routledge.
Hegelheimer, V., & O'Bryan, A. (2009). Mobile technologies and language education. In M. Thomas (Ed.), Handbook of research on Web 2.0 and second language learning (pp. 331–349). Hershey, PA: Information Science Reference.
Ho, I. T. (2001). Are Chinese teachers authoritarian?. In D. A. Watkins & J. B. Biggs (Eds), Teaching the Chinese learner: Psychological and pedagogical perspectives (pp. 99–114). Hong Kong & Melbourne: CERC & ACER.
Hoven, D., & Palalas, A. (2011). (Re) Conceptualizing design approaches for mobile language learning. *CALICO Journal, 28*(3), 699–720.
Hu, Y. (2007). China's foreign language policy on primary English education: What's behind it? *Language Policy, 6*, 359–376.
Hu, Z., & McGrath, I. (2011). Integrating ICT into College English: An implementation study of a national reform. *Education and Information Technologies, 17*(2), 147–165. doi:10.1007/s10639-011-9153-0.
Hurd, S., & Xiao, J. (2006). Open and distance language learning at the Shantou Radio and TV University, China, and the Open University, United Kingdom: A cross-cultural perspective. *Open Learning, 21*(3), 205–219. doi:10.1080/02680510600953161.
Klopfer, E., & Squire, K. (2008). Environmental detectives: The development of an augmented reality platform for environmental simulations. In Press for *Educational Technology Research and Development, 56* (2), 203–228. 52903.html.
Kukulska-Hulme, A., & Shield, L. (2008). An overview of mobile assisted language learning: From content delivery to supported collaboration and interaction. *ReCALL, 20*(3), 271–289.
Lam, A. (2002). English in education in China: Policy changes and learners' experiences. *World Englishes, 21*(2), 245–256.
Lan, Y., Sung, Y., & Chang, K. (2007). A mobile-device-supported peer-assisted learning. *Language Learning & Technology, 11*(3), 130–151.
Li, J. (2003). The core of Confucian learning. *American Psychologist, 58*(2), 146–147.
Li, C. (2014). Development, problems and solutions: A critical review of current situation of college English language education in Mainland China. *Arab World English Journal, 5*(3), 291–303.
Liu, D., Huang, Y., & Yang, Z. (2013). The application of mobile learning in college English education. *Huazhang, 11*, 187–187. doi:10.3969/j.issn.1009-5489.2014.11.158.
Meng, J., & Tajaroensuk, S. (2013). An investigation of tertiary EFL teachers' problems in their in-service professional development. *Journal of Language Teaching and Research, 4*(6), 1356–1364. doi:10.4304/jltr.4.6.1356-1364.
Miyakoda, H., Kaneko, K., & Ishikawa, M. (2011). Effective learning materials for mobile devices: Images vs. sound. In S. Barton et al. (Eds.). Proceedings of Global Learn Asia Pacific 2011 (pp. 1683–1690). Cheaspeake, VA: AACE.
Motallebzadeh, K., & Ganjali, R. (2011). SMS: Tool for l2 vocabulary retention and reading comprehension ability. *Journal of Language Teaching and Research, 2*(5), 1111–1115. doi:10.4304/jltr.2.5.1111-1115.
Netease News. (2013). The survey said sixty percent of urban Chinese have a smart phone. Retrieved from http://news.163.com/13/0522/15/8VG6KKEG00014JB6.html.
People's Daily Online. (2013). Eighty percent of college students own a smartphone is not to be cool to be practical. Retrieved on March, 2015 from http://tc.people.com.cn/n/2013/0328/c183008-209.
O'Mahony,C. (2003). Getting the information and communication technology formula right: access + ability = confident use. *Technology, Pedagogy and Education, 12*(2), 295–311.
Pyrczak, F. (2003). *Evaluating research in academic journals*. Los Angeles, CA: Pyrczak Publishing.

Quaglia, R., Marion, S., & McIntire, W. (1991). The relationship of teacher satisfaction to perceptions of school organization, teacher empowerment, work conditions, and community status. *Education, 112*, 206–217.
Rogers, E. (2003). *Diffusion of innovation* (5th ed.). New York: Free Press.
Ruan, Y., & Jacob, W. J. (2009). The transformation of college English in China. *Front. Educ China, 4*(3), 466–487.
Ruan, Y., & Ma, Y. (2014). Intelligent mobile phone environment English mobile learning system design and implementation. *Electronic Test, 2014*(9), 13–15.
Selwyn, N. (2007). Screw blackboard…do it on Facebook!: An investigation of students' educational use of Facebook. In: Presented at the Poke 1.0 – Facebook Social Research Symposium. London: University of London.
Spolsky, B., & Shohamy, E. (1999). *The languages of Israel: Policy, ideology, and practice.* Clevedon, UK: Multilingual Matters.
Tang, E. (2009). A cultural framework of "Chinese learn English": A critical review of and reflections on research. *English as International Language Journal, 4*(August), 7–44.
Wang, Q. (2002). Reasons for ineffective College English teaching and relevant countermeasures. *Foreign Language World, 4*, 27–35.
Wang, S., & Wang, H. (2011). On the state of college English teaching in China and its future development. *China Academic Journal Electronic Publishing House, 8*(5), 4–17.
Wang, W., Zhong, S., & Lv, S. (2009). Empirical research on mobile learning of college students. *Open Education Research, 15*(2).
Wong, L., Chin, C., Tan, C., & Liu, M. (2010). Students' personal and social meaning making in a Chinese idiom mobile learning environment. *Educational Technology & Society, 13*(4), 15–26.
Xie, B. (2013). Reflection on the position of English in the education and life of the Chinese today. *US-China Foreign Language, 11*(9), 716–722.
Xue, J. (2014). Research on construction and development of m-learning mode of College English. *Experimental Technology and Management, 31*(3), 176–179.
Yin, Y. (2013). Using WeChat to support undergraduates' preparation for CET 4. *Science & Technology Vision, 25*, 56–57.
Yu, H., & Zhong, X. (2008). A review of 2006-2007 Shanghai College student development report. *Fudan Education Forum, 6*(1), 19–25.
Zhang, R. X. (2006). *Report on College English reform (Series reports on quality of higher education)* (in Chinese). Beijing: China Education.
Zou, B., & Yan, X. (2014). Chinese students' perceptions of using mobile devices for English learning. *International Journal of Computer-Assisted Language Learning and Teaching, 4*(3), 20–33. doi:10.4018/ijcallt.2014070102

Author Biographies

Zhuo Wang is a student researcher pursuing her Doctorate degree in Instructional Technology at Towson University, United States. She is mostly keen on teaching English to speakers of other languages with an integrated use of mobile technology. Her most recent project involves designing a creative vocabulary learning mobile application for language learners and instructors.

Yang Cui holds a Master's degree in Applied Information Technology, and is now an app developer that specializes in programming hands-on and interactive mobile apps for diverse purposes. He is also an MBA student enrolled at University of Baltimore, aiming to enhance his app development and promotion with professional business skills.

Chapter 21
Enacting App-Based Learning Activities with Viewing and Representing Skills in Preschool Mathematics Lessons

Kay Yong Khoo

Abstract This chapter comprises discussion on research findings of this study on how apps can be used in the classroom to promote children's construction of mathematical knowledge by setting up specific learning contexts in ways that fundamentally transform the instructional environment. The study results identify how children enact viewing and representing skills through digital texts to acquire new strategies in their addition and subtraction learning. These skills enable children to externalise their understanding and internalise new meaning-making when interacting with peers. However, these dual reciprocal learning approaches require due consideration of the elements of the learners' learning styles, the standard of the game designs and the community settings of the classroom, all of which are crucial in determining the learners' engagement in a learning activity and active involvement in associated learning processes. With the appropriate level of autonomy and opportunity for choice, learner engagement will contribute to subsequent learning, with behavioural intensity and emotional quality at optimal levels. A detailed examination of the meaning-making processes through which viewing and representing skills mediate children's knowledge acquisition while seamlessly switching between individual and social interactions has led to the development of the framework in the preschool classroom's learner-centred mathematics learning model presented here.

21.1 Introduction

Mathematics competencies are cumulative over time (Jordan et al. 2009; National Mathematics Advisory Panel 2008). If not properly addressed and overcome, difficulties encountered at any stage of learning will lead to poor achievement in subsequent mathematics learning. For example, competencies in whole numbers are

K.Y. Khoo (✉)
E-Learn Dot Com Limited, Hong Kong SAR, People's Republic of China
e-mail: kykhoo@yahoo.com

D. Churchill et al. (eds.), *Mobile Learning Design*, Lecture Notes in Educational Technology, DOI 10.1007/978-981-10-0027-0_21

essential to arithmetic. Therefore, the effectiveness of the classroom teaching pedagogy during early education in developing fundamental numeric skills is crucial (Egan and Hengst 2012). Problem solving skills in addition and subtraction are dominant aspects of the fundamental competency domains in mathematics. When children practise solving problems, their underlying conceptual and procedural knowledge in addition and subtraction determines their competency (Canobi et al. 1998; Reeve et al. 2003). The causal relations between these two areas of knowledge have been found to be bidirectional; increases in conceptual knowledge will help to increase procedural knowledge and vice versa. The iterative relationship influences the development of conceptual–procedural knowledge, particularly in the competency domain of addition and subtraction (Bethany and Schneider 2015; Canobi 2009; Rittle-Johnson et al. 2001). Therefore, to develop an effective classroom teaching pedagogy, the integration of content knowledge (i.e. conceptual and procedural) into learning approaches is important (Chiu and Churchill 2015a, b). These approaches must engage children in the learning activities while leading to the process of meaning-making utilising the content knowledge (Hiebert and Wearne 1996; Star et al. 2011).

The conceptual and procedural knowledge of learners are observed explicitly via strategies applied during their routes to problem solutions. Some of the strategies applied in addition problem solving are direct modelling (represented by objects, which are all counted), counting on from first, counting on from larger and recalled with no apparent counting; those applied in subtraction problem solving are direct modelling (counting objects by separating from the total and counting those remaining), counting down from (a backward counting sequence from bigger numbers) and counting up from (a forward counting sequence from smaller numbers) (Carpenter and Moser 1984).

Problem solving is central to mathematics. One of the challenges in mathematics education is to help children to become skilled problem solvers rather than rote learners. Even after 30 years of reform, rote thinking is still common in classroom mathematics problem solving practices (Lithner 2008). Students often complete exercises in their textbook in which similar tasks are provided as exemplified in the book (Granberg and Olsson 2015). Rote learners are imitative; learners imitate a solution procedure memorised from the textbook. Conversely, creative reasoning engages students by allowing them to develop well-founded mathematically anchored arguments for their choice of methods in non-routine problem solving processes. Studies have shown that in most problem solving attempts students who engaged with creative reasoning performed significantly better than students who used imitative reasoning (Boaler 1998; Jonsson et al. 2014; Kapur 2011). In conjunction with challenging non-routine problems, collaboration is often suggested, since it can improve students' conceptual understanding (Boaler and GReeno 2000; Stahl et al. 2011). However, to accomplish collaborative creative reasoning, a suitable learning environment needs to be established. In this learning environment, students need to apply new strategies repeatedly with the objective being the advancement of their competency in addition and subtraction problem solving. Collaboration on a challenging problem cannot be automatically initiated within groups. The process of negotiation to seek the new knowledge (i.e. the correct strategies) must be made visible to the learners during their collaborative efforts. Therefore, mathematics learning that is solely print based and structured by content printed in a book is therefore inadequate (Clausen-May 2013).

Research evidence over the last 40 years regarding the impact of digital technology on learning consistently identifies positive benefits. In terms of teaching and learning technology resources, there are a number of free online mathematical problem solving digital artefacts. These tools mainly afford opportunities to learn interactively with ideas, content and modalities that were not previously possible (Yelland 2015). Research results also have indicated that the integration of digital devices in a classroom learning context facilitates cooperative participation of young learners with other classmates and teachers (Lindahl and Folkesson 2012; Wakefield and Smith 2012). That said, teaching could create and facilitate learning contexts, but not the actual learning. Learning mathematics and acquiring the competency to solve problems have been largely understood as a rational cognitive process (Chiu and Churchill 2015b; Zan et al. 2006). The actual learning takes place when learners make sense of mathematics through a meaning-making process.

The process of meaning-making is implicit and indirect (Seeger 2011). This interaction with digital text incorporates the four macro-skills of listening, speaking, reading and writing, but requires additional skills including frequent use of visuals, dynamic information and interaction (i.e. viewing and representing) (Khoo and Churchill 2013; Kress 2010; MOE Singapore 2010). Therefore, the focus of this study is not students' learning, but rather how, during the children's collaboration, the technology facilitates instant immersion in mathematical problem solving practices.

It is increasingly commonplace today for preschoolers to make use of computers in their out-of-school activities (MDG Advertising 2012). Although most children aged 2–5 years are more competent in interacting with a tablet computer than tying their shoe laces (Lunn 2012), the place of ICT in formal education in kindergarten has been contentious (Zaranis et al. 2013).

There is an emerging gap between the capabilities of digital learning in meaning-making and how preschoolers appropriate computers in their mathematics learning. Based on this concise review, we may conclude that the skills applied when preschoolers interact with digital texts are important in revealing the meaning-making processes. These interactions may facilitate the users externalising their understanding and internalising new knowledge. In order to develop a sustainable pedagogy utilising digital technology, it is imperative to investigate how these skills combine with collaborative interactions and result in the gaining of new knowledge. Moreover, we need a more explicit framework to integrate elements in the digital-based learning context of these sustainable practices in institutionalised education.

21.2 Literature Review

21.2.1 Designing a Creative Reasoning and Collaborative Learning Environment

Creative reasoning promotes the development of conceptual understanding in mathematical knowledge (Lithner 2008). In the creative reasoning learning environment, children construct their solving strategies and in this aspect, they are required to struggle with mathematics problems that are somewhat new to them. As they encounter challenging and non-routine problems, they undergo the process of testing and developing their solving strategies, visualising and verifying their arguments as to why their idea does or does not work. Brousseau suggested a didactic design that allows students to be responsible for arriving at solutions. In this design, teachers should not interfere or guide children (Brousseau 1997). Children's autonomous engagement in a teaching and learning activity is particularly important because it functions as a behavioural pathway through which their motivational processes contribute to their subsequent learning and development (Connell and Wellborn 1991; Hyungshim et al. 2010). However, in this learning design, if the students do not receive proper supportive activities, stagnation will likely result, hindering them from moving forward in their problem solving process (Ploetzner et al. 2009).

Therefore, an appropriate level of feedback in response to the students' actions should be introduced: collaborative engagements such as discussions, mutual explanations and elaborations are often suggested as means to assist children to improve their understanding (Hoffkamp 2011). However, having students work in groups do not automatically initiate collaboration. To design a creative reasoning and collaborative learning environment, the problem solving activity must be able to facilitate sharing. That is, students must be able to visualise the meaning-making processes and their representations (Rakes et al. 2010). Moreover, the activity must also allow them to effectively distribute their collaborative efforts through verbalization of mathematical concepts, referencing, testing, visualising, etc. One of the suggested methods is the use of dynamic software (Granberg and Olsson 2015). The proposition that digital texts may support problem solving activities brings us to the question as to how these texts facilitate creative reasoning and a collaborative learning environment.

21.2.2 Viewing and Representing Skills with Digital Texts

The notion of literacy in the twenty-first century has changed with the emergence of digital texts. The advancement of technology has led to some fundamental changes in the ways we receive and produce texts on screens. Digital interfaces can support user input and system output, with multi-mode capabilities including touch, eye gaze, speech, movement and hand gesture as a means of input or synthetic speech, graphical displays, and gesture as output. Thus, these texts are multimodal in nature; the users are required to design their path of engagement actively and continuously on screens both spatially and temporally to make sense of texts and interact when called for. However, the process of meaning-making is always about the interactions between one's perception and the mediums of communicative exchange. Perception is defined as a process of collecting information from the environment based on vision, touch, hearing and muscle to construct an internal representation (Gibson 1979). Therefore, the fundamental skills of the users are crucial in this aspect.

To make meaning with digital texts, one needs to understand various digital functionalities. For example, to search, translate, utilise the affordances of different modes in effective meaning-making, navigate digitally in different ways, and make meaning by placing elements of information with different modes in appropriate spatial/temporal positions. These profound changes brought about by digital texts have led to the development of emerging skills used to interact with digital texts—i.e. viewing and representing.

First, the viewer's interests draw attention to an element that is then selected; via the same process, another element is then selected and so on. In between the selections, an attempt is made by the viewer to integrate the selected elements to form meaning (Kress 2010). In the process of selections and integrations, the meaning will be translated from one mode into another (Mills 2011). In so doing, the viewer alters the meaning of the elements along the lines of their interest (Khoo 2012).

To represent messages with digital text, the producer must have an objective regarding what to show, what message to convey and what he wants to achieve socially, culturally or for other purposes (Kress 2010). The process of composing digital texts incorporates the competence of making meaning with multimodal elements, utilising the affordances of mode, creating meaning by contextually linking elements of different modes and utilising digital functions in meaning-making and navigation. Further, the composer's designs are derived from physical structures in real world settings. The skill of composing with an objective in mind is termed 'representing' (Khoo 2012). Table 21.1 summary of viewing and representing skills includes two levels of engagement with at least five aspects of competencies (Khoo and Churchill 2013). When the children interacted with digital texts in the mathematical problem solving apps, they applied different strategies to arrive at the answers. Viewing and representing skills are necessary when interacting with digital texts on screens to externalise what appears in the user's mind or to internalise the information.

Table 21.1 Viewing and representing skills framework (from Khoo and Churchill 2013)

Macro process	Element selection	Element integration
Skills	*Multi-mode*	*Contextual link*
	The skill to interpret or create elements of different modes to form information	The skill to interpret and create contextual links (in spatial/temporal layouts) with different elements to form information
	Affordances of mode	*Navigation*
	The skill to apply and engage with the affordances of different modes in elements to form information	The skill to move around a screen to integrate different elements to form information
	Digital functionality	
	The skill to assimilate digital functionalities in elements to form information	The skill to assimilate digital functionalities to integrate elements to form information

21.3 Research Design

The current study applies the Activity Theory (Engestrom 1987, 1999) as the theoretical framework in the current mobile technology-related contexts of learning. A close examination was made of the relations between the students (subjects), objectives and tools used in the learning activities (see Fig. 21.1). The study included observations, video recording and interviews, and employed an inductive research strategy that intuitively developed abstractions from the research (Merriam 1988).

The research design is qualitative (Merriam 1988; Yin 1994). Two research questions emerging from the literature reviews guided the data collection and analysis of the current study:

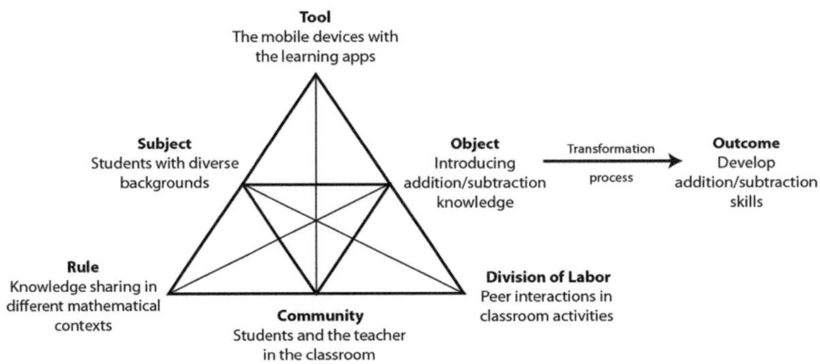

Fig. 21.1 The research framework of the study

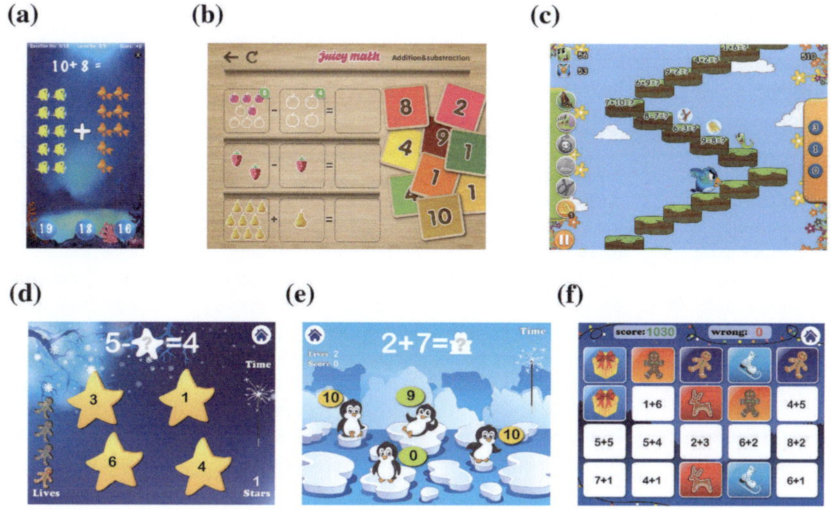

A: Addition & Subtraction for kids
B: Juicy Math
C: Worm Jump
D: Math Adventure (Penguin Craze)
E: Math Adventure (Catch A Star)
F: Math Adventure (Winter Match)

Fig. 21.2 Screen captures of the apps understudied **a**: Addition & Subtraction for kids, **b** Juicy Math, **c** Worm Jump, **d** Math Adventure (Penguin Craze), **e** Math Adventure (Catch A Star), **f** Math Adventure (Winter Match)

- How do digital texts facilitate collaboration in mathematics problem solving activities?
- What elements of digital-based problem solving activities might contribute to or obstruct students' collaborative creative reasoning practices?

Data were collected over a period of 8 months. During the two-level study (see Table 21.2), the study was explorative and a constant comparative method of data analysis was employed.

The study commenced by selecting suitable participants (the sampling units). The selection was terminated when no new information was forthcoming from new sampled units during the research period. Four participants were identified according to their learning styles and their community settings in the classrooms: their profiles are presented in Table 21.3. Pseudonyms are used to preserve their anonymity.

During the 4 months' study at the kindergarten, the teacher used different mathematics apps in children's practices in the classroom. The apps were introduced after the formal mathematics lessons. Each of the apps was used repeatedly two to four times per week. All the children were given the opportunity to use the

Table 21.2 Research methodology procedure and aims

Level 1	Level 2		
Observation	Classroom observations	Interviews with the teachers	Interviews with the children (before and after the lessons)
To find out whether: • The kindergartens were teaching mathematics with mobile devices • The children used the devices individually, and received appropriate feedback in the process of learning, resulting in acquiring new knowledge	• To observe how the participants learn in the classroom with mobile devices • To study how the participants interacted with the artefacts, their peers, the community and the rules in the lessons • To video record the lessons	Unstructured interviews • To determine the objectives of using the mobile devices in the classroom • To identify the addition and subtraction strategies taught by the teacher to the participants	Structured interviews To determine how well the participants understood the questions in the apps • To study how the participants interacted with their peers in lessons and understood the rules, and the maths concepts in the artefacts
• Kindergarten(s) that applied mobile learning in mathematics resulting in authentic learning being observed were shortlisted	• To determine how the students used viewing and representing skills • To ascertain the social learning skills emerging from use of mobile devices	• To verify the researcher's observations on how the teachers mediated mobile technology in learning	• To investigate how the emerging skills mediate digital text in learning addition/subtraction. • To confirm the observational data during the lessons
1st–4th month	5th–8th month		

Table 21.3 Details derived from the selection criteria for the four cases

Participants	The lessons				Age	Gender
	Addition and subtraction for kids (Session 1)	Adventure math (session 2)	Juicy math (Session 3)	Worm jump (Session 4)		
Peter	Team A	Each participant was provided with one device. They practised individually	Each participant was provided with one device. They practised individually	Each participant performed in front of the group until they had completed their practice	5	Male
Mary	Team A				5	Female
Ben	Team B				5	Male
Nicole	Team B				5	Female

apps during practices. Four apps were selected for the current study: "Addition & Subtraction for Kids", "Math Adventure", "Juicy Math" and "Worm Jump". These apps were selected based on the unique characteristics of the instructions they applied in learning mathematics (see Table 21.4).

Table 21.4 The learning apps

Apps	The Nature of the apps
Addition and subtraction for kids (The first app)	The app started with one question at a time. There are 10 questions in a set. The numbers featuring in the questions were from 1 to 20 and were randomly set. Each question was either addition or subtraction. Each provided three possible multi-choice answers. There was no time limit set for the questions. The accumulated scores (10 points for each correct answer) and the number of wrong attempts appeared on the top of the screen (See Fig. 21.2a)
Math adventure (The second app)	There were five different themes (Penguin Craze, Winter Match, Catch a Star, Snowman Hunter & Math Bingo). Different themes provide different instructions in problem solving. For each of the themes selected, the player must further select the type of questions (addition or subtraction) and the number range (1–10 or 1–20). In all themes, the questions had to be completed within a time limit."
	Penguin Craze addition allows the participants two lives (two chances to make mistakes before the game is terminated) and lasts for a maximum of three minutes. The question is an equation (i.e. 3 + 2 = ?) with four choices of answer. For every ten questions attempted, one free life is awarded (see Fig. 21.2e).
	Catch A Star subtraction has two lives and also lasts for a maximum of three minutes. The questions are in equation form and require the finding of an unknown in equations (i.e. "? – 4 = 2" or "8 − ? = 2"), with four choices of answer. For every ten questions attempted, one free life is awarded (see Fig. 21.2d).
	Winter Match addition has 20 boxes, 10 containing numbers and another 10 an incomplete addition equation (e.g. 3 + 2=). The player clicks a box and subsequently selects an answer by clicking another box. Once a question has been correctly solved, both boxes are closed. The score reduces as time advances, reaching a zero score in 110 s. The numbers of incorrect answers are displayed (see Fig. 21.2f)
Juicy math (The third app)	The game provides three choices: addition only, subtraction only or both. There is no set time limit for completion of the questions. The screen displays three questions at any one time. Each of the questions shows different objects in two boxes. The objects can be clicked. For addition, for each of the objects that the player clicks, the number counter increases by one. Once all the objects are clicked the total will appear in each box. The player will drag the answer to the answer box. It will bounce if the answer is wrong. Only when an answer is correct, will it be accepted and the two boxes on the left of the answer box change from objects to numbers. For subtraction, once an object in the second box is clicked, the object in the first box will disappear. The object remaining in the first box is the answer (see Fig. 21.2b)

(continued)

Table 21.4 (continued)

Apps	The Nature of the apps
Worm jump (The fourth app)	The bird in the game moves forward a step at a time. The worm is ahead of the bird and once a question is answered correctly, the worm will move forward a step. If the answer is answered wrongly, the worm will stay put. Once the bird reaches the worm, it will be eaten. To keep the game going, the player must answer promptly to stay ahead of the bird. Addition and subtraction questions are displayed and there are three choices per question. Some obstacles appear after the 40th–45th steps to slow down the player's moves (see Fig. 21.2c), thereby increasing tension and elevating the degree of difficulty in completing the game successfully

21.4 Observation of Participants' Enacting the App with Viewing and Representing Skills

21.4.1 Participant One: Peter

Peter was an outgoing child, asking questions spontaneously of his classmates in the classroom whenever doubts arose in his mind. His recent maths assessment result was about average. This study provides evidence of Peter's competency in enacting the viewing and representing skills (see Table 21.1). In Fig. 21.3, he enacted the viewing skills by interpreting meaning made by elements of different modes and integrating them to form contextual information. He read the question in numeric symbols "6 − 1=" (see Element 1, 3.1 in Fig. 21.3), counted the fish in the picture (see Element 2, 3.1 in Fig. 21.3) below the question and selected the numeric answer at the bottom of the screen (see Elements 3, 3.1 in Fig. 21.3). In Fig. 21.3, Peter selected and integrated the information contextually—i.e. in both spatial and temporal layouts. He chose "5 + 5" on one button and subsequently chose "10" on

3.1 3.2 3.3

Fig. 21.3 Peter enacted the viewing and representing skills

another (see Element 1, 3.2 in Fig. 21.3) to complete the question, while simultaneously observing the countdown score (as in Element 2 & 3, 3.2 of Fig. 21.3). In the fourth app, he read the questions and chose the answer while simultaneously striving to solve the question fast to maintain distance from the bird (see Elements 1, 2 & 3 in Fig. 21.3). He was highly aware of the movement of the bird seeking to eat the worm as it came closer. The game used the movement of the bird as a metaphor for the time limit. In the three apps, Peter's abilities in enacting the viewing and representing skills with digital texts were observed (Khoo and Churchill 2013).

In Peter's engagement with the first app (see Table 21.3), two students were assigned one mobile device and took turns to answer the questions. The teacher briefed all the children in the class before they started with the strategies of "counting on from first" and "counting down from" for solving both addition and subtraction questions. Peter started the game before Mary, with the task being to complete ten consecutive questions. He counted all the objects and chose the answer for two of the ten questions. The addition strategy of modelling (Goldin 1998) was observed. Peter's teammate, Mary, explained the strategies as briefed by the teacher. Subsequently, he answered the remaining questions with the new strategies. He turned to Mary to confirm his answers each time he had made a tentative choice of answers. Mary nodded to confirm her agreement with his choices. The situation demonstrated the process of peer learning (Hwang and Hu 2013; Liu and Carless 2006) where it was mediated in the context of using a mobile device as a learning tool.

For the second app, Peter was assigned a personalised device. He selected the Winter Match in his attempt to solve the set problems. There were two criteria for monitoring the participant's learning outcomes: the speed of solving the questions and the maximum number of wrongs allowed. A countdown timer limited to 110 s the time Peter was given to finish the questions. After the 110 s had elapsed, the game could still be continued but the score was always zero (see Table 21.4). The teacher set a rule that each of the children had to score 200 to finish the activity. "Engaging learners in thinking about achieving outcomes to certain agreed criteria is a learning process" (Liu and Carless 2006, p. 280). Peter started to count using the strategy of modelling. Slowly, he switched to "counting on from the first". He sought feedback from the teacher each time he was in doubt. The teacher guided Peter from time to time. Peter demonstrated his artfulness in engaging with the learning context, while his imprompt interactions with the teacher and the teacher's feedback regulated his learning. His ability to seek feedback was observed.

In the third app, Peter was also given a personalised mobile device, and the instructions of the app provided neither scores nor time limit. Peter attempted the questions using the strategy of "counting all" on the objects in the two boxes. He applied the same strategy to the rest of the questions. It was also observed that Peter could not obtain the answers for some of the questions in his initial attempts.

In the fourth app, Peter played the game in front of the teacher and the group of classmates. It was a 40-min lesson and the lesson was repeated on 3 consecutive days. The teacher set the condition that each child should answer at least ten questions before the termination of the game. Otherwise, he/she had to repeat. Peter started his first attempt with the "counting on from" strategy: e.g. he would count, "4 [pause], 5, 6, 7, 8", then clicked 8. The game was terminated after two questions. Peter observed how other children completed their attempts. The group was noisy. Some would speak out the numbers and answers without any apparent attempt at counting, e.g. 3, 5 [pause]… 8. Peter succeeded on his third attempt with a score of 120 (twelve correct answers). He acquired the number fact strategy by retrieving the recalled number fact from his memory (Carpenter and Moser 1984). His abilities to learn a new strategy along a learning hierarchy from acquisition to fluency were observed. His observational skill in learning was evidenced (Browder et al. 1986).

The interviews were conducted with Peter immediately before and after his lessons. The purpose of the interviews was to confirm his understanding of the strategies observed in the research. The researcher found that he applied the strategy of modelling before the lesson with the first app; he could correctly answer addition/subtraction questions involving small numbers (i.e. $2 + 3 =$, $4 - 3 =$ etc.). After the lessons, the results of the interview indicated that he could answer addition questions involving regrouping two one-digit addends (i.e. $5 + 8 = 13$); he could also answer one and two-digit subtraction questions (i.e. $18 - 4 =$). The interview revealed that he applied new counting strategies of "counting on from" (for addition) and "counting down from given" (for subtraction). In the second app, his ability to adopt the new strategy from "counting all" to "counting on from the first" was also observed.

The third app did not evidence his ability to gain any new strategies, but in the interview his ability to learn a new strategy from the "counting on from" to the "number fact" was evidenced in the fourth app.

21.4.2 Participant Two: Mary

Mary had an easy-going manner, and demonstrated good social assertive skills in the class. She scored good results in her schoolwork in mathematics and was very helpful to her classmates when asked for assistance. As with Peter, Mary demonstrated competencies in engaging with digital texts on screens in meaning-making by applying appropriate viewing and representing skills.

In her engagement with the first app, she had a vicarious experience from teaming with Peter and guiding him to complete the questions. Thus, when it was her turn, she completed the same app exercise quickly and with a full score. In her second app attempt, she was provided with an individualised mobile device to operate on her own. She clicked two consecutive buttons with the same values to close the two buttons (i.e. she clicked 4 & 4; 5 & 5, etc., instead of $1 + 3$ then 4; $2 + 2$ then 4; $1 + 4$ then 5; $2 + 3$ then 5). The remaining buttons panicked her because the problem solving questions were unfamiliar to her (i.e. $4 + 2$, $1 + 5$, $2 + 2$, $1 + 3$ etc.). She paused and pondered, then sought assistance from her teacher, who popped in and clued her up. The teacher introduced a new strategy, the "recall number fact" with no apparent counting (Carpenter and Moser 1984), and repeated the same strategy in the two subsequent questions. Thus, she coached Mary to the point where she was able to handle the problem solving on her own. Mary then independently solved the rest of the questions. The scaffolding process was evidenced (Beishuizen et al. 2010), which Mary embraced, learning the new strategy. In the fourth app, with the assistance of answer hints from the teacher and her peers, she demonstrated her ability to answer the two-digit and one-digit questions.

The interviews with Mary before and after the lesson with the first app activity revealed that she had learnt the new strategy from "counting all" to "counting on from the first". Observational learning was evidenced (Orelove 1982). The pre- and post-lesson interviews revealed that in the fourth app activity she applied the "recall number fact" strategy to addition and subtraction questions with no apparent counting for numbers in the equations less than 10. However, she applied the "counting-on from larger" strategy (Carpenter and Moser 1984) for additions involving "two digits with one digit" (e.g. $11 + 4$). After the lessons, she managed to acquire a new strategy, the decomposition strategy (e.g. $13 + 5 = 10 + 3 + 5 = 18$) (Canobi et al. 1998), learning through reflection from the teacher and peer hints.

21.4.3 Participant Three: Ben

Ben was a brilliant child. He had achieved full scores in most of his schoolwork. He was observant, quiet and able to learn quickly with less practice and repetition than typical of his peers. When dealing with the apps on the screens, he demonstrated competency in engaging with digital texts for meaning-making using viewing and representing skills.

In the second app, Ben chose "Catch a Star" with subtraction $1 - 10$. The questions were in the form of $x - 5 = 2$ and $5 - x = 2$, the task being to determine the value of x. Ben was not able to tackle the form of $x - 5 = 2$, and after two wrong attempts sought help from his teacher. The teacher showed him a trick to sum the numbers of 5 and 2. He managed to acquire knowledge of the new strategy and practised for the next few similar questions. When the question "$5 - x = 2$" appeared, the teacher showed him the strategy of subtracting 2 from 5.

The studies of Ben's classroom activities with the first, third and fourth apps were also examined, and the researcher found that observational learning and peer feedback had occurred.

The pre- and post-lesson interviews on the use of the second app disclosed that Ben had learnt the strategy of "addition to solve subtraction problems" (Peters et al. 2013). When he encountered uncertainty, he inquired in an attempt to find a solution. He also applied his prior knowledge in addition and subtraction to generate new understandings. His ability to learn reflectively was observed (Boud et al. 1985; Koong et al. 2014).

21.4.4 Participant Four: Nicole

Nicole was a shy but obedient student. Although her self-expression was awkward, she was able to learn mathematics, although her mastery of skills was relatively slow. Utilising the data collection methods as per Table 21.1, the researcher observed that Nicole had acquired viewing and representing skills in engaging with information on the screens. In her first app practice, she teamed up with Ben, observing how he completed his ten questions (i.e. counting the items and saying the numbers softly as he was solving the questions: e.g. for "9-1 = ?" he would say "nine minus one", and would then count the objects aloud, "1, 2, 3, …"; for the additional question, he pointed at the first number "9" and spoke it aloud, then counted the second pictorial quantity aloud, "10,11,12, …15"). Nicole observed and imitated what Ben had done. The imitation led to Nicole's learning to apply the new strategy of "the counting on from". She imitated the steps to start from the first number. She then counted the objects on the right and since each time the questions gave a different set of numbers, she practiced and learnt the new strategy. The imitation skill emerged from the current research with Nicole.

The interviews with Nicole were conducted before and after her lessons on the use of the first app. Before the lesson, she applied the strategy of modelling. She demonstrated that she had learnt the new strategy of "counting on from" after using the app in the lesson.

21.5 Discussions

In investigating the first research question, the current study has identified six collaborative skills. Five of these skills were identified in the literature before the data collection (Beishuizen et al. 2010; Goldin 1998; Hwang and Hu 2013; Liu and Carless 2006; Mezirow 1990; Orelove 1982; Pardo 2004): observational learning, embracing the process of scaffolding, reflection, peer learning and seeking feedback. The collaborative skill of imitation emerged in the current study. These collaborative skills take place at the social level and were enacted through interaction with digital texts. They are shown in Table 21.5 below.

In the study, activity theory was applied in categorising into four dimensions how the participants came to know new addition and subtraction learning strategies. The first such dimension is subject–tool–object. During Peter's practice with the first app, he enacted the app with the viewing and representing skills and his mathematical knowledge in addition, i.e. the strategy of modelling. Mary perceived the strategy used by Peter (the externalisation of Peter's understanding was observed) and she applied verbal and non-verbal responses with the alternative approach of pointing at the first number then continuing with the objects on the right (the strategy of counting on from), subsequently obtaining the answer. Her action led to Peter's new meaning-making and he used the new strategy and

Table 21.5 The collaborative skills enacted through interaction with digital texts

	Collaborative skills	
Skills	*Peer learning*	*Reflection*
	Verbal and non-verbal responses of peers to a person's actions or behaviour that lead to new meaning-making for that person	Intellectual activities in which one's own experiences are explored in order to generate new meaning-making
	Embracing scaffolding	*Observational learning*
	The ability to elicit information and reactions during the scaffolding processes that leads to new meaning-making	The learning skill that one develops along a learning hierarchy from acquisition and fluency to generalisation of initiative behaviour
	Seeking solutions	*Emulating*
	The ability to seek information from various sources or channels through social interaction in a learning context	The act of imitating someone else's successful steps in completing a task, while also applying one's own prior knowledge

Table 21.6 The new strategies acquired by participants compared with their social level

Community	Articulated interplay level	The new learning skills adapted			
		Peter	Mary	Ben	Nicole
Team of two (1st app)	Medium	✓	0	0	✓
Individual (2nd app)	Low	✓	✓	✓	✓
Individual (3rd app)	Low	0	0	0	0
Performed one by one in the group (4th app)	High	✓	✓	✓	✓

Note "✓": the frequency of a new strategy learnt

produced the correct answers for the next questions. Peter's internalisation of the new strategy was observed. Mary had also nodded in agreement with Peter's attempts in using the new strategy. The same dual conscious mental processes—i.e. cognitive processes—were also observed in the second and fourth apps for Peter, Mary and Ben and also first, second and fourth for Nicole (Table 21.6 refers).

The participants designed their path of engagement actively and continuously on screens, both spatially and temporally, to make sense of texts. In Peter's case, with the first app as an example, he read the question "6-1", counted the fish in the picture below the question and selected the answer at the bottom of the screen. First, Peter's interests drew attention to an element that was then selected; via the same process, another element was then selected and so on. In between the selections, an attempt was made by Peter to interact with the selected elements to form meaning (see Table 21.1). The processes were visualised by peers as shared context for their joint problem space. In Peter's case, Mary noted the process (numbers, images, spatial and temporal information) by which he answered his questions. They constructed a shared conception of the given problem by applying different

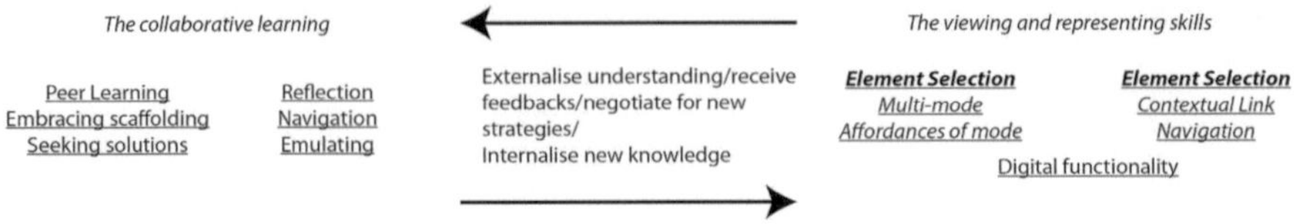

Fig. 21.4 The viewing skills facilitate collaboration amongst peers

collaborative skills as Mary explained the strategies as briefed by the teacher, emphasising parts that Peter might have missed (see Table 21.5). The crossed case study has revealed how digital texts, through viewing and representing skills, merged with the collaboration of the participants iteratively to enable the meaning-making process. These interactions facilitated the participants externalising their understanding and internalising new knowledge. The idea is demonstrated in Fig. 21.4. Research question one is answered. However, as we have observed in Table 21.6, some of the interactions might not result in the gaining of new knowledge. Results from the current research indicate that others elements must be further investigate.

In the second dimension, subject–community–object, the community setting provides the context in which the individual's learning interaction (articulated interplay) takes place through the activity (see Table 21.6). Although the findings of previous studies are that children's feelings of isolation and a sense of presence were positively correlated with the effectiveness of their learning (Cereijo et al. 2001; Rovai and Wighting 2005), the current research results show that the community setting does not have an absolute effect in new meaning-making. In the second app, all the participants gained new knowledge, but in the third app, none of the four participants gained any new knowledge.

In the third dimension, subject–division of labour–object, the individual's roles within the community differ in terms of the degree of engagement in the activities. The participants exhibited different learning styles: Peter and Mary preferred to deal with people, although Mary was smarter than Peter. Ben was excellent in inductive reasoning and Nicole was likely to solve problems in an intuitive trial-and-error manner (Carpenter et al. 1978; Kolb 1981). However, both Ben and Nicole were shy and neither was assertive when they encountered problems in answering their questions (see Table 21.7). Nevertheless, there were no significant results indicating that the different learning styles prevented them from acquiring new knowledge (see Table 21.7).

In the fourth dimension, subject–rule–object, three participants were observed acquiring new strategies in the course of attempting to adhere to the instructional requirement of the scores and speed limit when completing tasks (i.e. Peter and Mary with the second app, and Ben with the fourth). Intellectual engagement with outcomes and standards are focuses of participant involvement in solving the questions and led to clear standards in achieving high quality learning outcomes. The rules were different in different apps (see Table 21.8). The results demonstrated that the rules of the games were important in helping the participants to learn the

Table 21.7 The combination of different styles versus the new knowledge learnt

Learning style/Performance	Result is above average	Result is average
Assertive	Mary ✓✓✓	Peter ✓✓✓
Shy and not outspoken	Ben ✓✓	Nicole ✓

Note "✓": the frequency of a new strategy learnt

Table 21.8 The standard of different apps

The apps	The standard	The new strategy adopted by the four participants
Addition and subtraction for kids	Result score	✓✓
Math adventure	Time limit and result score	✓✓✓✓
Juicy math	–	–
Worm jump	Time limit	✓✓✓✓

Note "✓": the frequency of a new strategy learnt

new strategies. The third app was not designed with any rules that might help the participants to learn, although the app might have the potential to guide participants to new knowledge.

The analysis of the four dimensions also resulted in the following summaries.

- The externalisation and internalisation of the understandings might not result in new meaning-making, as observed in the first app with Mary and Ben (Table 21.6 refers).
- The differences in learning style of the participants did not significantly hinder them from learning new strategies (Table 21.7 refers).
- Community settings have no absolute effect on the meaning-making, whereas the type of rules embedded in the game designs might play important roles in this regard (see Tables 21.6 and 21.8).

The current study shows that digital technology enables students to enact creative reasoning with digital texts. They receive positive and negative feedback relative to their actions, which allows them to modify their subsequent actions, mostly without guidance from the teachers. They collaboratively constructed and shared conceptions through visualisation of their created and enacted solving strategies with digital texts. However, as discovered in I & III, in addition to the previous studies (Granberg and Olsson 2015; Lithner 2008), the application of dynamic software to engage students in a collaborative and creative reasoning learning environment might need to consider the interplay among elements that took place at the social level of the participants—i.e. the personal learning style, the community setting and the standard of the software design. The standard of the software design has an absolute effect in determining the new meaning-making that resulted in gaining new knowledge (see Tables 21.6 and 21.8). Learning style plays a significant role in determining the intensity of gaining new knowledge. However, the community setting does not appear to play a significant role in this aspect. The findings of research question 2 were, indeed, analogous to a jigsaw puzzle. Through the activity theory, the four dimensions of analysis served as fundamental pieces of the puzzle. Once these pieces are fitted together, the answer to research question 2 was answered in totality.

The summarised framework is presented in Fig. 21.5. The interactions between the participants and the digital artefacts transcend their knowledge levels through a positive meaning-making path that might not have been possible if the contextual setting was not aligned along the path (see Fig. 21.5). This conceptual and pedagogical framework serves as a reference to classroom instructors seeking to incorporate digital apps into classroom mathematical lessons.

A pedagogical implication arises from the current research. Over the past 50 years, research findings on addition and subtraction problem solving strategies have been very well defined and consistent (Carpenter and Moser 1984). Movements are currently are underway to reform the practices of mathematics problem solving by focusing on the flexible use of appropriate strategies, rather than standard school-taught approaches (De Corte et al. 2007; Peters et al. 2013). The finding implies that the conventional theories regarding young children's learning of addition and subtraction need to be reconceptualised, and flexible use of strategies in classroom activities with digital technology promoted and adopted to facilitate collaborative creative reasoning within the proposed framework (see Fig. 21.5).

The current research has several limitations. Although the study focuses on the implementation of classroom mathematics learning through activities, the investigation was confined to the topics of addition and subtraction. Thus, findings may

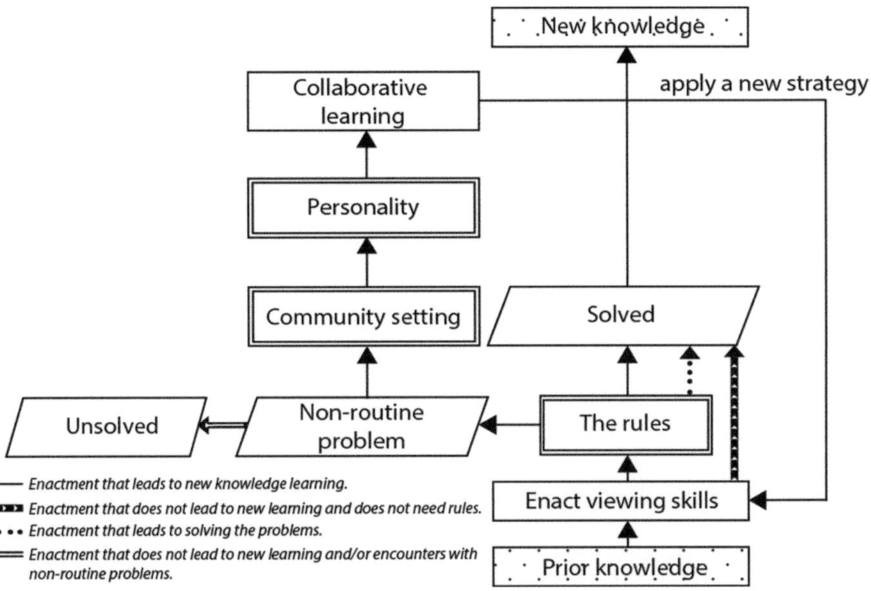

Fig. 21.5 The framework for enacting viewing/representing skills to acquire new knowledge

not be generalizable to all mathematics learning in the classroom. Future research could focus on more topics with a view to developing the current framework.

Further, the kindergarten encountered difficulties in selecting suitable apps for learning purposes. The apps used in the classroom were limited by the participants' age-related cognitive abilities (e.g. the level of written texts used and the aspects of virtual reality of the app content). Doubt must be acknowledged as to the validity and availability of apps suitable for children's classroom learning.

21.6 Conclusions

This paper synthesises relevant literature on collaborative learning and draws on viewing and representing skills to make a case for integrating digital devices in classroom mathematics learning. The study illuminates existing practices of learning mathematics from a new perspective of learning. A central concern of this new perspective is the ways in which children artfully engage with their peers and surroundings to create impromptu sites of learning. The results showed in this study demonstrated that learning apps were useful to facilitate children's classroom learning. The activities under study are learner, new knowledge, assessment and community-centred. Through digital technologies, these activities allow participants to enact autonomous tasks to construct their own meanings. The need for learners to externalise understanding is central to the activities. All parties focus on a common external representation of a subject that allows them to identify and discuss the topics (Laurillard 2002; Pask 1976).

Learning takes place effectively when learners are in control of the learning activity, able to assess and experiment with his/her ideas in the course of pursuing results, and to enquire by working with people in seeking new knowledge, then plan for new actions (Ravenscroft 2000). The current multiple case studies provide a framework for the integration of digital texts into activity-based classroom mathematics learning, along with specific recommendations that emerged from the research findings and implications.

References

Beishuizen, J., Volman, M., & Pol, J. V. D. (2010). Scaffolding in teacher-student interaction: A decade of research. *Education Psychology Review, 22*(1), 271–296.

Bethany, R.-J., & Schneider, M. (2015). Developing conceptual and procedural knowledge of mathematics. In R. C. Kadosh & A. Dowker (Eds.), *Oxford handbook of numerical cognition*. Oxford, UK: Oxford University Press.

Boaler, J. (1998). Open and closed mathematics: Student experiences and understandings. *Journal for Research in Mathematics Education*, 41–62.

Boaler, J., & Greeno, J. G. (2000). *Identify, agency, and knowing in mathematics worlds*. Multiple perspectives on mathematics teaching and learning.

Boud, D., Keogh, R., & Walker, D. (1985). *Reflection: Turning experience into learning*. London: Kogan Page.
Brousseau, G. (1997). *The theory of didactic situations* (N. Balacheff, M. Cooper, R. Sutherland, & V. Warfield, Trans.). Dordreht: Kluwer Academic Publishers.
Browder, D. M., Schoen, S. F., & Lentz, F. E. (1986). Learning to learn through observation. *Journal of Special Education, 20*(4), 447–461.
Canobi, K. H. (2009). Concept-procedure interactions in children's addition and subtraction. *Journal of Experimental Child Psychology, 102*(2009), 131–149.
Canobi, K. H., Reeves, R. A., & Pattison, P. E. (1998). The role of conceptual understanding in children's addition problem solving. *Developmental Psychology, 34*(5), 882–891.
Carpenter, T. P., McLeod, D. B., & McCornack, R. L. (1978). Cognitive Style and Mathematics Learning: The Interaction of Field Independence and Instructional Treatment in Numeration Systems. *Journal for Research in Mathematics Education, 9*(3), 163–174.
Carpenter, T. P., & Moser, J. M. (1984). The acquisition of addition and subtraction concepts in grades one through three. *Journal for Research in Mathematics Education, 15*(3), 179–202.
Cereijo, M. V. P., Young, J., & Wilhelm, R. W. (2001). Factors facilitating student participation in asynchronous web-based courses. *The Journal of Computing in Teacher Education, 18*(1), 32–39.
Chiu, T. K. F., & Churchill, D. (2015a). Design of learning objects for concept learning: Effects of multimedia learning principles and an instructional approach. *Interactive Learning Environments, ahead of print*, 1–16. doi:10.1080/10494820.2015.1006237.
Chiu, T. K. F., & Churchill, D. (2015b). Exploring the characteristics of an optimal design of digital materials for concept learning in mathematics: Multimedia learning and variation theory. *Computers & Education, 82*, 280–291.
Clausen-May, T. (2013). *Teaching mathematics visually & actively*. London: Sage.
Connell, J. P., & Wellborn, J. G. (1991). Competence, autonomy, and relatedness: A motivational analysis of self-system processes. In M. R. Gunnar & L. A. Sroufe (Eds.), *Self processes in development: Minnesota symposium on child psychology* (Vol. 23, pp. 167–216). Chicago, IL: University of Chicago Press.
De Corte, E., Verschaffel, L., & Greer, B. (2007). Whole number concepts and operations. In F. K. Lester (Ed.), *Second handbook of research in mathematics teaching and learning*. New York: MacMillian.
Egan, M., & Hengst, R. (2012). Software on demand: An early childhood numeracy partnership. *Contemporary Issues in Technology and Teacher Education, 12*, 189–213.
Engeström, Y. (1987). *Learning by expanding: An activity theoretical approach to developmental research*. Helsinki: Orienta-Konsultit.
Engestrom, Y. (1999). Activity theory and individual and social transformation. In Y. Engestrom, R. Miettinen, & R.-L. Punamaki (Eds.), *Perspectives on activity theory* (pp. 19–38). Cambridge: Cambridge University Press.
Gibson, J. J. (1979). *the ecology approach to visual perception Boston*. MA: Houghton-Mifflin.
Goldin, G. A. (1998). Representational systems, learning, and problem solving in mathematics. *The Journal of Mathematical Behavior, 17*(2), 136–165.
Granberg, C., & Olsson, J. (2015). ICT-supported problem solving and collaborative creative reasoning: Exploring linear functions using dynamic mathematics software. *The Journal of Mathematical Behavior, 37*(2015), 48–52.
Hiebert, J., & Wearne, D. (1996). Instruction, Understanding, and skill in Multidigit Addition and Subtraction. *Cognitive and instruction, 14*(3), 251–283.
Hoffkamp, A. (2011). The use of interactive visualizations to foster the understanding of concepts of calculus: Design principles and empirical results. *ZMD Mathematics Education, 43*(3), 359–372.
Hwang, W.-Y., & Hu, S.-S. (2013). Analysis of peer learning behaviors using multiple representations in virtual reality and their impacts on geometry problem solving.
Hyungshim, J., Hohnmarshall, R., & Edward, L. D. (2010). Engaging Students in learning activities: It is not autonomy support or structure but autonomy support an structure. *Journal of Educational Psychology, 102*(3), 588–600.

Jonsson, B., Norqvist, M., Liljekvist, Y., & Lithner, J. (2014). Learning mathematics through algorithmic and creative reasoning. *Journal of Mathematical Behavior, 36*, 63–68.

Jordan, N. C., Kaplan, D., Remineni, C., & Locuniak, N. (2009). Early math matters: Kindergarten number competence and later mathematics outcomes. *Developmental Psychology, 45*(3), 850–867.

Kapur, M. (2011). A further study productive failure in mathematical problem solving: Unpacking the design components. *Instructional Science, 39*(4), 561–579.

Khoo, K. Y. (2012). *A case study of children's viewing and representing practices through digital text.* Doctor of Education, The University of Hong Kong, Hong Kong.

Khoo, K. Y., & Churchill, D. (2013). The framework of viewing and representing skills through digital text. *Educational Technology & Society, 16*(3), 13.

Kolb, D. A. (1981). *Learning styles and disciplinary differences.* California: Jossey-Bass Inc., Publishers.

Koong, C.-S., Yang, T.-I., Wu, C.-C., Li, H.-T., & Tseng, C.-C. (2014). An investigation into effectiveness of different reflective learning strategies for learning operational software. *Computers & Education, 72*(1), 167–186.

Kress, G. (2010). *Multimodality- A social semiotic approach to contemporary communication.* Abingdon, Oxon: Routledge.

Laurillard, D. (2002). *Rethinking university teaching: A framework for the effective use of learning technologies* (2nd ed.). London: Routledge Falmer.

Lindahl, G., & Folkesson, A. (2012). ICT in Preschool: Friend or foe? The significance of norms in a changing practice. *International Journal of Early Years Education, 20*, 422–436.

Lithner, J. (2008). A research framework for creative and imitative reasoning. *Educational Studies in Mathematics, 67*(3), 255–276.

Liu, N.-F., & Carless, D. (2006). Peer feedback: The learning element of peer assessment. *Teaching in Higher Education, 11*(3), 279–290.

Lunn, S. (2012). Toddlers, touch screens and the parents' dilemma. *The Australian.* Retrieved from http://www.theaustralian.com.au/news/features/toddlers-touch-screens-and-the-parents-dilemma/story-e6frg6z6-1226238471750.

MDG Advertising. (2012). Kid Tech according to Apple [Infographic]. Retrieved from http://www.mdgadvertising.com/blog/kid-tech-according-to-apple-infographic/.

Merriam, S. B. (1988). *Case study research in education- A qualitative approach.* San Francisco: Jossey-Bass Publishers.

Mezirow, J. (1990). *Fostering critical reflection in adulthood.* San Francisco: Jossey Bass.

Mills, K. (2011). Transmediation in young children's multimodal and digital texts. *Australasian Journal of Early Childhood, 36*(3), 11.

MOE Singapore. (2010). English language syllabus 2010 (Primary & Secondary). Retrieved from http://www.moe.gov.sg/education/syllabuses/languages-and-literature/files/english-primary-secondary-express-normal-academic.pdf.

National Mathematics Advisory Panel. (2008). *Foundations for success: The final report of the National Mathematics Advisory Panel.* Washington, D.C.: Department of Education.

Orelove, F. P. (1982). Acquisition of incidental learning in moderately and severely handicapped adults. *Education and Training of the Mentally Retarded, 17*(2), 131–136.

Pardo, L. S. (2004). What ever teacher needs to know about comprehension. *Reading Teacher, 58*(1), 272–280.

Pask, G. (1976). *Conversation theory: Applications in education and epistemology.* Amsterdam and New York: Elsevier.

Peters, G., Smedt, B. D., Torbeyns, J., Verchaffel, L., & Ghesquière, P. (2013). Children's use of addition to solve two-digit subtraction problems. *British Journal of Psychology, 104*(1), 495–511.

Ploetzner, R., Lippitsch, S., Galmbacher, M., Heuer, D., & Scherrer, S. (2009). Students' difficulties in learning from dynamic visualizations and how they may be overcome. *Computer in Human Behavior, 25*, 56–65.

Rakes, C. R., Valentine, J. C., McGatha, M. B., & Ronau, R. N. (2010). Methods of instructional improvement in algebra: A systematic review and meta-analysis. *Review of Educational Reseach, 80*(3), 56–65.
Ravenscroft, A. (2000). Designing Argumentation for Conceptual Development. *Computer & Education, 34*(2000), 241–255.
Reeve, R. A., Canobi, K. H., & Pattison, P. E. (2003). Patterns of knowledge in children's addition. *Developmental Psychology, 39*(3), 521–534.
Rittle-Johnson, B., Siegler, R. S., & Alibali, M. W. (2001). Developing conceptual understanding and procedural skill in mathematics: An iterative processs. *Journal of Educational Psychology, 93*(2), 346–362.
Rovai, A. P., & Wighting, M. J. (2005). Feelings of alienation and community among higher education students in a virtual classroom. *The Internet and Higher Education, 8*(2), 97–110.
Seeger, F. (2011). On meaning making in mathematics education: Social, emotional, semiotic. *Educational Studies in Mathematics, 77*(2–3), 207–226.
Stahl, G., Rosé, C. P., & Goggins, S. (2011). Analyzing the discourse of GeoGebra collaborations. *Essays in Computer-Supported Collaborative Learning* (186), 33–40.
Star, J. R., Schneider, M., & Rittle-Johnson, B. (2011). Relations among conceptual knowledge, procedural knowledge, and procedural flexibility in two samples differencing in prior knowledge. *Developmental Psychology, 47*(6), 1525–1538.
Wakefield, J., & Smith, D. (2012). From socrates to satellites: iPad learning in an undergraduate course. *Creative Education, 3*, 643–648.
Yelland, N. (2015). Young children as multimodal learners in the information age. In K. L. Heider, & K.L. Renck Jalongo (Eds.), *Young children and families in the information age*. London: Springer.
Yin, R. K. (1994). *Case study research: Design and methods* (2nd ed.). Thousand Oaks, CA: Sage.
Zan, R., Brown, L., Evans, J., & Hannula, M. S. (2006). Affect in mathematics education: An introduction. *Educational studies in mathematics, 63*(2), 113–121.
Zaranis, N., Brown, L., Evans, J., & Hannula, M. (2013). Using mobile devices for teaching realistic mathematics in kindergarten education. *Creative Education, 4*(7), 1–10.

Author Biography

Kay Yong Khoo received his Ed.D from the Faculty of Education, University of Hong Kong. He has been the founding director of child education product development company E-Learn Dot Com Limited, Hong Kong since 2001. He has co-authored and edited 10 uniquely designed book series for kindergartens featuring toys and apps that enhance multimode classroom learning, and has had a leading role in developing classroom-learning apps (teaching and assessment) to promote interactive smart teaching in kindergartens. He has also been active in publishing in his academic field, authoring journal papers, book chapters and conference papers.

Chapter 22
Effects of Prior Knowledge on Mathematics Different Order Thinking Skills in Mobile Multimedia Environments

Thomas K.F. Chiu

Abstract This chapter presents a study that examined the effects of prior knowledge and multimedia design on developing mathematical conceptual understanding in a mobile learning environment. Two different approaches—instructional and noninstructional—were used in the design of the multimedia representation to facilitate students learning for a more complete understanding. Seventy students with different levels of prior knowledge in a secondary school participated in the experiment. Participants were assigned to the 2 (high vs. low prior knowledge group) × 2 (instructional vs. noninstructional) factorial groups to receive the 100-min treatment. The results revealed that the low prior knowledge group outperformed the high prior knowledge group in conceptual knowledge of low order thinking; the instructional group outperformed than the noninstructional group in conceptual knowledge of high order thinking and procedural knowledge; and there was no interaction of prior knowledge and design approach. These findings suggest that mobile multimedia environment enhancing viewing is sufficient for the low order thinking skill development, but not for the high order in mathematics concept learning and procedural skill. Finally, recommendations for future research were suggested.

22.1 Introduction

Mobile technologies are the most widely used information and communication technologies; 90 % of the world population has available access to mobile networks (International Telecommunication Union 2012). Using the mobile devices in classrooms can offer diverse opportunities for teachers and students (Boticki et al. 2015; Gedik et al. 2012). Many schools have been introducing mobile devices in learning and teaching. Most of the devices are small screens. Therefore, one of the

T.K.F. Chiu (✉)
Faculty of Education, The University of Hong Kong, Hong Kong, SAR China
e-mail: tchiu@hku.hk

D. Churchill et al. (eds.), *Mobile Learning Design*, Lecture Notes in Educational Technology, DOI 10.1007/978-981-10-0027-0_22

main challenges in mobile learning is how to design multimedia representations of digital materials for the small screens (Chen et al. 2008; Churchill 2011; Gay et al. 2001); design of multimedia representation should be studied more in mobile environments (Churchill 2011).

Literature addresses that there are limitations on designing effective multimedia presentation of educational materials in mobile environments (see Albers and Kim 2001; Churchill 2011; Churchill and Hedberg 2008; Lee and Bahn 2005). Due to the limited display space, reading multimedia messages on mobile devices is more difficult than that on paper (Albers and Kim 2001). Besides, interactive feature should be the main focus in mobile learning design (Churchill 2011). Churchill (2011) gives some recommendations for designing mobile devices based on the results of his study with Churchill and Hedberg (2008) regarding investigating different design of learning objects via handheld devices. For example, landscape presentation of learning material should be provided to increase the learning space that improved learner experience; no scrolling is recommended; the learning content should be task-oriented; the interaction should be single (one-step), which allows learners to notice immediately responses after manipulation; and zooming function should be provided, which allows learners to be able to enlarge learning content for better reading. Their studies indicated that the suggestions for mobile environments can enhance viewing that lead to better learning. Although these suggestions sound promising, the effects of prior knowledge on design of the multimedia presentation on learning in mobile environments are unclear. This present study aims to investigate the effects of prior knowledge and multimedia design on mathematics concept learning in mobile environments.

22.2 Literature Review

22.2.1 Multimedia Learning and Prior Knowledge

How to design multimedia representation influences the processes and outcomes of learning (Ainsworth 1999; Chiu and Churchill 2015a, b; Mayer 2009). Multimedia learning as learning with both visual and audio representation: visual representation is defined as pictures, graphs or video, audio representation as written or spoken words (Mayer 2009). Multimedia learning is designed to foster meaningful learning that helps learners to construct their knowledge organized in an integrated representation. The goals of multimedia learning are to develop learners' abilities to reproduce and apply the learning content presented in the materials—i.e., facilitating remembering and better understanding. Learning with images and words are more effective than learning with words alone—learners remember more (Chiu and Churchill 2015a, b; Doolittle 2002; Levin et al. 1987; Sankey et al. 2012). Images and words can complement each other, resulting in a better representation than either images or words used in isolation (Bodemer et al. 2004; Clark 1994; Lusk et al. 2009; Fletcher and Tobias 2005; Low and Sweller 2005). Moreover, while words are the basic form of representation in learning environments, Stokes (2002) notes that many studies claim that teaching with visuals leads to a greater degree of learning. Memories of learners are improved when information is supplemented by the use of images (Lusk et al. 2009), which suggests that most people might be expected to retain more information using learning materials including appropriate visual content.

Learner prior knowledge has effects on multimedia learning (Lust et al. 2009; Mayer 2009) in mobile environments (Liu et al. 2013). Students with different levels of prior knowledge responded differently to a multimedia design. Schnotz and Bannert (2003) suggest that images facilitate learning of learners with low prior knowledge; interfere learning of learners with high prior knowledge when the subject matter is visualized. Mayer (1997) suggest that learners with low prior knowledge benefit more from images and words than those with high prior knowledge since learners with higher prior knowledge can construct their mental understanding by reading text only. However, Schnotz and Lowe (2003) suggest that well-designed images and text are important for both low and high prior knowledge learners. Low prior knowledge learners need the images support in developing their mental understanding; and high prior knowledge learners simplify processing for developing mental understanding. Therefore, prior knowledge and multimedia representation have interaction effects on learning.

Many experimental studies have further demonstrated that prior knowledge and multimedia representation have close relationships on two different order thinking skills—remembering and understanding (Kalyuga 2014; Kalyuga et al. 2000; Leslie et al. 2012; Potelle and Rouet 2003; Rey and Fischer 2013; Spanjers et al. 2011), in mathematics learning (Guo et al. 2013; Lee and Chen 2009; Rittle-Johnson and Star 2007, 2009). For example, the design that presented steps to learn with images presented on screen worked best for weak students, but not for strong students (Kalyuga et al. 2000); images helped younger (less prior knowledge) students learn *science better* (*Leslie* et al. 2012)*; and continuous animations were more effective than* segmented animations for experienced students (Spanjers et al. 2011). These studies measured remembering and understanding skills in their experiments (Leslie et al. 2012; Rey and Fischer 2013; Spanjers et al. 2011). The results indicated that the multimedia designs are beneficial to both weak and strong students on remembering, but the designs were only beneficial for weak students on understanding. In conclusion, there are causal relationships between prior knowledge and order thinking skill in multimedia learning. It also seems that low prior knowledge students performed better when developing higher order thinking skills.

22.2.2 Instructional Design in Mathematics

Learning outcomes can be improved significantly when applying instructional strategies in design (Clark 1994; Riffell and Sibley 2005; Wiredu 2005). Learning mathematics concepts, unlike learning language, involved massive amount of learning information and thinking. Specific-discipline instructional strategies should be considered in designing content representation for mathematics concept learning. In algebra teaching, many researches on presenting various forms of learning information for students learning have been conducted. Rittle-johnson and Star (2007, 2009) suggest that comparing and contrasting solution method—student learning better by comparing an equation and its different solution methods or comparing different forms of an equation and their solution method. Moreover, Mok (2009) conducted a study on teaching techniques evolved from variation theory. By applying variation to algebra teaching, students have better understanding of concepts by experiencing different solving methods and algebraic forms at the same time on the same screen. In order to facilitate concept learning, teaching, and learning activities should be intended to help students build relationships among different forms of the same problem (Gu et al. 2004; Mok and Lopez-Real 2006). Mok and Lopez-Real (2006) suggest that variation in equation or solving method should be adopted by teachers for teaching algebra. However, in their studies, either different equations or solving method was varied for students learning, but not equation and solving method together. Moreover, Rittle-Johnson and his colleagues (2009) found that prior knowledge had impact on the effectiveness of the variation in content representation. Their results showed that lower prior knowledge students benefited more when learned by comparing various problems, and high prior knowledge students benefited more when learned by comparing solution methods. Apart from teaching strategies evolved from the variation, algebraic, numerical, graphical, and descriptive should be shown simultaneously when mathematics concepts was taught (NCTM 2000) to ensure effective algebra learning and teaching. The representation intends to help students in perceiving the relationships and associations between conceptual and procedural knowledge. Therefore, the teaching strategies—variation method and four-section representation—should be taken account into designing content representation for algebra learning.

22.2.3 The Present Study

The aim of the present study was to investigate the effects of prior knowledge (low vs. high) and multimedia representation incorporating instructional strategies (noninstructional vs. instructional design approach) in a mobile environment on development of conceptual understanding emerging conceptual knowledge that includes low order thinking skill—graphical property and concept association and high order thinking skill—evaluation of solutions and written explanation (CDC and HKEAA 2007; Kastberg 2003; Schneider and Stern 2005; Thompson 2008; Usiskin 1999), and procedural knowledge of low order thinking skill—graphical representation skills. We focused on developing student conceptual understanding through manipulating learning objects (Chiu and Churchill 2015a; Wagner 2002). More specifically, a conceptual model, a type of learning object, was used in the experiment (Chiu and Churchill 2015a, b; Churchill 2007, 2011, 2013). The representations of the conceptual model are interactive and visual mediated experiment (Chiu and Churchill 2015a, b; Churchill 2007, 2011, 2013). The model that presented property, parameters and relationships of discipline-specific concepts in an interactive and audio and visual way intends to improve conceptual understanding experiment (Churchill 2007, 2011). The learning activity in the experiment was self-learning.

The study examined the following questions. In a mobile environment using multimedia representation, (1) Do students with low prior knowledge outperform those with high prior knowledge on their conceptual understanding? (2) Does the instructional design approach have more positive effect than noninstructional design approach on students' conceptual understanding? (3) Does the combination of design approach and prior knowledge have an effect on students' conceptual understanding?. Hence, we explored three hypotheses: In a mobile multimedia environment, (1) low prior knowledge students will achieve better performance on conceptual knowledge that requires low order thinking skill, (2) Students who learn with instructional design approach will achieve better performance on conceptual knowledge that requires high order thinking skill than those who would learn with noninstructional design approach. (3) Students who would learn with instructional design approach will achieve better performance on procedural knowledge than those who would learn with noninstructional design approach.

22.3 Method

22.3.1 Participants

Seventy students of a secondary school in Hong Kong participated and completed in this study. All the students had been taught the essential concepts about quadratic equation—solving and forming equation skills—by the same teacher approximately 2 weeks before the start of the experiment. The students were divided into two groups (high prior knowledge and low prior knowledge) according to their scores of a recent mathematics quiz on the essential concept taught. The students scored less than 2/3 of full mark in the quiz were assigned to the low prior knowledge group; and those scored more than or equal to 2/3 of full mark in the quiz were assigned to the high prior knowledge group.

22.3.2 Design

A 2 (different design approach: instructional vs. noninstructional) × 2 (prior knowledge: low vs. high) between-subjects design was used to address the hypotheses in this study. The low and high prior knowledge students were assigned to the two situations with different design approaches of a conceptual model. This resulted in the four experimental conditions—18 students with low prior knowledge learning with multimedia representation using instructional design approach, 18 students with high prior knowledge learning with multimedia representation using instructional design approach, 17 students with low prior knowledge learning with multimedia representation using non-instructional design approach, 17 students with lower prior knowledge learning with multimedia representation using non-instructional design approach. Moreover, this study used pre- and post- conceptual and procedural knowledge performance tests to measure the improvement of conceptual understanding in the experiment (Schneider and Stern 2005).

22.3.3 Materials

Materials in the experiment included learning materials, pre and posttests. Learning materials used in the experiment was conceptual models. The conceptual model was adopted from the studies of Chiu and Churchill (2015a, b). Two different designs were non-instructional and instructional approaches. The instructional approach adopted the four-form representation suggested by NCTM and variations of equations and solving method together; and the noninstructional approach showed an equation and its graphical representation. Since well-designed images and text are important for both high and low prior knowledge learners, five multimedia learning design principles—coherence, signaling, spatial contiguity, temporal contiguity, and segmenting—were applied to the two approaches to reducing cognitive load of students. These can make the image and text presented better. The topic was secondary school-level quadratic equations. The design of the conceptual model in the experiment followed the recommendations of the Churchill' mobile multimedia environment design. For example, presentation of the learning content was full-screen and landscape; and information of the conceptual model was tasked-centered; and control slides offered students to control for instant responses (one-step).

In the prior knowledge test, ten questions of multiple choices were given to students before the experiment and each of the questions was scored 1. In the conceptual knowledge performance tests, the four different thinking order skills were measured. They were graphical property, concept association, evaluation of solutions, and written explanations. Total score of each measure was 12.

22.3.4 Procedure

We conducted the experiment in the students' school. The students finished pretests in 40 and 20 min, respectively before the experiment in their classroom. In the experiment, the students conducted self-learning with the conceptual models in two lessons. The students followed a worksheet to learn in the first lesson and conducted their own self-learning by manipulating the models without having the worksheets in the second lesson. After the experiment, the students finished posttests in their lesson.

22.4 Result

The data of gain score in the pre- and post- conceptual knowledge performance tests were analyzed using multivariate analyses of variance (MANOVAs) with graphical property and concept association as dependent variables and evaluation of solution and written explanation as dependent variables. Moreover, the data of gain score in the pre- and post- procedural knowledge performance tests were analyzed using univariate analyses of variance (ANOVAs) with the dependent variables—solving equation skill, forming equation skill, and graphical representation skill. The data of the conceptual and procedural knowledge performance tests are presented in Table 22.1.

22.4.1 Graphical Property and Concept Association

For the dependent variables graphical property and concept association, the MANOVA revealed a significant main effect of prior knowledge, Wilk's $\lambda = 0.81$, F (2,65) = 7.65, $p < 0.001$, partial $\eta^2 = 0.19$, with low prior knowledge students performing better than the high low prior knowledge students. No significant main effect of design approach was found, Wilk's $\lambda = 0.94$, $F(2,65) = 1.97$, $p > 0.05$, partial $\eta^2 = 0.06$. No significant interaction effect between prior knowledge and design approach was found, Wilk's $\lambda = 0.98$, $F(2,65) = 0.61$, $p > 0.005$, partial $\eta^2 = 0.018$.

Table 22.1 Means and SDs of the dependent variable

Measures	Instructional design approach				Noninstructional design approach			
	High prior knowledge		Low prior knowledge		High prior knowledge		Low prior knowledge	
	M	SD	M	SD	M	SD	M	SD
Conceptual knowledge performance test								
GP	3.72	3.74	5.83	3.73	1.47	2.12	5.35	3.50
CA	1.67	1.91	2.56	2.53	0.76	1.99	2.00	2.32
ES	3.28	240	1.56	2.43	0.59	3.20	0.59	1.50
WE	3.39	3.24	2.00	3.41	0.53	1.59	0.82	1.67
Procedural knowledge performance test								
GR	2.33	4.06	4.89	5.40	1.29	4.22	1.53	4.49

Note GP Graphical property; *CA* Concept association; *ES* Evaluation of solution; *WE* Written explanation; *GR* Graphical representation

The follow-up ANOVAs' results showed the main effect of prior knowledge yielded an F ratio of $F(1, 66) = 13.96$, $p < 0.001$, partial $\eta^2 = 0.18$, indicating that the mean gain score was significantly greater for low prior knowledge ($M = 5.6$, $SD = 3.57$) than for high prior knowledge ($M = 2.63$, $SD = 3.23$).The main effect of design approach yielded an F ratio of $F(1, 66) = 2.90$, $p > 0.05$, partial $\eta^2 = 0.042$, indicating that the mean change score was not significant. The interaction effect was not significant, $F(1, 66) = 1.22$, $p > 0.05$, partial $\eta^2 = 0.018$.

With regard to concept association, a significant main effect of prior knowledge was found, $F(1, 66) = 4.08$, $p < 0.05$, partial $\eta^2 = 0.058$. The mean gain score was significantly greater for low prior knowledge ($M = 2.29$, $SD = 2.41$) than for high prior knowledge ($M = 1.23$, $SD = 1.97$). No significant main effect of design approach was found, $F(1, 66) = 1.92$, $p > 0.5$, partial $\eta^2 = 0.28$. The interaction effect was not significant, $F(1, 66) = 0.11$, $p > 0.05$, partial $\eta^2 = 0.002$.

For the dependent variables graphical property and concept association, the significant main effect was found for the independent factor prior knowledge but not the design approach. This result revealed that low prior knowledge students achieved higher conceptual knowledge performance that requires low order thinking skill when they learned with interactive multimedia representation.

22.4.2 Evaluation of Solution and Written Explanation

For the dependent variables evaluation of solution and written explanation, the MANOVAs' results showed a significant main effect of design approach was found, Wilk's $\lambda = 0.81$, $F(2,65) = 7.51$, $p < 0.001$, partial $\eta^2 = 0.19$, with the instructional design approach group performed better than the noninstructional design group. No significant main effect of prior knowledge was found, Wilk's $\lambda = 0.97$, $F(2,65) = 1.15$, $p > 0.05$, partial $\eta^2 = 0.03$. No significant interaction effect between prior knowledge and design approach was found, Wilk's $\lambda = 0.96$, $F(2,65) = 1.48$, $p > 0.005$, partial $\eta^2 = 0.04$.

The follow-up univariate ANOVAs with dependent variable evaluation of solutions yielded a main effect for the design approach, $F(1, 66) = 9.69$, $p < 0.01$, partial $\eta^2 = 0.13$, such that the mean gain score was significantly greater for instructional design approach ($M = 2.42$, $SD = 2.53$) than for noninstructional design approach ($M = 0.59$, $SD = 2.46$). The main effect of prior knowledge was nonsignificant, $F(1, 66) = 2.15$, $p > 0.05$, partial $\eta^2 = 0.032$. The interaction effect was nonsignificant, $F(1, 66) = 2.15$, $p > 0.05$, partial $\eta^2 = 0.03$.

With regard to the dependent variable written explanation, univariate ANOVAs yielded a main effect for the design approach, $F(1, 66) = 10.20$, $p < 0.01$, partial $\eta^2 = 0.13$, such that the mean gain score was significantly greater for experimental design ($M = 2.69$, $SD = 3.35$) than for control design ($M = 0.68$, $SD = 1.61$). The main effect of prior knowledge was nonsignificant, $F(1, 66) = 0.75$, $p > 0.05$, partial $\eta^2 = 0.01$. The interaction effect was not significant, $F(1, 66) = 1.77$, $p > 0.05$, partial $\eta^2 = 0.03$.

For the dependent variables evaluation of solution and written explanation, the significant main effect was found for the independent factor design approach but not the prior knowledge. This result revealed that students learned with the instructional design approach achieved higher conceptual knowledge performance that requires high order thinking skill than those learned with the noninstructional design approach.

22.4.3 *Procedural Knowledge*

For the dependent variable graphical representation, an univariate ANOVA yielded a main effect for the design approach, $F(1, 66) = 4.04$, $p < 0.05$, partial $\eta^2 = 0.06$, such that the mean gain score was significantly greater for instructional design approach ($M = 3.61$, $SD = 4.88$) than for noninstructional design approach ($M = 1.41$, $SD = 4.29$). The main effect of prior knowledge was nonsignificant, $F(1, 66) = 1.62$, $p > 0.05$, partial $\eta^2 = 0.02$. The interaction effect was not significant, $F(1, 66) = 1.12$, $p > 0.05$, partial $\eta^2 = 0.02$.

The significant main effect was found in the dependent variable graphical representation, for the independent factor design approach but not the prior knowledge. This result revealed that students learned with the instructional design approach achieved higher performance on the question of graphical representation than those learned with the noninstructional design approach.

22.5 Discussions

As predicted, students' prior knowledge had impact on the effectiveness of design approach in the mobile multimedia environment. The results of this experiment indicated that there were no interactions between design approach of multimedia representation and level of prior knowledge on concept learning. That is, only one factor—either design approach or prior knowledge—affected the effectiveness of development of conceptual understanding. These results were in line with the results of Guo et al. (2013) study.

22.5.1 *Prior Knowledge and Low Order Thinking Skill in Conceptual Knowledge*

The results of the present study indicated that the effectiveness of design approach was absent for students on improving conceptual understanding that was easily transferred to conceptual knowledge that requires low order thinking skill—graphical property and concept association. Those students did not show any preferences for either design approach of multimedia representation. However, students' prior knowledge effect was significant. Low prior knowledge students benefited more from the multimedia representation than high prior knowledge students, which is in line with the research of Mampadi et al. (2009). That is, prior knowledge of students are the main effect of the improvement of their conceptual understanding that were easily transferred to the conceptual knowledge of graphical property and concept association when a well-designed mobile learning environment were provided.

22.5.2 Design Approach and High Order Thinking Skill in Conceptual Knowledge

The findings showed that design approach of multimedia representation was the main effect on improving students' conceptual understanding that was transferred to conceptual knowledge that requires higher order thinking skill—evaluation of solutions and written explanation. These findings support the teaching techniques evolved from variation theory (Gu et al. 2004; Mok and Lopez-Real 2006) and the four-form representation (NCTM 2000). The effect of prior knowledge was not significant. The improvement of conceptual understanding depended on the design approach rather than prior knowledge. This suggests effective multimedia representation was the significant predictors on predicting improvement of conceptual understanding developed through manipulation. Students learned better in a specific design using teaching strategies rather than a general design. That is, how to design multimedia representation in a mobile learning environment affects the effectiveness of high order thinking skill in algebra.

22.5.3 Design Approach and Procedural Skill

The findings suggested that design approach of multimedia representation was the main effect on developing students' conceptual understanding for answering the questions regarding to procedural knowledge. The specific multimedia representation—using instructional approach—had a more promising effect in the experiment, suggesting the instructional approach is likely to help students to develop a more complete conceptual understanding that was translated to graphical representation skill than the noninstructional approach. This also suggests that the interactive multimedia representation was not sufficient in designing mobile learning environment even though the graphical representation skill only requires conceptual understanding regarding to the relationships between equations and their graphs.

22.6 Conclusions

This study addresses the issue of multimedia representation in designing mobile learning environment and how different level of prior knowledge responses the environment. To measure the improvement of conceptual understanding, conceptual and procedural knowledge performances were measured. Several pedagogical implications for mobile design can be drawn from this study. First, mobile multimedia environment designed using recommendations enhance viewing can provide different level of prior knowledge students with learning opportunities that improve on conceptual knowledge requires low order thinking skill; and low prior knowledge students can benefit more. This also implicates that learning opportunities for mathematics low order thinking skill offered by mobile environment is as effective as other environment as long as it is designed for enhancing effective viewing. Second, the results suggest instructional design approach was more beneficial to students than that noninstructional approach in improving conceptual knowledge that requires high order thinking skill. This indicates that although enhancing viewing in mobile design recommendations is important, it can be insufficient for higher order thinking skill development. It can be better when instructional design approach was applied to provide more pedagogical aspect—representations of relationships among, equations, graphs, solving methods and descriptions. Therefore, multimedia representation design is very important in mobile environment when it comes to improving high order thinking skills. Students likely required a more structured and pedagogical multimedia representation to develop a more complete understanding. Finally, the combination of prior knowledge and multimedia representation could not significantly predict student conceptual and procedural knowledge performances. Therefore, in mobile multimedia learning environment, ways to incorporate students' prior knowledge was the key to promoting their conceptual understanding that leads to low order thinking skills; and focusing on the multimedia representation design was important to development of conceptual understanding that easily transferred to conceptual knowledge of high order thinking skill and procedural knowledge.

The present study shows good results, but there are some limitations. First, only one procedural skill was measured in the knowledge performance test. More research efforts are necessary to consider other types of procedural skills. Finally, future research could explore other factors that may have effects on the mobile multimedia learning environment design.

References

Ainsworth, S. (1999). The functions of multiple representations. *Computers & Education, 33*, 131–152.

Albers, M., & Kim, L. (2001). Information design for the small-screen interface: An overview of web design issues for personal digital assistants. *Technical Communications, 49*(1), 45–60.

Bodemer, D., Ploetzner, R., Feuerlein, I., & Spada, H. (2004). The active integration of information during learning with dynamic and interactive visualisations. *Learning and Instruction, 14*(3), 325–341.

Boticki, I., Baksa, J., Seow, P., & Looi, C. K. (2015). Usage of a mobile social learning platform with virtual badges in a primary school. *Computers & Education, 86*, 120–136.

Chen, N. S., Hsieh, S. W., & Kinshuk, K. (2008). Effects of short-term memory and content representation type on mobile language learning. *Language Learning & Technology, 12*(3), 93–113.

Chiu, T. K. F., & Churchill, D. (2015a). Design of learning objects for concept learning: Effects of multimedia learning principles and an instructional approach. *Interactive Learning Environment, 1*, 1–16. doi:10.1080/10494820.2015.1006237.

Chiu, T. K. F., & Churchill, D. (2015b). Exploring the characteristics of an optimal design of digital materials for concept learning in mathematics: Multimedia learning and variation theory. *Computers & Education, 82*, 280–291. doi:10.1016/j.compedu.2014.12.

Churchill, D. (2007). Towards a useful classification of learning objects. *Education Technology Research and Development, 55*, 479–497.

Churchill, D. (2011). Conceptual model learning objects and design recommendations for small screens. *Educational Technology & Society, 14*, 203–216.

Churchill, D. (2013). Conceptual model design and learning uses. *Interactive Learning Environments, 21*, 54–67.

Churchill, D., & Hedberg, G. (2008). Learning object design considerations for small-screen handheld devices. *Computers & Education, 50*(3), 881–893.

Clark, R. E. (1994). Media will never influence learning. *Education Technology Research and Development, 42*, 21–29.

Curriculum Development Council, & Hong Kong Examinations and Assessment Authority. (2007). *The new senior secondary mathematics curriculum and assessment guide (secondary 4–6)*. Hong Kong, China: The Government Printer.

Doolittle, P. E. (2002, August). Multimedia learning: Empirical results and practical applications. In *The proceedings of the Irish Educational Technology Users' Conference*.

Fletcher, J. D., & Tobias, S. (2005). The multimedia principle. *The Cambridge Handbook of Multimedia Learning, 117*, 133.

Gay, G., Stefanone, M., Grace-Martin, M., & Hembrooke, H. (2001). The effects of wireless computing in collaborative learning environments. *International Journal of Human-Computer Interaction, 13*(2), 257–276.

Gedik, N., Hanci-Karademirci, A., Kursun, E., & Cagiltay, K. (2012). Key instructional design issues in a cellular phone-based mobile learning project. *Computers & Education, 58*(4), 1149–1159.

Gu, L., Huang, R., & Marton, F. (2004). Teaching with variation: A Chinese way of promoting effective mathematics learning. In L. Fan, N. Y. Wong, J. Cai, & S. Li (Eds.), *How Chinese learn mathematics: Perspectives from insiders* (pp. 309–347). Singapore: World Scientific.

Guo, J. P., Yang, L. Y. & Ding, Y. (2013). Effects of example variability and prior knowledge in how students learn to solve equations. *European Journal of Psychology of Education*, *29*, 1–22.

International Telecommunication Union. (2012). *Measuring the Information Society*. Retrieved May 15, 2014, from http://www.itu.int/en/ITU-/Statistics/Documents/publications/mis2012/MIS2012_without_Annex_4.pdf.

Kalyuga, S. (2014). The expertise reversal principle in multimedia learning. In R. E. Mayer (Ed.), *The Cambridge handbook of multimedia learning* (pp. 576–597). New York: Cambridge University Press.

Kalyuga, S., Chandler, P., & Sweller, J. (2000). Incorporating learner experience into the design of multimedia instruction. *Journal of Educational Psychology, 92*(1), 126.

Kastberg, S. (2003). Using Bloom's taxonomy as a framework for classroom assessment. *Mathematics Teacher, 96*, 402–405.

Lee, C. Y., & Chen, M. P. (2009). A computer game as a context for non-routine mathematical problem solving: The effects of type of question prompt and level of prior knowledge. *Computers & Education, 52*(3), 530–542.

Lee, K., & Bahn, H. (2005). A compressed display technique for 240 x 320 resolution personal digital assistants (PDAs). *IEEE Transactions on Consumer Electronics, 51*(4), 1268–1272.

Leslie, K. C., Low, R., Jin, P., & Sweller, J. (2012). Redundancy and expertise reversal effects when using educational technology to learn primary school science. *Educational Technology Research and Development, 60*(1), 1–13.

Levin, J. R., Anglin, G. J., & Carney, R. N. (1987). On empirically validating functions of pictures in prose. *The Psychology of Illustration, 1*, 51–85.

Liu, T. C., Lin, Y. C., & Paas, F. (2013). Effects of prior knowledge on learning from different compositions of representations in a mobile learning environment. *Computers & Education, 72*, 328–338

Low, R., & Sweller, J. (2005). The modality principle in multimedia learning. *The Cambridge handbook of multimedia learning, 147*, 158.

Lusk, D. L., Evans, A. D., Jeffrey, T. R., Palmer, K. R., Wikstrom, C. S., & Doolittle, P. E. (2009). Multimedia learning and individual differences: Mediating the effects of working memory capacity with segmentation. *British Journal of Educational Technology, 40*(4), 636–651.

Mampadi, F., Chen, S., & Ghinea, G. (2009). The effects of prior knowledge on the use of adaptive hypermedia learning systems. In *Human-computer interaction. Interacting in various application domains* (pp. 156–165). Springer, Berlin

Mayer, R. E. (1997). Multimedia learning: Are we asking the right questions? *Educational psychologist, 32*(1), 1–19.

Mayer, R. E. (2009). *Multimedia learning*. New York: Cambridge Press.

Mok, I. A. C. (2009). *Learning of algebra inspiration from students' understanding of the distributive law*. Hong Kong, China: Hong Kong Association for Mathematics Education.

Mok, I. A. C., & Lopez-Real, F. (2006). A tale of two cities: A comparison of six teachers in Hong Kong and Shanghai. In D. Clarke, C. Keitel, & Y. Shimizu (Eds.), *Mathematics classrooms in 12 countries: The insiders' perspective* (pp. 237–246). Rotterdam, Netherlands: Sense Publishers B.V.

National Council of Teachers of Mathematics. (2000). *Principles and standards for school mathematics*. Reston, VA: Author.

Potelle, H., & Rouet, J. F. (2003). Effects of content representation and readers' prior knowledge on the comprehension of hypertext. *International Journal of Human-Computer Studies, 58*, 327–345.

Rey, G. D., & Fischer, A. (2013). The expertise reversal effect concerning instructional explanations. *Instructional Science, 41*(2), 407–429.

Riffell, S., & Sibley, D. (2005). Using web-based instruction to improve large undergraduate biology courses: An evaluation of a hybrid course format. *Computers & Education, 44*(3), 217–235.

Rittle-Johnson, B., & Star, J. R. (2007). Does comparing solution methods facilitate conceptual and procedural knowledge? An experimental study on learning to solve equations. *Journal of Educational Psychology, 99*(3), 561–574.

Rittle-Johnson, B., & Star, J. R. (2009). Compared with what? The effects of different comparisons on conceptual knowledge and procedural flexibility for equation solving. *Journal of Educational Psychology, 101*(3), 529–544.

Sankey, M., Birch, D., & Gardiner, M. (2012). The impact of multiple representations of content using multimedia on learning outcomes across learning styles and modal preferences. *International Journal of Education and Development using ICT, 7*(3), 18–35.

Schneider, M., & Stern, E. (2005). Conceptual and procedural knowledge of a mathematics problem: Their measurement and their causal interrelations. In *Proceedings of the 27th Annual Conference of the Cognitive Science Society.*

Schnotz, W., & Bannert, M. (2003). Construction and interference in learning from multiple representation. *Learning and Instruction, 13*(2), 141–156.

Schnotz, W., & Lowe, R. (2003). External and internal representations in multimedia learning. *Learning and Instruction, 13*(2), 117–123.

Spanjers, I. A. E., Wouters, P., Van Gog, T., & Van Merrienboer, J. J. G. (2011). An expertise reversal effect of segmentation in learning from animated worked-out examples. *Computers in Human Behavior, 27*(1), 46–52.

Stokes, S. (2002). Visual literacy in teaching and learning: A literature perspective. *Electronic Journal for the Integration of technology in Education, 1*(1), 10–19.

Thompson, T. (2008). Mathematics teachers' interpretation of higher-order thinking in Bloom's taxonomy. *International Electronic Journal of Mathematics Education, 3*, 96–109.

Usiskin, Z. (1999). Conceptions of school algebra and uses of variables. In B. Moses (Ed.), *Algebraic thinking, grades k–12: Readings from NTCM's school-based journals and other publications* (pp. 7–13). Reston, VA: National Council of Teachers of Mathematics.

Wagner, E. D. (2002). The new frontier of learning object design. *The E-Learning Developers' Journal.*

Wiredu, G. (2005, August). *The reconstruction of portable computers: On the flexibility of mobile computing in mobile activities.* Paper presented at IFIP 8.2 Conference on Designing Ubiquitous Information Environments, Cleveland, OH.

Chapter 23
An Investigation of the Effects of Individual Differences on Mobile-Based Augmented Reality English Vocabulary Learning

Tong-Ann Sytwu and Chien-Hwa Wang

Abstract Mobile devices are now widely owned and available to the majority of people. While the affordances of mobile learning include supporting a more personalized, authentic, situated learning based on the findings of many studies, it is crucial and urgent to start rethink pedagogy and learning using mobile devices. Additionally, as the concept of augmented reality (AR) enables learners to receive additional, valuable information in a real setting, this study, thus, aims to investigate the effects of a mobile-based augmented reality simulation learning system for English vocabulary acquisition on learners of different learning styles (field independence/dependence, FI/FD) and English proficiency (high/low) in terms of learning outcome and motivation. An experimental research design was used in this study to identify any differences between FI, FD students and high/low English proficiency learners. The results showed that FD learners benefitted significantly from the mobile AR instruction on learning outcome; there was a borderline significant difference between high and low English proficiency learners on learning outcome; and neither learning styles nor English proficiency affected learning motivation after the mobile AR instruction was applied. From the findings of the present study, individual differences should be considered when a new instructional approach is applied in order to make learning more effective and motivating.

T.-A. Sytwu (✉) · C.-H. Wang
National Taiwan Normal University, Taipei, Taiwan
e-mail: sytwu510@gmail.com

C.-H. Wang
e-mail: wangc@ntnu.edu.tw

D. Churchill et al. (eds.), *Mobile Learning Design*, Lecture Notes in Educational Technology, DOI 10.1007/978-981-10-0027-0_23

23.1 Introduction

English is significant when it comes to cross-cultural, international exchanges, ranging from political, business, scientific, communications, and academic issues (Chang 2011). Accordingly, learning English as a foreign language (EFL) has become a critical matter for both educators and learners. Vocabulary learning, in particular, has always played an important role in laying a solid foundation for the acquisition of a foreign language (Beck et al. 2002; Bormuth 1966; Davis 1944, 1968). As the basic building blocks of English sentences, vocabulary acquisition is necessary for second language (L2) learners to make correct inferences or to understand the content, or even to avoid being diagnosed as learning disabled (Gu 2003; Huang 2007; Nation 2001; August et al. 2005). Furthermore, as Wilkins (1972) noted, "without vocabulary nothing can be conveyed (p. 111)." Thus, constant, numerous researches into vocabulary learning for English as a foreign language have been conducted, showing a keen, urgent interest in finding out how words can best be learned.

Vocabulary learning is crucial for EFL learners to master English; however, both instructors and L2 learners are now encountering difficulties among English vocabulary learning. Aside from the problems of formal English education in Taiwan, such as lack of teaching hours, teacher shortage, and different required vocabulary size resulting in a huge gap between learners' English proficiency (Chang 2011), Barab (2002) stated the main problems in traditional schooling practices are that information becomes decontextualized, knowledge appears to be more indirect, abstract, and experiences are second-handed confined in classroom context. That is, instructions tend to be more fragmented, teacher-centered and irrelevant to student's needs and interests (Cullen 1994). To be more specific, acquiring vocabulary from abstract, textual definitions from a dictionary results in problems when using language in real situations (Brown et al. 1989).

Learners' individual differences also play an essential role in the effectiveness of learning. For example, Oxford and Anderson (1995) suggests that there is a need for language instructors to understand students' learning style to achieve optimal language progress. Furthermore, in the field of learning styles, field independence/dependence (FI/D), in particular, has been considered potentially important in second language acquisition (Chapelle and Green 1992). In addition to learning styles, learners' initial English proficiency may also make a difference in terms of learning outcome. However, previous studies that treated learners' English proficiency as an independent variable were mostly about their usage of learning strategies rather than the effect of a particular instructional method (Su 2005; Lai 2009).

The effects of individual differences, such as learning style, prior knowledge, and competencies, are also widely investigated in technology-supported learning, as the aid of technology in educational settings enables learning to be personalized and creates learning environments that support learners' diversity and individual needs (Chen and Tsai 2012; Wade and Ashman 2007; Kraus et al. 2001); however, the results whether individual differences affect learning outcome with an technology-enhanced instruction are unclear.

It is also noticeable that there has been a gradual shift of learning approach from behaviorist to contextualized, situated approaches (Chuo 2004). Situated learning theory, proposed by Lave and Wenger in the 1990s, posits that knowledge should be constructed in an authentic context and that learning requires social interaction and collaboration (1990). As technology advances, augmented reality (AR) incorporating with the use of mobile devices then provides a solution to support situated learning theory, since AR has the affordances of the real world setting by offering additional and contextual information to support learning, blending learner's learning environment into their real-life contexts (Squire and Klopfer 2007). That is, with mobile devices, wireless connection and location-based technology, a mobile AR learning system then enables and enhances learning by making it ubiquitous, collaborative, personalized, and situated while at the same time bridging formal and informal learning (Wu et al. 2013).

23.2 Literature Review

23.2.1 English Vocabulary Learning

Vocabulary learning is a primary, endless, and indispensable task for English learners (Schmitt 2008). When it comes to the acquisition of L2 vocabulary, Nation (2001) categorized the methods of learning and teaching high frequency words as four main approaches, "direct teaching," "direct learning," "incidental learning," and "planned encounters." To be more specific, "high frequency words" with a commonly agreed coverage of 2,000 English words, refer to vocabulary other than that of academic, technical, and low-frequency words (Nation 2001). As for the learning strategies which learners use to acquire vocabulary, Gu and Johnson (1996) listed six major strategies commonly employed by EFL learners–guessing, dictionary, note-taking, rehearsal, encoding, and activation strategies.

Among the abovementioned four ways of learning/teaching high-frequency words and six vocabulary learning strategies, incidental learning, or the "guessing strategies," which means to guess word meanings from context, has been acknowledged to result in vocabulary growth (Krashen 1989). However, drawbacks of incidental vocabulary learning are also addressed. According to Hunt and Beglar (2002), guessing strategies may only be more beneficial to learners of higher proficiency and it may also be time-consuming. Furthermore, there are risks that learner take wrong guesses or make incorrect inference resulted from the ambiguous information presented in contexts (Shahrokni 2009; Yoshii and Flaitz 2002).

Aside from the drawbacks such as wrong guessing and incorrect inferences, there is a certain degree of difficulty to practice incidental vocabulary learning, since many learners do not have the environment that are needed for this kind of learning to happen (Nation 2001). And while incidental vocabulary learning is often regarded as opposed to the direct intentional learning and teaching, Nation (2001) proposed that the two should be complementary activities, enhancing each other simultaneously, and that a well-designed language learning program should have a proper balance between meaning-focused activities (e.g., incidental learning through reading and speaking activities) and language-focused activities (e.g., the direct study of language items).

As for "dictionary strategies," Brown et al. (1989) pointed out that dictionary-based learning might result in problems when learners try to use the language in real situations. Barab (2002) also argued that the main problems while practicing traditional teaching methods are that information becomes decontextualized, knowledge becomes more indirect, abstract, and experience are second-handed confined in classroom context. Thus, the present study proposed a learning system that aimed to realize the idea of "recognizing a word automatically in natural contexts" (Gu and Johnson 1996, p. 660) by enabling learners to learn in a physical context, which is the actual environment, while at the same time, also providing learners with glosses (i.e., the direct study of language items) through the aid of augmented reality technology.

When it comes to learning styles, there is no one universally acknowledged definition. Keefe (1979, p. 4) described learning styles as "cognitive, affective, and physiological traits that are relatively stable indicators of how learners perceive, interact with, and respond to the learning environment (as cited in Park et al. 2006)." It is necessary for language instructors to understand how students perceive and approach learning tasks, that is, students' distinct learning styles, in order to achieve optimal language learning progress (Oxford and Anderson 1995). Among different classifications of learning styles, field independence/dependence (FI/D) in particular, has been extensively investigated and acknowledged to be potentially important in second language acquisition (Alptekin and Atakan 1990; Chapelle and Green 1992; Wyss 2002; Dörnyei and Skehan 2003). Nonetheless, the results of the empirical studies motivated by the FI/D conducted in the field of second language acquisition showed that the correlation between FI/D and language learning achievement is usually low, and that the FI/D interpretation is simply a measure of intelligence in disguise, which in turn means FI stylists are often the ones showing significant positive correlations (Dörnyei and Skehan 2003).

Regardless of the criticisms, Chapelle and Green (1992) provided a powerful defense that knowing learners' FI/D is still significant to offer a better L2 learning experience, since the measure of FI/D of previous studies only tackles one of the three major constructs, the "cognitive restructuring skills," and ignoring the two other components, "interpersonal competencies" and "reliance on internal versus external referents" (Witkin and Goodenough 1981, p. 54).

The success of second language acquisition is associated with both ends of the FI/D continuum. For instance, FI learners are claimed to be more intense in focusing "on the language stimuli relevant to the language learning task at hand" (Naiman et al. 1978, p. 30) and thus excel in classroom learning which involves analysis, attention to details and mastering in exercises such as being better at tracking grammatical correctness, acquiring linguistic rules, and scoring better on classroom-oriented language tests like cloze test (Chapelle and Green 1992; Wyss 2002). In terms of the personality dimension, Seliger (1977) and Day (1984) denoted that FI learners tend to be the more confident language learners as they depend more on internal reference, and thus may speak out actively and take risks in class. As for FD learners, Chapelle and Roberts (1986) and Brown (1987) suggested that the preference for social interaction of field dependents assists them to acquire language through contextualized practice with native speakers. That is, according to Wyss (2002), FD learners tend to achieve a higher degree of success in everyday language situations beyond the constraints of the classroom, and tackle problems by utilizing interpersonal communication skills. The theory of FI/D, as described by Witkin, is "an ever-changing framework, continuously incorporating new discoveries and new insights about the nature of the dimension" (Goodenough 1986, p. 6).

There are several differences between learners of different English proficiency levels when it comes to English learning. For example, students with higher English proficiency are generally more confident, and hold relatively positive language learning beliefs in language acquisition, while learners of lower English proficiency show passive attitudes, shyness and fear due to students' smaller vocabulary size (Hong-Nam and Leavell 2006; Huang and Tsai 2003). Studies have found that explicitly teaching English vocabulary, in other words, direct teaching benefitted both high and low English proficiency learners (Huang and Tsai 2003; Carlo et al. 2004). Furthermore, to reach a prevalent success in both high and low English proficiency learners, August et al. (2005) suggested that vocabulary definitions and contextual information should be provided and actively involving students through student-directed activity would help reach the goal. Moreover, learners of higher English proficiency tend to use a greater amount of vocabulary learning strategies and they prefer to learn vocabulary in its natural contexts (O'Malley et al. 1985; Gu and Johnson 1996; Lai 2009; Su 2005). Accordingly, the present study, based on situated learning theory, aims at investigating how FI/FD learners differ in learning performance and motivation when vocabulary is acquired in a real setting.

23.2.2 Situated Learning

Situated learning theory, or situated cognition, which has a significant impact on educational thinking, was first proposed by Brown et al. in 1989, asserting that knowledge is constructed in an authentic context requiring social interaction and collaboration and that learning is the outcome of interactions among the people, places, objects, processes, and culture within the given context (Brown et al. 1989; Lave and Wenger 1990). Implementing these ideas in instructional design can cause difficulties as the model of situated learning continue to evolve and develop with new research and technology. In order to incorporate new technology into situated instructional design, it is necessary to pay careful attention to some of the critical characteristics, such as providing authentic context that reflects the way knowledge would be used in reality, providing authentic activities, access to expert performances, multiple roles, perspectives, supporting collaborative construction of knowledge, and so on (Herrington and Oliver 1995).

Augmented reality, AR, refers to the concept to augment virtual information to the reality. Azuma (1997) defined AR to be able to reveal the three following features: a combination of real and virtual world, real-time interaction, and accurate 3D registration of virtual and real objects. As for the implementation of AR, varied technologies can be used, such as PC, handheld devices, head-mounted computers, and so on. According to Squire and Klopfer (2007), AR has the affordances of the real-world setting by offering additional and contextual information to support situated learning.

A number of empirical studies have implemented AR in educational settings and have been proved to enable situated learning. However, most of the proposed AR-facilitated learning systems to date are developed for science and mathematics education, because learning such subjects require visualization of abstract concepts (Wu et al. 2013). Still, there are a few learning systems that incorporated AR or other contextual technology (e.g., mobile technology, RFID, GPS) to support situated learning for other disciplines like language education. The following are three empirical studies of context-aware learning applied specifically in vocabulary acquisition (Wong and Looi 2010; Chen and Li 2010; Ogata and Yano 2004). Wong and Looi (2010) conducted a learner-created, design-oriented, mobile-assisted language learning (MALL) study that required primary students to take photos outside the classroom using mobile devices to demonstrate their knowledge of English prepositions (in, on, over, under, in front, and behind), and were subsequently asked to share, describe their photos in the classroom to illustrate the spatial relationship of the prepositions by making sentences. The researchers discovered that students were excited, engaged, and became “active knowledge builders” rather than passively receiving knowledge in a formal learning setting during the activities (Wong and Looi 2010). According to the teachers, the photo-taking and sentence-making activity helped students to “internalize and enhance the ability to apply the prepositions” with the aid of mobile technology that made learning seamless and thus bridging the gap between formal and informal learning (Wong and Looi 2010). Chen and Li (2010) proposed the idea that ‘context’ is an essential factor in vocabulary learning which also enhances learners’ learning interest and efficiency, and thus came up with an English vocabulary learning system called PCULS (personalized context-aware ubiquitous learning system) that personalizes learning by sending learners location-based English vocabulary through positioning techniques. The results indicated that incorporating context-awareness into the learning system increases learning performance. One of the studies conducted by Ogata and Yano (2004) used a system called TANGO (Tag Added learNinG Objects) that enables learners to acquire vocabulary through authentic objects in the environment with their mobile devices and with the aid of RFID (radio frequency identification) technology. Learners were asked to complete tasks assigned by the system through scanning the RFID tag attached to a specific object, and they reported that relating vocabulary to authentic objects helps them understand the words with greater ease, interest, and engagement. The significance of Ogata and Yano's study lies in its corporation of context-awareness and self-pacedness into vocabulary learning.

23.2.3 The Present Study

Based on the theoretical foundation of situated learning, the present study proposes a learning system that incorporates the technology of augmented reality into English vocabulary learning with the use of mobile devices, enabling learners to acquire vocabulary in an authentic context by actually seeing and interacting with the environment. Learning motivation and learning outcome will then be measured and analyzed to find out whether there is a difference between FI and FD learners. This study aims to describe early research into augmented reality-based mobile learning that attempts to assess its effect on students of different learning styles' learning outcome and perceived motivation in English vocabulary learning of elementary students, by enabling students to actually see, touch and interact with the "vocabulary" in a real setting. The research objective of this study is to investigate whether there is a difference among learning motivation and learning outcome of students of different learning styles exposed to a mobile-based AR simulations learning system proposed in this study. Accordingly, the three primary research questions are:

RQ1 Is there a significant difference between FD and FI learners in the mobile augmented reality-supported English vocabulary instruction?

RQ2 Is there a significant difference between FD and FI learners in learning motivation while the mobile augmented reality instruction was applied?

RQ3 Is there a significant difference between learners of high and low English proficiency levels in the mobile augmented reality-supported English vocabulary instruction?

RQ4 Is there a significant difference between learners of high and low English proficiency levels in learning motivation while the mobile augmented reality instruction was applied?

23.3 Methods

This study employed a quasi-experimental design, a pretest–posttest design, to examine the difference in learning motivation and outcome of participants of two different learning styles (field independence/dependence) and two English proficiency levels (high/low). Before the experiment, a pilot test was conducted to discover problems before the main experiment and thus to ensure the validity of the experimental design. Then, all the participants would take the Group Embedded

Figures Test to distinguish field independent and field dependent learners, followed by a pretest on English vocabulary. After the pretest, participants would then use mobile devices to learn vocabulary in a real setting, which would be their classroom. Finally, they were then given questionnaires and vocabulary tests to measure their learning motivation and learning outcome, respectively. A semi-structured student interviews were also conducted at the end of the experiment in order to provide an in-depth understanding of the lived experience of the third graders regarding their opinions and learning attitude toward the usage of the proposed learning system.

23.3.1 Participants

In order to align the learning system with the existing curriculum guidelines, the participants in this study were 52 third-grade students, from two different classes, class 303 and class 308, at an elementary school in XiZhi District, New Taipei City. Students from both classes used the mobile-based AR simulations learning system in a real setting. As for the participants in the pilot test, a total of 26 third-grade students from the same elementary school were assigned using a convenience sampling method. These students, aged between 9 and 10 year-old have at least received 2 years of formal English education at school for 80–120 min a week. As for the usage of mobile devices, 84 % of the students have the experience of using mobile devices, such as tablets and smartphones or iPods; while 44 % possess their own mobile devices; and 44 % of the participants indicate that they have used mobile devices as learning tools. In brief, from the survey, it can be concluded that participants are fairly familiar with mobile devices to a certain degree, and would not have encountered major problems in terms of operation.

23.3.2 Mobile-Based AR Simulations Learning System Overview

The mobile-based AR simulations learning system requires learners to collaborate with teammates to complete the assigned task using an augmented reality platform called *Aurasma*, a free mobile application enabling users to generate their own augmented reality content and is available for both iPhones and Android phones. With wireless Internet connection and the built-in video camera on, additional information in the form of images, videos, animations, and audio can be shown on users' screen when recognizing the objects one wishes to augment more information to, see Fig. 23.1 for a demonstration of this concept. When participants turn on the *Aurasma* application with their tablets, SAMSUNG Note 2, and found the assigned items in the classroom, relevant digital content would appear on their screen, showing the English word, Chinese equivalent, and English pronunciation, which was previously recorded by a native English speaker who is also a professional EFL teacher.

Fig. 23.1 User screen

23.3.3 *Procedure*

Before the activity using the proposed mobile AR English vocabulary learning system, students were given pretests on their prior knowledge, namely, their English vocabulary knowledge, see Fig. 23.2. Group Embedded Figures Test (GEFT) was also administrated to all 52 participants to research into their learning styles, field dependence or field independence. An instruction on how the test works was given before participants took it, as shown in Fig. 23.3, depicting one of the participants demonstrating how to locate a simple geometric shape embedded in a more complex one. The activity began with the anticipatory set, where students were asked if they had hands-on experience of using a particular item in a classroom, and to contribute to a discussion about their personal experiences and knowledge of a classroom vocabulary, which they had showed great interest in sharing their own thoughts (as depicted in Fig. 23.3), since the topic was closely related to every student's daily life.

After a pre-information and instruction of the activity explaining what they need to do in the classroom, students were randomly divided into nine sub groups with three members in one group, given a mobile device, and began with the activity. In the classroom, they were assigned to a task: First, they were given a clue for the first item. Second, when they successfully found out the first item, additional

Fig. 23.2 Participants taking pretests on English vocabulary (on the *left*) and a participant demonstrating how GEFT works (on the *right*)

Fig. 23.3 Students actively participating

information of the item, the English vocabulary, Chinese equivalent, and audio pronunciation would appear on their screen when scanning the right item. Then, they had to return to the control center to show the screenshot proving they had successfully discovered the assigned item and utter the Chinese and English pronunciation of the vocabulary to get the next clue. In order to accomplish the task, participants need to collect every required item. After all groups had finished the task, the group who spent the least time possible would be awarded.

To ensure grouping (i.e., the composition of group members) would not have an effect on students' learning outcome, learners were asked to take turns using the tablets to eliminate possible confounding factors caused by grouping. Finally, after the activity, they were asked to answer questions on the motivation questionnaire and English vocabulary posttest to gauge their learning outcome

23.3.4 Variables and Instruments

Two independent and two dependent variables were examined in this study, which are learning styles (FI/FD), English proficiency level, learning outcome, and learning motivation, respectively. As for the two independent variables, learning styles and English proficiency levels, students' previous English (midterm and final) exam scores were used to indicate their proficiency in English, dividing them into two groups: high English proficiency (top 27 % among 52 students) and low English proficiency (lowest 27 %); while the Group Embedded Figures Test (GEFT) was used to measure their learning style, field dependence/independence. As for the two dependent variables, learning outcome and learning motivation, to investigate participants' learning outcome, an English vocabulary pre and posttests were used; while a motivation survey was conducted to measure the learning motivation.

The Group Embedded Figures Test (GEFT), developed by Witkin and his associates in 1971, is one of the most widely used measures of field independence/dependence (FI/D), especially in second language acquisition research (Khatib and Hosseinpur 2011). The test requires subjects to locate and trace simple geometric figures embedded in a more complex figure. For example, the participants are asked to identify the simple figure labeled "x," see Fig. 23.4 (Witkin et al. 1971), from a more complex one below, and outline the shape out of it.

The Group Embedded Figures Test used in this study, based on Witkin et al. (1971), was a modified, Chinese version with Mandarin phonetic symbols, zhuyin, to ensure the third-grade participants could fully comprehend the instructions on the test paper. Several geometric figures were also re-illustrated due to some of its undistinguishable features. The test consists of three sections/pages: the first page comprising seven simpler geometric questions serves as a warm-up exercise for students to get familiarize with the test; while the second and third part containing nine questions each with more complex figures are the part where students are given a limited time and get scored to determined whether they are FI or FD stylist.

In general, FI/D is determined by the numbers of the correct answers given by the test takers. That is, those who scored higher are labeled as FI, while those who score lower are branded as FD learning stylists. There are two common ways of

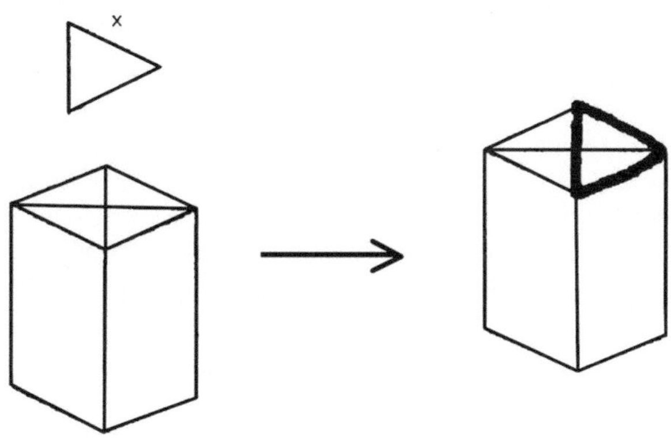

Fig. 23.4 An example of the inquiries in the GEFT (Adapted from "A Manual of Embedded Figures Tests," by Witkin et al. 1971)

determining FD and FI learners. The first method assumes learners who score above the median of the overall scores as FI learners, while those below as FD learners. The second method labels the upper 25 % of the subject as FI learners, while the lower 25 % as FD learners.

Taking the present study's relatively smaller sample size into account, the first method was adopted to determine FI and FD learners. In the present study, a total number of 52 students were examined. The median score was four, and those who scored four and below were categorized as FD learners, consisted of 30 persons; while students who scored above four were deemed to be FI learners, consisted of 22 persons.

The questionnaire (see Appendix A) used in the present study to measure learning motivation was adapted from Liu and Chu (2010), which the two researchers aimed to investigate if their proposed ubiquitous English instructional approach would affect learning outcome and motivation, a research topic similar to the present study. That is, both studies probed into the impact of a more innovative, mobile, context-aware instructional approach on learning outcome and motivation, and particularly in the field of English as foreign language learning. Thus, this motivation scale is best suited for this study since the two studies shared similar learning method, learning topic and variables.

The motivation questionnaire developed by Liu and Chu (2010) applied Keller's attention, relevance, confidence and satisfaction (ARCS) motivation model, which is a model well used in student-centric instructional tasks (Keller 1987; Keller and Suzuki 2004). The core values of this motivation model were also practiced in the design of the proposed mobile AR instruction. That is, with an eye to building an effective learning environment with interesting activities, the present study aimed at stimulating students' visual and auditory senses to draw their *attention*, providing opportunities for self-learning, and cooperation with teammates at the same time. As the proposed mobile AR instruction was designed to be closely related to students' daily life, such as "items one encounters in a classroom," students could also be able to find *relevance* when they were involved in this learning activity. What is more, the activity offered students opportunities to accomplish challenging tasks, and thus providing the possibility of building self-*confidence* and gaining a sense of *satisfaction* afterwards.

The motivation questionnaire used in the present study was composed of 16 questions in total, 4 questions for each dimension, *attention*, *relevance*, *confidence,* and *satisfaction.* The level of motivation will be indicated on a five-point Likert scale from (1) strongly disagree, to (5) strongly agree. English vocabulary tests for the pre and posttests used to examine students after the experiment were designed with a focus on word recognition. In the first section, students were asked to match the English vocabulary and the pictures of that item. For the second section, students had to match the English vocabulary with its Chinese equivalents. The target words of this learning activity were selected from the basic 2000 English word list at elementary school level, issued by Taiwan's Ministry of Education. The English vocabulary pre and posttests were also validated in advance by three experts in related field, mobile-assisted learning and EFL education.

At the end of the experiment, semi-structured interviews with the students were conducted in order to gain in-depth understanding of the participants' live experience in using the proposed mobile AR vocabulary learning system. The interviews contained a predetermined set of questions as follows:

- Do you think the method of English learning employed in this course is interesting? Why or Why not?
- Do you think the method of English learning employed in this course is attractive? Why or Why not?
- Do you think the method of English learning employed in this course is useful? Why or Why not?
- Do you think this course improved your confidence in learning English? Why or Why not?
- Are you satisfied with your English learning achievement? Why or Why not?

23.3.5 Data Analysis

The overall performance, students' English vocabulary test scores, and scores of the questionnaire used to measure learning motivation were collected and analyzed using ANCOVA and independent *t*-test to identify any significant differences between learners of the two different learning styles, FI/FD, and those of high/low English proficiency levels. Pre and posttest designs are widely adopted and considered well suited to investigate the effects of educational innovations. ANCOVA on posttest and pretest as a covariance is a more appropriate and informative analysis (Dugard and Todman 1995).

As for the qualitative data, all interviews were audio-taped and transcribed by the researcher and analyzed with the procedure by first organize the data, generate categories, themes, and patterns; search for alternative explanation for the data, and write the report, as proposed by Marshall and Rossman (1989). Five interviewees from three different classes were chosen and coded according to the class they belong; capital letters, A, B, C for the class code, and the numeric numbers, 1–5 representing their identity.

23.4 Results

23.4.1 Research Question 1

To investigate whether there is a significant difference in the improvement of scores between students of Field Dependence and Field Independence on learning outcome while the proposed mobile AR instruction was applied, analysis of covariance

(ANCOVA) was conducted to analyze the data. The descriptive statistics of FI and FD learners' performance are showed in Table 23.1.

Before ANCOVA, the first step is to analyze the homogeneity of regression coefficients. The result, $F = 0.464$, $p = 0.499 > 0.05$, does not reach the significant level, thus meaning the regression slope of FI and FD groups is equivalent. This result confirms the assumption of homogeneity of regression coefficients; ANCOVA could then be further executed. The ANCOVA evaluation results of FI and FD learners' learning outcome are presented in Table 23.2. The results, $F = 10.010$, $p = 0.003 < 0.005$, indicate that learning styles, FI and FD, do make a difference when students received the proposed mobile AR instruction.

Table 23.3 displayed the estimated posttest score after removing the effect of covariance, showing FD learners' posttest scores (9.602) higher than that of FI learners (8.543). Thus, it can be concluded that FD learners benefitted more than FI learners while a mobile augmented reality English vocabulary learning approach was applied.

23.4.2 *Research Question 2*

To examine whether there is a statistical difference in learning motivation between FI and FD learners, independent samples *t*-test was conducted. The mean and SD of FI were 70.68 and 13.40, respectively; the mean and SD of FD were 68.80 and 6.91, respectively. The t-test results further showed that there is no statistical difference, $p = 0.51$, in motivation between learners of two different learning styles.

Table 23.1 Descriptive statistics of FI and FD's learning outcome

Learning styles	Mean		Std. deviation		*N*
	Pretest	Posttest	Pretest	Posttest	
FI	7.68	8.86	3.153	1.457	22
FD	5.41	9.50	3.232	1.102	30

Table 23.2 ANCOVA evaluation results

Source	SS	df	MS	*F*	*p*
Styles (FI/FD)	12.732	1	12.732	10.010	0.003
Error	62.322	49	1.272		
Total	4456.000	52			

Table 23.3 The estimated score of FI and FD learners after adjusting the dependent effect with respect to the covariance

	Mean	SD	Lower bound	Upper bound
FI group ($N = 22$)	8.543	0.248	8.044	9.042
FD group ($N = 30$)	9.602	0.211	9.178	10.062

That is, learning motivation of both FI and FD learners is more or less similar after they received the proposed mobile AR English vocabulary instruction.

23.4.3 Research Question 3

To investigate whether there is a difference between students of different English proficiency levels after receiving the mobile AR English vocabulary instruction, students were first classified as two groups, higher English proficiency level and lower English proficiency level. Students whose average exam scores among the top 27 % ($N = 14$) were categorized as high English proficiency group; while the lowest 27 % ($N = 14$) were deemed as low English proficiency group. The descriptive statistics of the two groups' learning performance are showed in Table 23.4.

Before the ANCOVA test to explore whether English proficiency level has an impact on learners' learning outcome, the first step is to analyze the homogeneity of regression coefficients. The result, $F = 0.789$, $p = 0.383 > 0.05$, does not reach the significant level, thus meaning the regression slope of high English proficiency and low English proficiency groups is equivalent. In other words, this result confirms the assumption of homogeneity of coefficients; ANCOVA could then be further executed. The ANCOVA evaluation results of high and low English proficiency learners' learning outcome are presented in Table 23.5. The results, $F = 4.179, p = 0.052 > 0.005$, indicate there is no statistical difference in English proficiency in terms of learning outcome. However, the p value = 0.052 has reached the brink of significance. In other words, learners' prior English ability still made a difference in the effectiveness of learning when students received the proposed mobile AR instruction.

23.4.4 Research Question 4

To examine whether there is a statistical difference in learning motivation between high and low English proficiency learners, independent samples t-test was

Table 23.4 Descriptive statistics of high and low English proficiency learners' learning outcome

English proficiency	Mean		Std. deviation		N
	Pretest	Posttest	Pretest	Posttest	
High Eng. proficiency	9.36	9.93	1.393	0.267	14
Low Eng. proficiency	4.36	7.86	3.249	1.657	14

Table 23.5 ANCOVA test on high and low English proficiency learners' learning outcome

Source	SS	df	MS	F	p
Eng. Proficiency	5.425	1	5.425	4.179	0.052
Error	32.458	25	1.298		
Total	2281.000	28			

conducted. The mean and SD of High proficiency were 72.07 and 8.18, respectively; the mean and SD of low proficiency were 71.21 and 4.93, respectively. The results of the *t*-test further inferred that there is no statistical difference, $p = 0.74$, in motivation between learners of two different English proficiency levels. That is, learning motivation of both high and low English proficiency learners is more or less similar after they received the proposed mobile AR English vocabulary instruction.

23.4.5 *Findings from the Interviews*

The main focus of the qualitative part of the study centered on students' self-perceptions upon their experience in the mobile AR learning activity. The findings are presented in clusters that described the five third-grade students' self-reported perceptions of using mobile technologies in learning English vocabulary. Specifically, four categories emerged according to the elements the interviewees brought out, which are, *fun*, *effectiveness*, *confidence,* and *satisfaction*.

23.4.5.1 Fun

One of the main findings from the interviews was that all five interviewees found the learning method fun, interesting, special, and different from the way they used to acquire English vocabulary.

> I never knew English vocabulary could be learned this way, with tablets and with teamwork. It's so different from how our teacher teaches us in the classroom. This activity impressed me a lot. (A1)
> Learning with tablets made everything so fun and exciting! (C4)

The idea of learning with tablets excites most of the third graders, giving them an impression of fun, untraditional, novel learning experience that they are willing to actively take part in.

23.4.5.2 Effectiveness

Effectiveness was what all five interviewees found in the instructional approach. According to the students, the learning activity enabled them to better memorize those vocabulary by actually seeing the object.

> Clearly seeing, touching, and feeling the actual objects enables me to memorize the words. Although there are pictures for the vocabulary in the textbook, you can't see the side or back of that object from a flat image. (C5)
>
> Seeing the real objects helps memorizing it. Helping teammates to pronounce the word when we check with the instructor to get the next clue also helps with my own learning. (B3)

I can better memorize these words because they are important, needed, and widely seen in my daily life. (A2)

Hearing the pronunciation of that word from the tablet helps me pronounce and memorize the word. (C4)

From interviewees' viewpoint, it can be concluded that seeing the authentic object in front of them, finding relevance in the vocabulary, helping teamwork by teaching them, and hearing the pronunciation enabled them to fully understand and retain the word.

23.4.5.3 Satisfaction

Interviews expressed the fun and effectiveness of such learning method satisfied them, because they find it interesting and useful. However, there were some technical problems that unsatisfied one of the interviewees:

I encountered problem like failing to detect the object, problems with saving the screenshot and having the camera to focus. These made me upset because I really wanted to find all of the assigned items. (C5)

The technical problems need to be solved in order to provide learners with a more pleasant, effective learning experience.

23.4.5.4 Confidence

Interviewees gained confidence from knowing they could fully memorize the words:

I felt confident when I knew I would not forget these words for a while. (A2)

Confidence may be derived from the abovementioned effectiveness and satisfaction of the learning activity. In other words, all four factors, *fun, effectiveness, satisfaction,* and *confidence* are all correlated. For example, *fun* of the activity motivated learners to give it a try; while trying, they found it to be an effective way of acquiring vocabulary. Satisfaction and confidence were then subsequently enhanced.

23.5 Conclusion

23.5.1 Summary and Discussion of Major Findings

The researcher first investigates the first independent variable, learning styles, examining whether there is a significant difference between FI and FD learners in learning outcome and learning motivation while the proposed mobile AR English vocabulary instruction was applied. The second part of the results probes into the other independent variable, English proficiency, divided into two groups, high/low English proficiency levels, and examines whether learners' initial English proficiency has an impact on their learning outcome and learning motivation while the proposed mobile AR English vocabulary instruction was applied.

There was a significant difference in FI and FD's learning outcome. Both FI and FD learners improved greatly after the instruction. However, FD learners appear to benefit a lot more than FI learners when this mobile AR instruction was applied. The result showing FD learners benefitting more from the mobile AR instruction which allowed learners to learn in an authentic context is consistent with what Chapelle and Roberts (1986), Brown (1987) and Wyss (2002) suggested that FD learners tend to learn language through contextualized practice and achieve greater success beyond the constraints of traditional teaching method in the classroom. The results also debunked the claims that the significant positive correlations are always in favor of FI learners (Dörnyei and Skehan 2003). Instead, FD learners can be the significant positive correlations given an instructional approach that relates to their learning style.

The results showed that there was no significant difference between FI and FD learners in learning motivation after the learning activity. That is, it can be concluded that both FI and FD learners found this mobile AR learning activity to be motivating at a similar level, which proves the findings of Ogata and Yano (2004) that when vocabulary learning is associated with authentic objects and context, learning interest, and engagement are then increased.

There was a borderline significant difference in high and low English proficiency learners' learning outcome, showing high English proficiency benefits slightly more than low English proficiency learners while the mobile AR instruction was applied. In other words, students of high English proficiency did statistically outperform those who have lower initial English proficiency after removing the effect of covariance. However, low English proficiency learners showed significant improvement in scores after the proposed mobile AR instruction. It can be inferred that a learning method that incorporates context-awareness into daily life with the help of augmented reality and mobile devices may be an effective way for lower English proficiency learners to acquire L2 vocabulary, as it is more related to the learner, enabling learners to find relevance in learning and at the same time providing them with hands-on understanding of the vocabulary by actually seeing and feeling it. Aside from the possibility of high relevance and hands-on learning experience offered by the proposed learning method, learning motivation may be another critical key that leads to ideal learning performance, especially for those lower English proficiency learners, who have more difficulties finding motivation in a subject which they are already poor at. As one interviewee who has lower English proficiency learner suggested, being able to use tablets in learning enhanced his motivation in English learning, which he used to have no interest and confidence in, and in turn showing an improved learning result.

The results further showed that there was no significant difference between high and low English proficiency learners in learning motivation after the learning activity. In other words, both high and low English proficiency learners share more or less the same perception concerning learning motivation, which echoes with the findings of August et al. (2005) that the use of technology in vocabulary learning can generally provide learners with incentives, and thus be motivated regardless of learners' individual differences.

23.5.2 Pedagogical Implications

This study aims to investigate whether individual differences, learning styles, and English proficiency make a difference when receiving the proposed mobile-facilitated augmented reality instruction for English vocabulary learning. The present study would like to shed light on the relations between individual difference and the mobile AR instruction. So the two main pedagogical implications of this study are as follow.

To begin with, teacher should fully recognize students' individual differences, such as their learning styles and prior knowledge. Learning styles can be easily revealed, in the present study, by taking a Group Embedded Figures Test that takes no longer than 20 minutes. After getting the information of students' learning styles, it is important to learn the difference of each learning style and come up with appropriate instructional methods. Students' prior knowledge should also be paid careful attention to, since students at different level may benefit from different instruction. From the results of the present study, students of lower achievement are more likely to benefit from an instructional approach that learners find related to and one that enhances motivation and confidence in learning; while FD learners on the other hand, may also perform outstandingly in a contextualized learning environment like this, rather than in the traditional classroom learning which involved more analysis, attention to details and mastering in exercises such as tracking grammatical correctness, acquiring linguistic rules, and taking classroom-oriented language tests like cloze tests.

Second, the idea of using mobile devices as a learning tool is fully appreciated by elementary school students, where they find enormous learning interest and motivation. As mobile devices usage and ownership have reached a new peak among students, there is great potential in mobile learning. With the help of technology such as augmented reality, learning can then take place everywhere, at any time, beyond the constraints of traditional classrooms. If mobile devices are well used in educational context, students' common belief that mobile devices are only for recreational purposes can be reversed. Instructors should also ponder how mobile learning can be taken advantages of inside the classroom and outside of the classroom, balancing formal and informal learning to the upmost.

In general, instructors should be aware of individual difference, come up with appropriate instruction that benefits students of different characters, and always welcome the idea of incorporating new concepts and technology into instruction.

23.5.3 Limitations of the Study and Suggestions for Future Research

Several limitations have been found in the process of the research, including small sample size, time constraint, limited teaching material, and constrained learning environment. Some suggestions for future studies are also brought out within the limitations.

First, limitations from this study mainly stem from its scope, which is the size and composition of the sample population, and lack of a control group, in particular. The present study involved only 52 third-grade students in elementary school. It would be inappropriate to generalize the results of the study and jump to a conclusion that the effectiveness of a mobile AR instruction applies to every student and that it successfully leads to students' improvement. There is a need for future empirical studies with a larger and more varied sample to clarify the present findings and investigate the relationship between individual differences and such learning approach. What is more, as this study applied random grouping arrangements, it is recommended that future studies look into the effects of different grouping methods, for example, heterogeneous or homogeneous, on the effectiveness of learning among different students.

Second, the time which the mobile AR instruction was carried out was not enough to ensure there is a steady improvement in English vocabulary acquisition. And when any technology is introduced in an educational context, there is always a problem with a novelty effect; implying students tend to get more motivated in the beginning as it is something new to them. However, as students get used to such innovative instruction, the novelty might wear off. Thus, it is recommended that future studies on education incorporated with new technology should include more longitudinal research to examine the effectiveness of learning and its relationship with individual differences.

Third, the target words in the present study consist of only ten vocabularies, which is about the amount students learn from one unit in their formal English classes. It would be more desirable if the number of target words increase. If so, the time for test and burden of the students should also be considered.

Fourth, learning was still constrained in a classroom setting due to practical matters and safety issues. Although the classroom setting counts as an authentic context, the effectiveness of learning English vocabulary outside of the classroom setting should also be investigated in future studies. As there is great potential in the present instructional approach to be utilized in self and informal learning, which learning mostly occurs outside of the classroom setting.

Acknowledgments This paper is dedicated to Dr. Chien-Hwa Wang, Dr. Huey-Kang Sytwu, Dr. Shuh-Jen Sheu, the Sytwu brothers, and Hegel Tsai, who helped and guided me all the way.

Appendix A: Motivation Questionnaire

學習動機的問卷

班級：_________　座號：_________　組別：_________

題號	注意力	非常同意	同意	普通	不同意	非常不同意
1	這個學習單字的方式能吸引我的注意力。					
2	螢幕上面並沒有出現太多東西，我可以抓住重點。					
3	這個學習活動很有趣。					
4	這個學習活動能引起我的好奇心。					
題號	關連	非常同意	同意	普通	不同意	非常不同意
5	這個活動中的內容值得學習。					
6	這個活動的學習內容跟我的日常生活有關。					
7	這個學習活動對我的英文學習有幫助。					
8	我可以將這次所學到的內容與我以前學到的東西做出連結。					
題號	信心	非常同意	同意	普通	不同意	非常不同意
9	活動進行的時候，我有信心可以學好這些內容。					
10	活動進行的時候，我可以自己控制學習的步調。					
11	活動進行的時候，我有把握自己可以通過等下的測驗。					
12	這個學習單字的方式能夠增加我的學習自信。					
題號	滿足	非常同意	同意	普通	不同意	非常不同意
13	我很喜歡這種學英語單字的方式。					
14	這個活動結束後，我很有成就感。					
15	我很滿意我在這個活動中所學到的東西。					
16	這個活動結束後，我感到滿足。					

References

Alptekin, C., & Atakan, S. (1990). Field dependence-independence and hemisphericity as variables in L2 achievement. *Second Language Research, 6*(2), 135–149.

August, D., Carlo, M., Dressler, C., & Snow, C. (2005). The critical role of vocabulary development for English language learners. *Learning Disabilities Research & Practice, 20*(1), 50–57.

Azuma, R. T. (1997). A survey of augmented reality. *Presence-Teleoperators and Virtual Environments, 6*(4), 355–385.

Barab, S. (2002). Human-field interaction as mediated by mobile computers. In T. Koschmann, R. Hall, & N. Miyake (Eds.), *CSCL 2: Carrying forward the conversation* (pp. 533–537). Mahwah, NJ: Lawrence Erlbaum.

Beck, I. L., McKeown, M. G., & Kucan, L. (2002). *Bringing words to life*. New York, NY: The Guilford Press.

Bormuth, J. R. (1966). Readability: A new approach. *Reading Research Quarterly, 1*(3), 79–132.

Brown, H. D. (1987). *Principles of language learning and teaching* (2nd ed.). Upper Saddle River, NJ: Prentice Hall.

Brown, J. S., Collins, A., & Duguid, P. (1989). Situated cognition and the culture of learning. *Educational Researcher, 18*(1), 32–41.

Carlo, M. S., August, D., McLaughlin, B., Snow, C. E., Dressler, C., Lippman, D. N., et al. (2004). Closing the gap: Addressing the vocabulary needs of English-language learners in bilingual and mainstream classrooms. *Reading Research Quarterly, 39*(2), 188–215.

Chang, V. W. (2011). English education in Taiwan: Current situation and examination. *Journal of Educational Resources and Research, 69*, 129–144.

Chapelle, C., & Green, P. (1992). Field Independence/Dependence in second-language acquisition research. *Language Learning, 42*, 47–83.

Chapelle, C., & Roberts, C. (1986). Ambiguity tolerance and field independence as predictors of proficiency in English as a second language. *Language Learning, 36*, 27–45.

Chen, C. M., & Li, Y. L. (2010). Personalised context-aware ubiquitous learning system for supporting effective English vocabulary learning. *Interactive Learning Environments, 18*(4), 341–364.

Chen, C. M., & Tsai, Y. N. (2012). Interactive augmented reality system for enhancing library instruction in elementary schools. *Computers & Education, 59*(2), 638–652.

Chuo, T.-W. I. (2004). The application of computer technology in foreign language instruction. *Languages, Literacy Studies and International Studies, 1*, 15–33.

Cullen, R. (1994). Incorporating a language improvement component in teacher training programmes. *English Language Teachers, 48*, 162–172.

Davis, F. (1944). Fundamental factors in comprehension in reading. *Psychometrika, 9*, 185–190.

Davis, F. (1968). Research in comprehension in reading. *Reading Research Quarterly, 4*, 499–545.

Day, R. (1984). Student participation in the ESL classroom or some imperfections in practice. *Language Learning, 34*(3), 69–102.

Dörnyei, Z., & Skehan, P. (2003). Individual differences in second language learning. In C. J. Doughty & M. H. L. Long (Eds.), *The handbook of second language acquisition* (pp. 589–630). Oxford, England: Blackwell.

Dugard, P., & Todman, J. (1995). Analysis of pre-test-post-test control group designs in educational research. *Educational Psychology, 15*(2), 181–198.

Goodenough, D. R. (1986). History of the field dependence construct. In M. Bertini, L. Pizzamiglio, & S. Wapner (Eds.), *Field dependence in psychological theory, research, and application-two symposia in memory of Herman A. Witkin* (pp. 5–13). Hillsdale, NJ: Lawrence Erlbaum.

Gu, Y. (2003). Fine brush and freehand: The vocabulary-learning art of two successful Chinese EFL learners. *TESOL Quarterly, 37*(1), 73–104.

Gu, Y., & Johnson, R. K. (1996). Vocabulary learning strategies and language learning outcomes. *Language Learning, 46*(4), 643–679.
Herrington, J., & Oliver, R. (1995). Critical characteristics of situated learning: Implications for the instructional design of multimedia.
Hong-Nam, K., & Leavell, A. G. (2006). Language learning strategy use of ESL students in an intensive English learning context. *System, 34*(3), 399–415.
Huang, H. T. (2007). Vocabulary learning in an automated graded reading program. *Language Learning & Technology, 11*(3), 64–82.
Huang, S. C., & Tsai, R. R. (2003). A comparison between high and low English proficiency learners' beliefs.
Hunt, A., & Beglar, D. (2002). Current research and practice in teaching vocabulary. *Methodology in language teaching: An anthology of current practice* (pp. 258–266). Cambridge, England: Cambridge University Press.
Keefe, J. W. (1979). Learning style overview. In *Student learning styles: Diagnosing and prescribing programs* (pp. 1–18). Reston, VA: National Association of Secondary School Principals.
Keller, J. M. (1987). Development and use of the ARCS model of motivational design. *Journal of Instructional Development, 10*(3), 2–10.
Keller, J. M., & Suzuki, K. (2004). October). Learner motivation and E-learning design: A multinationally validated process. *Journal of Educational Media, 29*(3), 229–239.
Khatib, M., & Hosseinpur, R. (2011). On the validity of the Group Embedded Figure Test (GEFT). *Journal Of Language Teaching And Research, 2*(3), 640–648.
Krashen, S. (1989). We acquire vocabulary and spelling by reading: Additional evidence for the input hypothesis. *Modern Language Journal, 73*, 445–464.
Kraus, L. A., Reed, W. M., & Fitzgerald, G. E. (2001). The effects of learning style and hypermedia prior experience on behavioral disorders knowledge and time on task: A case-based hypermedia environment. *Computers in Human Behavior, 17*(1), 125–140.
Lai, Y. C. (2009). Language learning strategy use and English proficiency of university freshmen in Taiwan. *TESOL Quarterly, 43*(2), 255–280.
Lave, J., & Wenger, E. (1990). *Situated learning: Legitimate peripheral participation*. Cambridge, England: Cambridge University Press.
Liu, T. Y., & Chu, Y. L. (2010). Using ubiquitous games in an English listening and speaking course: Impact on learning outcomes and motivation. *Computers & Education, 55*, 630–643.
Marshall, C., & Rossman, G. B. (1989). *Designing qualitative research*. Newbury Park, CA: Sage.
Naiman, N., Frohlich, M., Stern, H., & Todesco, A. (1978). *The good language learner* (Research in Education Series No. 7). Toronto, ON: The Ontario Institute for Studies in Education.
Nation, I. S. P. (2001). *Learning vocabulary in another language*. Cambridge, England: Cambridge University Press.
O'Malley, J. M., Chamot, A. U., Stewner-Manzanares, G., Kupper, L., & Russo, R. P. (1985). Learning strategies used by beginning and intermediate ESL students. *Language Learning, 35* (1), 21–46.
Ogata, H., & Yano, Y. (2004). *Context-aware support for computer-supported ubiquitous learning*. Paper presented at the 2nd IEEE international workshop on wireless and mobile technologies in education, JhongLi, Taiwan, March.
Oxford, R., & Anderson, N. (1995). A crosscultural view of learning styles. *Language Teaching, 28*, 201–221.
Park, C. C., Endo, R., & Goodwin, A. L. (Eds.). (2006). *Asian and Pacific American education: Learning, socialization and identity*. Charlotte, NC: Information Age Publishing.
Schmitt, N. (2008). Review article: Instructed second language vocabulary learning. *Language Teaching Research, 12*(3), 329–363.
Seliger, H. W. (1977). Does practice make perfect? A study of' interaction patterns and L2 competence. *Language Learning, 27*, 263–278.

Shahrokni, S. A. (2009). Second language incidental vocabulary learning: The effect of online textual, pictorial, and textual pictorial glosses. *TESL-EJ, 13*(3). Retrieved Jan 1, 2015 from http://www.tesl-ej.org/wordpress/issues/volume13/ej51/ej51a3/.

Squire, K., & Klopfer, E. (2007). Augmented reality simulations on handheld computers. *Journal of the Learning Sciences, 16*(3), 371–413.

Su, M. H. M. (2005). A study of EFL technological and vocational college students' language learning strategies and their self-perceived English proficiency. *Electronic Journal of Foreign Language Teaching, 2*(1), 44–56.

Wade, V. P., & Ashman, H. (2007). Guest editors' introduction: Evolving the infrastructure for technology-enhanced distance learning. *IEEE Internet Computing, 3*, 16–18.

Wilkins, D. A. (1972). *Linguistics in language teaching*. London, England: Edward Arnold.

Witkin, H. A., & Goodenough, D. R. (1981). Cognitive styles: Essence and origins- Field dependence and field independence. *Psychological Issues Monograph, 51*, 1–141.

Witkin, H. A., Oltman, P. K., Raskin, E., & Karp, S. A. (1971). *A Manual for the Embedded Figures Tests*. Palo Alto, CA: Consulting Psychologists Press.

Wong, L. H., & Looi, C. K. (2010). Vocabulary learning by mobile-assisted authentic content creation and social meaning-making: Two case studies. *Journal of Computer Assisted learning, 26*(5), 421–433.

Wu, H. K., Lee, S. W. Y., Chang, H. Y., & Liang, J. C. (2013). Current status, opportunities and challenges of augmented reality in education. *Computers & Education, 62*, 41–49.

Wyss, R. (2002). *Field independent/dependent learning styles and L2 acquisition.* Computer data on-line. *The weekly column article 102.* Retrieved from http://www.eltnewsletter.com/back/June2002/art1022002.htm

Yoshii, M., & Flaitz, J. (2002). Second language incidental vocabulary retention: The effect of picture and annotation types. *CALICO Journal, 20*(1), 33–58.

Author Biographies

Tong-Ann Sytwu Recently received her master's degree from the Department of Graphic Arts and Communications, National Taiwan Normal University, Taiwan. She majored in English as an undergraduate, during which she developed great interest in communication technology and benefitted greatly herself from technology-supported self learning. Subsequently, Sytwu combined her training and personal interest, and strived to construct an effective formula for English language learning through the benefit of technology. Currently fighting nasty autoimmune diseases, and she's going to win the battle.

Chien-Hwa Wang Professor at the department of Graphic Arts and Communications, National Taiwan Normal University, Taiwan. Wang researches into the application of communication technology in educational settings, currently probing into the effects of mobile-based augmented reality on learning of different disciplines.

Part VI
Future Developments

Chapter 24
Future Directions in Mobile Learning

Mark Pegrum

Abstract The shape of mobile learning (m-learning) depends very much on the complex cultural, social, political, economic and, above all, educational ecologies in which mobile technologies are embedded. Focusing primarily on the developed world, this chapter begins by surveying our contemporary technological context, highlighting new forms of hardware and emerging patterns of usage. It then turns to our contemporary educational context, outlining seven major trends—towards contextualisation, personalisation and diversification of learning; towards student support, engagement and creativity; and towards wider collaboration—which reflect aspects of the broader cultural, social, political and economic landscape. It is suggested that the future of digital learning generally, and m-learning in particular, will take shape at the point where ongoing technological developments intersect with ongoing educational trends.

24.1 Introduction

In any discussion of the future of m-learning, we must take into consideration the larger cultural, social, political, economic and educational ecologies in which mobile technologies, like all technologies, are embedded (Selwyn 2014; Warschauer 2011). In particular, we must consider dominant educational trends, since these will play a major role in determining which current and future technologies will find their way into education, and how they will or will not be deployed. In exploring this territory in this chapter, we will focus primarily on the developed world: while m-learning has extensive applications in the developing world, where it serves to increase educational access in line with a social justice agenda (GSMA 2010; UNESCO 2013), cutting-edge technologies and new educational strategies typically emerge from the developed world.

M. Pegrum (✉)
The University of Western Australia, Crawley, Australia
e-mail: mark.pegrum@uwa.edu.au

D. Churchill et al. (eds.), *Mobile Learning Design*, Lecture Notes in Educational Technology, DOI 10.1007/978-981-10-0027-0_24

After sketching out a definition of m-learning, we will survey today's *technological context,* with its growing mobile ecosystem nestled within a larger digital ecosystem, and consider innovations in the form of new hardware and emerging patterns of technology usage. We will then turn to the *educational context,* pinpointing seven contemporary educational trends, and considering how each intersects with new technological developments. Specifically, we will focus on educational shifts towards: contextualisation of learning (enabled, notably, through augmented reality), personalisation of learning (through big data and learning analytics), diversification of learning (through Massive Open Online Courses, or MOOCs), student support (through virtual assistants), student engagement (through gaming and gamification), student creativity (through makerspaces), and wider collaboration (through digital networking).

Ultimately, it is in the space where ongoing technological developments intersect with ongoing educational trends that the future of m-learning will take shape. Yet because unexpected technological developments may arise, and because the educational landscape, like the wider social, cultural, political and economic landscape, is a partially contested, evolving terrain, it is impossible to chart the future with certainty. With this caveat, this chapter extrapolates from current developments and trends to offer glimpses of a possible future.

24.2 Present and Future Mobile Learning

Mobile learning is of course governed by the principle of mobility, which can apply variously to the devices, the learners, and the learning experiences. The most fundamental of these categories is that of the devices since, unless we make mobile digital devices central to our definition of m-learning, we are obliged to widen it to include other kinds of learning-on-the-move, such as learning supported by toys or books (Pegrum 2014, in press b). It has been suggested that the category of mobile digital devices comprises those which can be used at Point A and Point B, as well as everywhere in between, without stopping (Puentedura 2012). This conception makes room for yesterday's personal digital assistants (PDAs) and digital music players like Apple's discontinued iPod, as well as still widely used feature phones; it makes room for today's multifunctional smart devices, notably smartphones and tablets, as well as increasingly niche products like digital cameras; and, perhaps most importantly, it makes room for emerging devices carried on or in the body, such as wearables and embeddables/implantables, as well as independently or semi-independently mobile technologies like drones or robots. In light of these newer developments, we might slightly modify Puentedura's definition to say that mobile devices are *digital devices which can operate* (rather than 'be used') *at Point A, Point B, and continuously everywhere in between.*

M-learning, then, would be *any learning involving mobile digital devices,* as defined above. Yet if the category of devices is the most important when it comes to the *mobile* part of mobile learning, it is not necessarily the most important when it comes to the *learning* part. It is only when the devices *and* the learners *and* ideally the learning experience are all mobile that the full educational potential of mobile devices can be realised, that is, where their potential to support collaborative, constructivist learning meets their potential to support situated, embodied learning in real-world contexts (Pegrum 2014, in press b). It is important to bear in mind this complex character of m-learning as we consider how it is likely to develop at the intersection of present and future technological and educational trends.

24.3 The Technological Context

The signs of our shift into a mobile, wireless era are clear. As of late 2014, some 60 % of all internet-connected devices were estimated to be smartphones or tablets, with some 70 % of all new computing devices running the Android mobile operating system (Blodget et al. 2014). Over time, internet connectivity and accessibility will gradually become a reality for the more than 50 % of the world's population who currently remain offline; this is thanks to a range of projects like Facebook Zero, Google Free Zone and Wikipedia Zero; cooperative alliances like the Facebook-led Internet.org; and experiments in beaming internet access from drones, as in Facebook's Aquila project, or balloons, as in Google's Project Loon. Yet for the foreseeable future the developed world will remain far ahead in terms of speed and bandwidth, and is thus likely to originate most new technological and related educational developments.

As the educational potential of smart devices is becoming more widely recognised in the developed world, we are seeing a shift towards Bring Your Own Device (BYOD) approaches and, particularly in Western institutions which promote student autonomy, technology-supported flipped classroom approaches (Johnson et al. 2015). Teachers and students have been working with educational apps (Oakley et al. 2012)—despite some concerns over their pedagogical limitations (Gardner and Davies 2013; Pegrum 2014)—as well as exploring the value of social media for facilitating multimodal communication and constructivist networking, of MOOCs for opening up access to diverse educational content and learning communities, of gaming platforms for engaging attention and reinforcing learning, and of new learning spaces for supporting blended educational interactions.

But m-learning is not all about phones and tablets: the mobile ecosystem is rapidly becoming larger and more diverse, as older devices mix with emerging devices that heighten the power of mobility. Leading the new category of *wearables,* most of which do not currently function as standalone tools but rather sync by Bluetooth to accompanying smart devices, are two device types: increasingly popular fitness trackers, often in the form of digital wristbands, which encourage users to make interventions in their lifestyles on the basis of tracked patterns of activity and sleep, visualised through mobile apps; and multipurpose smartwatches, which display notifications, messages and information, control other digital devices, and double as fitness trackers. Smart clothing, such as sensor-imbued shirts, socks or shoes, has to date been used largely to improve sporting performance, though wider applications may be in the pipeline, such as clothing that detects and responds to a user's mood (Bryner 2010).

Additionally in this category we find smart glasses. As distinct from virtual reality (VR) headsets, which provide an immersive experience of a fully simulated environment, smart glasses are *augmented reality* (*AR*) tools that superimpose digital information and communication channels on our existing view of the real world. The first publicly available product, Google Glass, carries a small screen mounted above the user's right eye which serves as an information display and a recording and communication interface synced to a smart device. Although, in the wake of limited commercial success, Google has halted sales to individuals and begun redevelopment work (Barr 2015), the device has already started to find a role in medical and other educational programmes (Johnson et al. 2015; Open Colleges, n.d. b). Microsoft's new HoloLens goggles, while bulkier, contain a small Windows 10 computer and appear to offer a more integrated AR approach where digital information will be literally overlaid on our view of our surroundings (Lee 2015). It is likely, however, that AR glasses and headsets are only the first stage on the road to smart contact lenses (Carmigniani and Furht 2011; Scoble and Israel 2014).

It is conceivable that multiple wearable devices could be connected into a 'body area network' or 'BAN', perhaps co-ordinated through a smartphone (Woodill 2015). A set of wearables that enable on-time, in-place access to digital information, communication channels and recording options, most likely controlled through an AR interface, has the potential to support many kinds of situated, embedded, embodied learning (Delgado 2014; Johnson et al. 2015), including just-in-time workplace learning (Baty 2014), and moreover to support those with special needs (ELI 2013c).

It has been suggested that 'the long-term survival for the keyboard, mouse and monitor suddenly seems precarious' (Lee 2015), due largely to the rise of *natural user interfaces,* which have proven to be intuitive for young children and accessible for many users with disabilities (Johnson et al. 2012; Kukulska-Hulme 2010). These may involve touch recognition, like the ubiquitous swiping and pinching motions used on today's touchscreens. They may involve gesture recognition, as seen in Microsoft's Kinect or the Leap Motion controller, 'enabling 3D input that involves users in the computing activity' (ELI 2014b), with likely implications for situated learning, notably in combination with AR (ibid.). Or they may involve voice recognition, as seen in virtual assistants like Apple's Siri, Google Now, or the open source Sirius (Hauswald et al. 2015), which represent early examples of a 'conversational user interface' (Kaplan 2013) that facilitates our interaction with digital data.

Much like AR glasses, such input mechanisms may be temporary stopping points on a longer journey towards more intimate human–machine interaction. Biometrics like fingerprint or iris scanning can already be used to unlock digital devices. Research is underway on brain–computer interfaces (Pegrum 2014; Scoble and Israel 2014), with applications ranging from neurogaming to supporting those with special needs, for instance through devices like the Indian 'brainphone' which allows users to navigate the web via brainwaves (Trivedi 2013). At the same time as input mechanisms are becoming more natural, so too are output mechanisms, as seen in the automated translations enabled by Google Translate or Microsoft's Skype (Orsini 2015), or indeed a recent Microsoft technology which can synthesise a speaker's voice translated into a foreign language he or she has never learned (Microsoft Research, n.d.). Work is also proceeding on new devices which will simulate smell and taste (Woodill and Udell 2015).

It is likely that 'the human body will be the next computer interface' (Goodman and Righetto 2013), a trend whose beginnings we can observe in today's gesture-based computing and in emerging hardware like smart contact lenses. Yet we can, and will, go much further in the 'embodiment of mobile devices' (Woodill 2015, Kindle location 2255). In time, what we might call *embeddables* or *implantables* will find their way both onto and under the skin, leading us into the territory of human–machine *cyborgs*. Early experiments include Motorola's passcode tattoo and an indigestible pill which, activated by stomach acids, effectively turns the body into a passcode transmitter (Gannes 2013), thus taking the idea of a BAN to a whole new level. Going beyond today's retinal and cochlear implants—arguably early cyborg technologies—humans could conceivably one day hear ultrasonic sounds or see ultraviolet light (Kaku 2014). There is speculation, too, about where brain implants might lead (Open Colleges, n.d., a); Google's Sergey Brin has suggested: 'Perhaps in the future, we can attach a little version of Google that you just plug into your brain' (cited in Carr 2008, p. 213). Meanwhile, research in *nanotechnology* is opening up the possibility of *nanobots* swimming through the bloodstream; futurists imagine that these tiny robots might eventually move beyond their envisaged medical and health applications and interact with neurons to extend human intelligence (Kurzweil 2013). Whatever the exact development trajectory of this technology as it extends into the far future, it seems that its immediate effect will be to further expand and blend the concepts of mobility and ubiquity of computing, thereby boosting the potential for situated, digitally enhanced learning already seen in today's wearables.

While the mobility of all the above devices depends on human wearers or carriers, we are also beginning to see the appearance of *independently or semi-independently mobile technologies*. These include *connected cars* with dashboards of apps projected from a synced smartphone using software like Apple's CarPlay or Google's Android Auto; but, like wearables, more and more cars will be manufactured with embedded connections which link them directly to the internet. Connected cars will in time give way to the fully autonomous *self-driving cars* being developed by technology companies like Google and Uber as well as car manufacturers like Ford. Here, the initial self-study possibilities inherent in an in-car app dashboard will explode into a plethora of informational and communicational resources available to 'drivers' who no longer have to focus on the road.

More direct educational applications are evident with camera- and sensor-carrying *drones,* which may be controlled by humans via mobile devices or may be programmed to fly autonomously. As they gradually move out of the shadow of their military origins, these helicopter-like machines are finding their way into areas like surveying and mapping; remote viewing of difficult-to-access terrain, flora and fauna; news-gathering; and even the filming of movies or creation of digital stories (ELI 2015).

Telepresence *robots*—which for now are controlled remotely through smart devices but may in the future allow gesture-based control, and which for now are grounded but in the future might fly like drones—permit distant teachers to show their faces on a screen on the robot's head, to view and move around a local space, and to engage in an embodied manner with participants in that space (ELI 2013b). Meanwhile, semi-independent robotic language teaching assistants have already found their way into Asian classrooms, where they can interact in simple ways with students (Han 2012; Pegrum 2014). Ongoing research into *affective technologies* (as in MIT's Personal Robots project at www.media.mit.edu/research/groups/personal-robots) and, more broadly, *empathic systems* (as in the EU's Empathic Products project at www.empathic.eu) is leading to robots and systems capable of evaluating human emotions and reacting appropriately, allowing 'a more organic relationship' (Isaías 2014) between humans and our (mobile) technologies.

As noted earlier, the mobile ecosystem is part of a larger hardware ecosystem. Indeed, the power of mobile technologies derives in part from their capacity to connect easily to more stationary devices, like desktop and laptop computers which offer superior computing power and enhanced input options; data projectors and smart TVs which facilitate sharing and collaboration via large displays; emerging 3D printers, which foster a 3D literacy linked to novel approaches to design thinking (Thompson 2013); and even 4D printers, which will produce objects that mutate over time (Marks 2013), fostering further sophisticated digital literacy skills. But even more than this, the power of mobiles derives from their connection to the network of devices, sensors and beacons that make up the evolving *internet of things*:

> The Internet of Things, or IoT, represents a major departure in the history of the Internet. The Internet is moving beyond the rectangular confines of smartphones and tablets and helping to power billions of everyday devices, from parking meters to home thermostats. (Rubin 2013, p. 2)

The use of low-power sensors and transmitters is permitting growing numbers of everyday devices to become 'smart', collecting and sharing data with each other and with us, often via smartphone apps that can be used for data visualisation as well as device management. Wearable, embeddable or implantable devices can hook into this same communicational network, as can independently or semi-independently mobile technologies like smart cars, drones or robots. Indeed, extending beyond the internet of things, we are arguably seeing the advent of an *internet of everything:*

> It is no longer far-fetched to envision a world where all people, objects and devices are connected to act in concert, regardless of brand or vendor. This idea is also known as The Internet of Everything (IoE), which is comprised of machine-to-machine (M2M), machine-to-person, and person-to-person networked technologies. (Johnson et al. 2015, p. 46)

By weaving our mobile and stationary devices together, the IoT (or IoE) amplifies the educational potential inherent in individual devices. Considerable implications follow from our devices' capacity to receive or retrieve detailed localised data from objects in our real-world environments, possibly displayed in AR interfaces. In this way they can support 'hypersituation', where learning is informed by 'a host of interdisciplinary information that is pushed to [us] from [our] surroundings' (Johnson et al. 2015, p. 47), or indeed pulled by us from those surroundings.

Similarly, considerable implications follow from our growing ability to generate, collect, access and analyse networked, up-to-the-minute *big data* (Mayer-Schönberger and Cukier 2013). Mobile apps already display our *quantified selves,* based on the tracking, evaluation and visualisation of our health, fitness and sleep patterns (Feinleib 2013; Mayer-Schönberger and Cukier 2013). These quantified selves will increasingly intersect with *learning analytics* (Johnson et al. 2014), where our learning patterns are analysed to predict our future successes and challenges (US Dept of Education 2012). As time goes on, our devices will provide us with even more tailored answers to our questions, along with even more tailored advice which we have not yet thought to request.

As our machines increasingly offer what appears to be intelligent advice, delivered in what appears to be the right emotional tone, we might feel we are witnessing the birth of *artificial intelligence* (*AI*). But this is at best weak AI, where machines simulate thinking in a narrowly defined area, often based largely on the rapid processing of big data, rather than strong AI, where machines actually think (Russell and Norvig 2010). This may change: in today's AI research there is a new interest in biological models and processes as well as the role of emotions and embodiment in shaping intelligence (Kaku 2014; Warwick 2012), and there is speculation on the eventual merging of human and technological intelligence in more fundamental ways than via implants (Kaku 2014; Kurzweil 2013; Warwick 2012). The consequences of such distant developments are difficult to anticipate from our current standpoint, though it is probable that along the way we will learn a great deal more about learning itself. In the meantime, it is important to keep a critical eye on these, and all, new technological developments, remembering that today's immediate concerns around privacy, surveillance and security (Mayer-Schönberger and Cukier 2013; Schneier 2015) are likely to be heightened by tomorrow's concerns around the control of AI and robots (Kaku 2014; Russell and Norvig 2010).

24.4 The Educational Context

In the foregoing account of our technological context, we have begun to sketch out the educational potential of new and emerging technological developments. But the realisation of that potential depends on whether, and how, these developments fit with today's major educational trends. We will turn now to these trends, seeking an example of the technological developments that might be enlisted to support each trend. Neither the list of educational trends, nor the list of supporting technologies, is exhaustive. Rather, they serve as pointers to the possible shape of future m-learning at the intersection of technology and pedagogy.

24.4.1 Contextualisation of Learning (E.G., Through Augmented Reality)

Recent years have seen a growing emphasis on contextualised learning, that is, learning which occurs in contexts as similar as possible to real-world contexts, and ideally in the actual real-world contexts themselves where the learning applies. Unlike desktop devices that 'operate in their own little world', mobile devices 'operate in *the* world' (Traxler 2010, p. 5, italics in original), meaning that m-learning reverses the move away from the real world inherent in much e-learning, and opens up the possibility of situated learning (Lave and Wenger 1991) and indeed embodied learning (Lipson Lawrence 2012) in everyday surroundings. Mobile devices can thus help break down the walls between the classroom and the world, thereby alleviating the problem of transfer distance, that is, the need to transfer learning across the gap between formal learning contexts and everyday contexts.

There is little doubt that today's smart devices permit ubiquitous learning or *u-learning* (Milrad et al. 2013), provided we treat them not as screens but as lenses:

> The mobile device as a *lens rather than a screen* is a critical design metaphor … it is critical that designers do not create experiences where the technology becomes a barrier to the environment. Rather the technology needs to drive the students deeper into the authentic observation and interaction with the environment and with each other… (Dunleavy 2014, p. 32)

Here, an appropriate AR interface could mediate between the digital and the real, immersing us in a ubiquitous, omnipresent *mixed reality* where digital material is directly overlaid on our view of the real world. In other words, AR could offer us a magnifying lens rather than an impermeable screen (even if, pending fuller development of AR wearables, the lens will be displayed for now on a mobile screen) to support and enhance situated, embodied learning.

Within the classroom, mixed reality technologies already offer students simulated environments with which they can physically interact (Lindgren and Johnson-Glenberg 2013). But the greatest promise may lie outside the classroom: on the AR Heritage Trails in Singapore, students learn about their city's history in surroundings whose significance is unlocked by digital overlays (Pegrum 2014); on the AR TIEs (Trails of Integrity and Ethics) in Hong Kong, students engage with ethical issues in the everyday university settings where they may arise (Chow et al. 2015); and in the MASELTOV project in Europe, a dashboard app provides students with context-aware language learning recommendations (Gaved et al. 2014).

Indeed, with the support of mobile, and ideally AR, technologies, students can turn any real-world context of their choice into a user-generated learning context (Cook 2010). While early educational forays into this area augur well for the future, we must bear in mind that challenges may arise for students in disentangling the real from its representations, and for teachers in deciding how best to guide, track and evaluate learning which ebbs and flows between the virtual and the real.

24.4.2 Personalisation of Learning (E.G., Through Big Data and Learning Analytics)

Much is heard nowadays about the personalisation of learning. This may be achieved in multiple, overlapping ways: for instance through BYOD approaches where students use individually customised hardware (Pegrum 2014), through flipped approaches which enhance student control over information delivery (Johnson et al. 2015), or through user-generated learning contexts which students create in their own everyday environments, as described above. Digital, including mobile, technologies also lend themselves to the differentiation of learning to suit varying student needs.

Such differentiation is now becoming possible at a large scale through the use of big data subjected to learning analytics. After collecting and analysing past performance data, an *adaptive learning system* can decide what and how to teach a student next (Feinleib 2013; Waters 2014), so that each student may see a different, tailored version of an online course (Johnson et al. 2014), with an early warning system serving to identify at-risk students (Aljohani and Davis 2012; de Freitas et al. 2014). Ideally, students should have access to an open learner dashboard, or 'personal dashboard' (Aljohani and Davis 2012), which displays an overview of their learning journey, their strengths, and their challenges. On the basis of group performance, course design as a whole may be adapted automatically, or adapted by teachers operating as informed *learning designers* (Feinleib 2013; Johnson et al. 2015; Mor et al. 2015). Because we carry our mobile devices with us at all times, they can gather far more data about us than our desktop or even laptop computers, especially in terms of our quantified selves. As briefly noted earlier, new insights are sure to emerge at the intersection of the quantified self and learning analytics, highlighting the link between lifestyles and learning outcomes (de Freitas et al. 2014; Johnson et al. 2014).

Explorations are underway in this space by private companies such as Knewton, for-profit universities like the University of Phoenix (with its patented Adaptive Activity Stream), public universities like the University of Michigan (with its GradeCraft learning management system) and, increasingly, private-public hybrids like Smart Sparrow (spun off from the University of New South Wales) and ventures like the ongoing collaboration between Arizona State University, Knewton and Pearson (Johnson et al. 2015; Waters 2014). There would seem to be potential for strengthening engagement by gamifying both the quantified self, notwithstanding certain risks (Whitson 2013), and adaptive learning systems, as seen in GradeCraft, which may open up future avenues of research. Important questions, however, remain about the accuracy and limitations of big data (boyd and Crawford 2012; note that danah boyd does not capitalise her name), the level of data literacy required of those who interpret it (Pegrum, in press a), and its privacy and surveillance implications (Mayer-Schönberger and Cukier 2013).

24.4.3 Diversification of Learning (E.G., Through MOOCs)

Diversification of learning correlates to some extent with personalisation of learning, in the sense that greater diversity opens up scope for greater personalisation. This can be seen for example in the growing diversification of hardware through BYOD, of courses through learning analytics, and of learning environments through user-generated contexts. The spread of MOOCs can also support the diversification of learning on a mass scale. Of course, MOOCs have been plagued with bad press: not unlike some first-generation e-learning innovations, many MOOCs simply wrap up old pedagogies in new technologies; most have startlingly low completion rates of between 5 and 16 % (Johnson et al. 2014); and nearly all appeal only to an educated elite rather than the once hoped-for broader demographic. On the other hand, there are MOOCs, sometimes called cMOOCs, which hold to the original connectivist ideal of autonomous, flexible, networked learning shaped by students' interests, in contrast with cognitivist–behaviourist xMOOCs whose instructor-determined videos, quizzes and discussion forums mirror more standard online courses (Hew 2014; Margaryan et al. 2015). At least within the boundaries of an upper secondary- and tertiary-educated demographic, MOOCs offer access to a wide range of content and teachers (especially, though not only, in the case of xMOOCs) and peers and learning communities (especially, though not only, in the case of cMOOCs). Once accreditation issues are settled, MOOCs may prove especially useful in diversifying flipped approaches (Hew 2014).

There is something of a trend towards MOOCs about mobile learning as well as a trend towards MOOCs intended to be accessible through mobile devices, with the latter sometimes called mMOOCs (de Waard 2013). These trends are seen for instance in the 2011 and 2012 mobiMOOC (mobimooc.wikispaces.com); the 2013 #IDML13 (Instructional Design for Mobile Learning; facultyecommons.org/instructional-design-for-mobile-learning-idml13-2/); and the 2015 #MOSOMELT (MObile SOcial MEdia Learning Technologies; mosomelt.wordpress.com). While only the last of these explicitly badges itself as a cMOOC, all have a connectivist orientation. mMOOCs which are simultaneously cMOOCs capitalise on the capacity of today's chief networking tool, mobile devices, to support networked learning, while mMOOCs in general facilitate contextualised learning in our everyday surroundings and personalised learning on our own devices. As such, mMOOCs are 'a step forward in realizing an m-learning format that fits the contemporary and future needs of education in this knowledge era' (de Waard 2013, p. 360).

In the European 3D World MOOC project, empathic technologies are being built into a virtual world-based MOOC with the aim of better immersing and supporting students in the learning process (Isaías 2014). While there is still much research to be done, we might ask whether there is future potential in the merging of empathic systems with mMOOCs that take the real world as their learning context. But before tackling such questions, we may need to address more basic concerns about the role of MOOCs in fostering a postmodern, even neoliberal, marketplace of learning, linked to the unbundling of today's educational institutions (Barber et al. 2013), and conceivably threatening the conceptual coherence of traditional qualifications.

24.4.4 Student Support (E.G., Through Virtual Assistants)

In an era of mass education, techniques for contextualising, personalising and diversifying learning at scale are of great importance. These processes can be reinforced by providing individualised support to students—not necessarily human support, which is difficult to scale, but the automated support of virtual assistants. Such virtual assistants depend on a convergence of three technological trends: conversational user interfaces, personal context awareness, and service delegation (Johnson et al. 2014). Conversational user interfaces, part of the trend towards natural user interfaces, allow us to interact with virtual assistants much as we would with human assistants. Personal context awareness involves understanding patterns in each individual's manner of speaking, but also ties into a broader analysis of big data: 'The most advanced versions of this software actually track user preferences and patterns so they can adapt over time to be more helpful to the individual' (ibid., p. 46). Service delegation entails virtual assistants accessing and managing our collections of apps.

In time, we may see virtual assistants that can tailor AR displays to our needs at a given moment, and guide us through adaptive learning experiences based on patterns they have extracted at the point where our quantified selves intersect with our learning analytics trails. Meanwhile, empathic technologies and AI may make our assistants seem ever more lifelike and ever more intelligent. Yet as Scoble and Israel (2014) remind us in a broader discussion of big data: 'The marvels of the contextual age are based on a tradeoff: the more the technology knows about you, the more benefits you will receive' (Kindle location 225). Convenience and privacy are thus negatively correlated and, in observing the evolution of our apparently obsequious helpers, we should bear in mind the dangers of surveillance, the limitations of big data, and even the risks of AI which may slip out of our control.

24.4.5 Student Engagement (E.G., Through Gaming and Gamification)

Just as educational institutions are seeking to bolster student support, so too are they seeking to promote student engagement. In this respect, digital gaming would seem to hold some promise. First, complex multiplayer games (as opposed to behaviourist drills or quizzes dressed up as games) are 'pedagogically rich' (Klopfer 2008, Kindle location 22) spaces that involve critical thinking and creativity, negotiation of meaning and collaborative problem-solving (Gee 2007; Pegrum 2014); and they provide a clear learning structure incorporating reasons, goals, assessment criteria and potential rewards for all activities undertaken (Hoyle 2012). Second, games are highly motivating (ELI 2014a; Klopfer 2008) as players engage in intensive learning to accomplish in-game tasks. While lacking the full impact of gaming, *gamification* essentially the process of using gaming elements like badges, levels and leaderboards to shape or reshape educational activities—can also offer motivational benefits (Johnson et al. 2014), as seen for example in the aforementioned GradeCraft (ELI 2014a).

We are now seeing a major shift towards mobile gaming. This can just mean app-based games played on small screens rather than big screens. But of much more interest is the use of networked, context-sensitive devices to engage in AR games played in real-world settings annotated with digital information (Klopfer 2008; Pegrum 2014). Early explorations are promising: in the US 'Alien Contact!' game, teams of students moved around their school grounds using GPS-enabled mobile devices to seek clues and develop an explanation as to why aliens had landed on earth (Dunleavy et al. 2009; Potter 2011); in 'Mentira', a US game for learners of Spanish, teams equipped with iPhones or iPod Touches seek clues to a historical murder mystery, which includes navigating a local Spanish-speaking neighbourhood (Holden and Sykes 2011; Pegrum 2014); and the MASELTOV app, discussed above, incorporates a competitive gaming element based on learners' real-world target language conversations (ibid.). As AR interfaces continue to improve in conjunction with wearables and embeddables, more and more situated learning options will open up, with Google's Ingress game (www.ingress.com) hinting at future possibilities. But educators will need to ensure that gamified learning is designed in line with sound pedagogical principles so that gaming does not overtake or trivialise learning.

24.4.6 Student Creativity (E.G., Through Makerspaces)

Much attention is currently focused on students' development of creativity and related 21st century skills (Khan 2012; Robinson 2011), in line with perceived national and international needs for innovative political leadership and entrepreneurial economic leadership. This notion is gaining considerable traction in Asia, where creativity is increasingly viewed as an essential complement to education systems which have traditionally produced strong performances on standardised international tests (Barber et al. 2012; Zhao 2012). One means of fostering creativity involves linking BYOD approaches with new learning spaces whose layouts and furnishings are crafted to support different kinds of learning, underpinned by different kinds of interactions, supported by students' own chosen mobile devices, and situated within an inviting, inspiring décor.

Among the new learning spaces emerging on campuses and in libraries are *makerspaces* (ELI 2013a; Johnson et al. 2014) equipped with a range of craft tools and digital tools–the latter sometimes including devices like 3D printers–which students can use as they actively engage in 'creative, higher-order problem solving through hands-on design, construction and iteration' (Johnson et al. 2015, p. 40). Makerspaces may also involve fully digital creation, as in the production of digital stories (ibid.) or the building of apps; indeed, as the Mozilla Webmaker slogan reminds us with reference to coding, 'making is learning' (Santo 2013). In makerspaces, students work both autonomously and collaboratively, cross disciplinary boundaries, and expand institutional limits as they focus on innovative approaches to real-world issues, and become 'creators rather than consumers' (Johnson et al. 2014, p. 7) of knowledge and design. Of course, perennial questions remain about how creativity is best guided and supported, and how it can be captured and assessed.

24.4.7 Wider Collaboration (E.G., Through Digital Networking)

There are great advantages in students learning about real-world issues alongside the real-world communities which are dealing with those issues. Participatory pedagogy, often underpinned by a problem-based or inquiry-based design, promotes engaging, hands-on, contextualised learning through which students come to understand that their learning can, and perhaps should, have an impact on the wider world (Pegrum, in press a):

> At the heart of the idea is to allow students to participate in knowledge-creating activities around shared objects and to share their efforts with the wider community for further knowledge building that is a legitimate part of civilization (Scardamalia and Bereiter 2006). (Vartiainen 2014, p. 109)

Such participation is very much facilitated by the shift towards a network society (Castells 2013; Rainie and Wellman 2012), where human networking is supported by digital networking thanks to pervasive social media accessed on mobile devices. Research shows a correlation between online networking and offline social and civic engagement (Castells 2013; Thompson 2013). The most dramatic results have been seen in protest movements like those surrounding the Arab Spring (Castells 2012), though naturally the online/offline nexus is likely to be promoted in more low-key ways in educational contexts. Within an appropriate educational design, digital networking with experts, peers and wider communities via mobile media allows students to situate their learning within real communities and contexts, develop their understanding with input from those communities, and share their emerging insights in the contexts where they might be of benefit. Again, as with creativity, and as with contextualised learning more broadly, questions remain about how best to capture and assess such learning.

24.5 Conclusion

Mobile hardware and software will continue to proliferate and mutate for the foreseeable future, but it is likely that much of it will sit within a cluster of ongoing technological trends, namely towards greater mobility of devices, greater ubiquity of computing, and more seamless integration of our devices with other devices and our wider environments. It may also link to subtrends towards smaller sizes (for wearable and embeddable technologies) and greater independence (for technologies like drones and robots). The relevance of these developments for education, however, will depend less on their fit with technological trends and more on their fit with ongoing educational trends, notably towards contextualisation, personalisation, and diversification of learning; towards student support, engagement, and creativity; and towards wider collaboration.

These educational trends are in turn influenced in complex ways by partially overlapping and partially conflicting cultural, social, political and economic trends. With the shift towards a network society comes an upswing in digital networking; with the shift towards a knowledge economy comes the promotion of transferrable 21st century skills; and with the shift towards neoliberalism comes an emphasis on marketplace diversification, not to mention personal entrepreneurship. At the same time the global emphasis on quantification of performance, and the standardised assessment of education, may fit neatly with big data and learning analytics, but it flies in the face of personalisation, diversification and notoriously difficult-to-assess 21st century skills.

As we noted at the outset, this chapter provides glimpses of a possible future for digital learning in general and mobile learning in particular. Exactly how the future plays out will depend on technological developments and, moreover, on how these fit with evolving educational—and cultural, social, political and economic—trends. In attempting to anticipate the future of m-learning, we need to keep an eye on both technological and educational trends, and ask ourselves constantly how the two are likely to intersect.

Acknowledgments I would like to thank Pedro Isaías and Julian Stodd for their help with specific technological queries that arose during the writing of this chapter.

References

Aljohani, N. R., & Davis, H. C. (2012). Learning analytics in mobile and ubiquitous learning environments. In M. Specht, M. Sharples, & J. Multisilta (Eds.), *mLearn 2012: Proceedings of the 11th International Conference on Mobile and Contextual Learning 2012*, Helsinki, Finland, October 16–18 (pp. 70–77). Retrieved from http://ceur-ws.org/Vol-955/papers/paper_70.pdf.

Barber, M., Donnelly, K., & Rizvi, S. (2012, August). *Oceans of innovation: The Atlantic, the Pacific, global leadership and the future of education.* London: Institute for Public Policy Research. Retrieved from http://www.ippr.org/images/media/files/publication/2012/09/oceans-of-innovation_Aug2012_9543.pdf.

Barber, M., Donnelly, K., & Rizvi, S. (2013, March). *An avalanche is coming: Higher education and the revolution ahead.* London: Institute for Public Policy Research. Retrieved from http://www.ippr.org/images/media/files/publication/2013/04/avalanche-is-coming_Mar2013_10432.pdf.

Barr, A. (2015, January 16). Google Glass gets a new direction. *The Wall Street Journal* (p. B1).

Baty, J. (2014, November 23). Can I really train my employees with an Apple Watch? *Medium.* Retrieved from https://medium.com/@jodybaty/is-there-a-place-for-the-apple-watch-in-employee-training-ed6e260ecd59.

Blodget, H., Ballvé, M., Danova, T., Smith, C., Heggestuen, J., et al. (2014, December 2). *The future of digital: 2014.* Retrieved from https://intelligence.businessinsider.com/the-future-of-digital-2014-edition-2014-12.

Boyd, D, & Crawford, K. (2012). Critical questions for big data: Provocations for a cultural, technological, and scholarly phenomenon. *Information, Communication & Society, 15*(5), 662–679.

Bryner, M. (2010, June 8). 'Smart' clothing responds to wearer's emotions. *NBC News.* Retrieved from http://www.nbcnews.com/id/37579367/ns/technology_and_science-innovation/t/smart-clothing-responds-wearers-emotions/.

Carmigniani, J., & Furht, B. (2011). Augmented reality: An overview. In B. Furht (Ed.), *Handbook of augmented reality* (pp. 3–46). New York: Springer.

Carr, N. (2008). *The big switch: Rewiring the world, from Edison to Google*. New York: W.W. Norton.

Castells, M. (2012). Networks of outrage and hope: Social movements in the internet age. Cambridge: Polity.

Castells, M. (2013). *Communication power* (2nd ed.). Oxford: Oxford University Press.

Chow, E. C. H., Pegrum, M., Thadani, D. R., & Wong, E. Y. W. (2015). *Learning about academic integrity and ethics using mobile technologies and augmented reality.* Paper Presented at eLearning Forum Asia, SIM University, Singapore, June 17–19

Cook, J. (2010). Mobile learner generated contexts: Research on the internalization of the world of cultural products. In B. Bachmair (Ed.), *Medienbildung in neuen Kulturräumen: Die deutschsprachige und britische Diskussion* (pp. 113–125). Wiesbaden: VS Verlag für Sozialwissenschaften.

de Freitas, S., Gibson, D., Du Plessis, C., Halloran, P., Williams, E., et al. (2014). Foundations of dynamic learning analytics: Using university student data to increase retention. *British Journal of Educational Technology* (Early view).

Delgado, R. (2014, April 20). Imagining the classroom of 2016, empowered by wearable technology. *EmergingEdTech.* Retrieved from http://www.emergingedtech.com/2014/04/imaging-the-classroom-of-2016-empowered-by-wearable-technology/.

de Waard, I. (2013). mMOOC design: Ubiquitous, open learning in the cloud. In Z. L. Berge & L. Y. Muilenburg (Eds.), *Handbook of mobile learning* (pp. 356–368). New York: Routledge.

Dunleavy, M. (2014). Design principles for augmented reality learning. *TechTrends, 58*(1), 28–34.

Dunleavy, M., Dede, C., & Mitchell, R. (2009). Affordances and limitations of immersive participatory augmented reality simulations for teaching and learning. *Journal of Science Education and Technology, 18*(1), 7–22.

ELI (EDUCAUSE Learning Initiative). (2013a). *7 things you should know about ... makerspaces*. EDUCAUSE. Retrieved from http://net.educause.edu/ir/library/pdf/eli7095.pdf.

ELI. (2013b). *7 things you should know about ... telepresence robots*. EDUCAUSE. Retrieved from https://net.educause.edu/ir/library/pdf/ELI7103.pdf.

ELI. (2013c). *7 things you should know about ... wearable technology*. EDUCAUSE. Retrieved from https://net.educause.edu/ir/library/pdf/ELI7102.pdf.

ELI. (2014a). *7 things you should know about ... games and learning*. EDUCAUSE. Retrieved from https://net.educause.edu/ir/library/pdf/ELI7106.pdf.

ELI. (2014b). *7 things you should know about ... gesture-based computing*. EDUCAUSE. Retrieved from https://net.educause.edu/ir/library/pdf/ELI7104.pdf.

ELI. (2015). *7 things you should know about ... drones*. EDUCAUSE. Retrieved from http://net.educause.edu/ir/library/pdf/ELI7116.pdf.

Feinleib, D. (2013). *Big data demystified: How big data is changing the way we live, love and learn*. San Francisco: The Big Data Group.

Gannes, L. (2013, June 3). Passwords on your skin and in your stomach: Inside Google's wild Motorola research projects (video). *All Things D*. Retrieved from http://allthingsd.com/20130603/passwords-on-your-skin-and-in-your-stomach-inside-googles-wild-motorola-research-projects-video/.

Gardner, H., & Davies, K. (2013). *The app generation: How today's youth navigate identity, intimacy, and imagination in a digital world*. New Haven, CT: Yale University Press.

Gaved, M., Luley, P., Efremidis, S., Georgiou, I., Kukulska-Hulme, A., et al. (2014). Challenges in context-aware mobile language learning: The MASELTOV approach. In M. Kalz, Y. Bayyurt, & M. Specht (Eds.), *Mobile as mainstream—towards future challenges in mobile learning. 13th World Conference on Mobile and Contextual Learning, mLearn 2014, Istanbul, Turkey, November 3–5, 2014, Proceedings* (pp. 351–364). Cham, Switzerland: Springer. Retrieved from http://www.academia.edu/9001887/.

Gee, J. P. (2007). *What video games have to teach us about learning and literacy* (new ed.). New York: Palgrave Macmillan.

Goodman, A., & Righetto, M. (2013, March 5). Why the human body will be the next computer interface. *Fast Company*. Retrieved from http://www.fastcodesign.com/1671960/why-the-human-body-will-be-the-next-computer-interface.

GSMA (Global System for Mobile Communications Association). (2010). *mLearning: A platform for educational opportunities at the base of the pyramid*. London: GSMA. Retrieved from http://www.gsma.com/mobilefordevelopment/wp-content/uploads/2012/04/mlearningaplatformforeducationalopportunitiesatthebaseofthepyramid.pdf.

Han, J. (2012). Robot assisted language learning. *Language Learning & Technology, 16*(3), 1–9. Retrieved from http://llt.msu.edu/issues/october2012/emerging.pdf.

Hauswald, J., Laurenzano, M. A., Zhang, Y., Li, C., Rovinksi, A., et al. (2015). *Sirius: An open end-to-end voice and vision personal assistant and its implications for future warehouse scale computers*. Presented at ASPLOS (Architectural Support for Programming Languages and Operating Systems) '15, Istanbul, Turkey, March 14–18. Retrieved from http://jasonmars.org/wp-content/papercite-data/pdf/hauswald15asplos.pdf.

Hew, K. F. (2014). Promoting engagement in online courses: What strategies can we learn from three highly rated MOOCs. *British Journal of Educational Technology* (Early view).

Holden, C., & Sykes, J. (2011). Mentira: Prototyping language-based locative gameplay. In S. Dikkers, J. Martin, & B. Coulter (Eds.), *Mobile media learning: Amazing uses of mobile devices for learning*. Pittsburgh, PA: ETC Press.

Hoyle, M. A. (2012, July 10). Learning in world of warcraft. *Classroom Aid*. Retrieved from http://classroom-aid.com/2012/10/07/learning-in-world-of-warcraft/.

Isaías, P. (2014). *Empathy, empathic information systems and new directions for learning.* Presented at ICEduTech (International Conference on Educational Technologies), Tamkang University, Taipei, Taiwan, December 10–12.

Johnson, L., Adams, S., & Cummins, M. (2012). *NMC Horizon Report: 2012 K-12 Edition.* Austin, TX: New Media Consortium. Retrieved from http://www.nmc.org/pdf/2012-horizon-report-K12.pdf.

Johnson, L., Adams Becker, S., Estrada, V., & Freeman, A. (2014). *NMC horizon report: 2014 higher education edition.* Austin, TX: The New Media Consortium. Retrieved from http://www.nmc.org/pdf/2014-nmc-horizon-report-he-EN.pdf.

Johnson, L., Adams Becker, S., Estrada, V., & Freeman, A. (2015). *NMC horizon report: 2015 higher education edition.* Austin, TX: The New Media Consortium. Retrieved from http://cdn.nmc.org/media/2015-nmc-horizon-report-HE-EN.pdf.

Kaku, M. (2014). *The future of the mind: The scientific quest to understand, enhance and empower the mind.* London: Penguin.

Kaplan, R. (2013, March 21). Beyond the GUI: It's time for a conversational user interface. *Wired.* Retrieved from http://www.wired.com/2013/03/conversational-user-interface/.

Khan, S. (2012). *The one world schoolhouse: Education reimagined.* London: Hodder & Stoughton.

Klopfer, E. (2008). *Augmented learning: Research and design of mobile educational games.* Cambridge, MA: MIT Press.

Kukulska-Hulme, A. (2010). *Mobile learning for quality education and social inclusion.* Policy Brief. Moscow: UNESCO Institute for Information Technologies in Education. Retrieved from http://iite.unesco.org/pics/publications/en/files/3214679.pdf.

Kurzweil, R. (2013). *How to create a mind: The secret of human thought revealed.* London: Duckworth Overlook.

Lave, J., & Wenger, E. (1991). *Situated learning: Legitimate peripheral participation.* Cambridge: Cambridge University Press.

Lee, A. (2015, January 22). The end may be nigh for keyboard, mouse and monitor, thanks to Microsoft. *ReadWrite.* Retrieved from http://readwrite.com/2015/01/21/hololens-windows-10-holographic-cortana-microsoft.

Lindgren, R., & Johnson-Glenberg, M. (2013). Emboldened by embodiment: Six precepts for research on embodied learning and mixed reality. *Educational Researcher, 42*(8), 445–452.

Lipson Lawrence, R. (Ed.). (2012). *Bodies of knowledge: Embodied learning in adult education. Special issue of New Directions for Adult and Continuing Education, 134.* San Francisco: Jossey-Bass.

Margaryan, A., Bianco, M., & Littlejohn, A. (2015). Instructional quality of Massive Open Online Courses (MOOCs). *Computers & Education, 80*, 77–83.

Marks, O. (2013, March 14). 4D printing: The new frontier. *ZDNet.* Retrieved from http://www.zdnet.com/article/4d-printing-the-new-frontier/.

Mayer-Schönberger, V., & Cukier, K. (2013). *Big data: A revolution that will transform how we live, work and think.* London: John Murray.

Microsoft Research. (n.d.). How technology can bridge language gaps. *Microsoft Research.* Retrieved from http://research.microsoft.com/en-us/research/stories/speech-to-speech.aspx.

Milrad, M., Wong, L.-H., Sharples, M., Hwang, G.-J., Looi, C.-K., & Ogata, H. (2013). Seamless learning: An international perspective on next-generation technology-enhanced learning. In Z. L. Berge & L. Y. Muilenburg (Eds.), *Handbook of mobile learning* (pp. 95–108). New York: Routledge.

Mor, Y., Ferguson, R., & Wasson, B. (2015). Editorial: Learning design, teacher inquiry into student learning and learning analytics: A call for action. *British Journal of Educational Technology, 46*(2), 221–229.

Oakley, G., Pegrum, M., Faulkner, R., & Striepe, M. (2012). *Exploring the pedagogical applications of mobile technologies for teaching literacy.* Report for the Association of Independent Schools of Western Australia. Retrieved from http://www.education.uwa.edu.au/research/?a=2195652.

Open Colleges. (n.d., a). *The evolution of learning technologies*. Retrieved from http://www.opencolleges.edu.au/informed/evolution-of-learning-technologies.

Open Colleges. (n.d., b). *How Google Glass might be used in education*. Retrieved from http://www.opencolleges.edu.au/informed/features/google-glasses-education.

Orsini, L. (2015, January 12). Microsoft, Google aim to break the language barrier. *ReadWrite*. Retrieved from http://readwrite.com/2015/01/12/google-microsoft-language-translation.

Pegrum, M. (2014). *Mobile learning: Languages, literacies and cultures*. Basingstoke: Palgrave Macmillan.

Pegrum, M. (in press a). Languages and literacies for digital lives. In E. Martín-Monje, I. Elorza, & B. García Riaza (Eds.), *Technological advances in specialized linguistic domains: Learning on the move*. London: Routledge.

Pegrum, M. (in press b). Three agendas for MALL. In A. Palalas, & M. Ally (Eds.), *The international handbook of mobile-assisted language learning*.

Potter, G. (2011). Augmented reality and mobile technologies. In A. Kitchenham (Ed.), *Models for interdisciplinary mobile learning: Delivering information to students* (pp. 212–230). Hershey, PA: Information Science Reference.

Puentedura, R. R. (2012). *Building upon SAMR*. Presented at Presbyterian Ladies' College, Perth, Australia, September 14

Rainie, L., & Wellman, B. (2012). *Networked: The new social operating system*. Cambridge, MA: MIT Press.

Robinson, K. (2011). *Out of our minds: Learning to be creative* (revised ed.). Chichester: Capstone.

Rubin, R. (2013, October 23). Here comes the Internet of things. *BI Intelligence*. Retrieved from https://intelligence.businessinsider.com/the-internet-of-things-2013-10.

Russell, S. J., & Norvig, P. (2010). *Artificial intelligence: A modern approach* (3rd ed.). Upper Saddle River, NJ: Pearson.

Santo, R. (2013, February 12). Is making learning? Considerations as education embraces the Maker Movement. *Empathetics: Integral Life*. Retrieved from http://empathetics.org/2013/02/12/is-making-learning-considerations-as-education-embraces-the-maker-movement/.

Scardamalia, M., & Bereiter, C. (2006). Knowledge building: Theory, pedagogy, and technology. In K. Sawyer (Ed.), *Cambridge handbook of the learning sciences* (pp. 97–118). New York: Cambridge University Press.

Schneier, B. (2015). *Data and Goliath: The hidden battles to collect your data and control your world*. New York: W.W. Norton.

Scoble, R., & Israel, S. (2014). *Age of context: Mobile, sensors, data and the future of privacy*. Lexington: Patrick Brewster Press.

Selwyn, N. (2014). *Digital technology and the contemporary university: Degrees of digitization*. Society for Research into Higher Education Series. London: Routledge

Thompson, C. (2013). *Smarter than you think: How technology is changing our minds for the better*. London: William Collins.

Traxler, J. (2010). Will student devices deliver innovation, inclusion, and transformation? *Journal of the Research Center for Educational Technology, 6*(1), 3–15.

Trivedi, K. R. (2013). *Brainphone for m-learning—brainwave enabled multifunctional, communication, controlling and speech signal generating device*. Presented at UNESCO Mobile Learning Week Symposium, Paris, France, February 18–19. Retrieved from http://www.unesco.org/new/en/unesco/themes/icts/m4ed/unesco-mobile-learning-week/speakers/kiran-trivedi/.

UNESCO (United Nations Educational, Scientific and Cultural Organization). (2013). *UNESCO policy guidelines for mobile learning*. Paris: UNESCO. Retrieved from http://unesdoc.unesco.org/images/0021/002196/219641E.pdf.

US Dept of Education. (2012). *Enhancing teaching and learning through educational data mining and learning analytics: An issue brief*. Washington, DC: US Dept of Education. Retrieved from http://tech.ed.gov/wp-content/uploads/2014/03/edm-la-brief.pdf.

Vartiainen, H. (2014). Designing participatory learning. In P. Kommers, T. Issa, T. Issa, D.-F. Chang, & P. Isaías (Eds.), *Proceedings of the International Conferences on Educational Technologies (ICEduTech 2014) and Sustainability, Technology and Education 2014 (STE 2014), New Taipei City, Taiwan, December 10–12, 2014* (pp. 105–112). IADIS Press.

Warschauer, M. (2011). *Learning in the cloud: How (and why) to transform schools with digital media*. New York: Teachers College Press.

Warwick, K. (2012). *Artificial intelligence: The basics*. London: Routledge.

Waters, J. K. (2014, April 16). The great adaptive learning experiment. *Campus Technology*. Retrieved from http://campustechnology.com/Articles/2014/04/16/The-Great-Adaptive-Learning-Experiment.aspx.

Whitson, J. R. (2013). Gaming the quantified self. *Surveillance & Society, 11*(1/2). Retrieved from http://library.queensu.ca/ojs/index.php/surveillance-and-society/article/view/gaming.

Woodill, G. (2015). Unique affordances of mobile learning. In C. Udell & G. Woodill (Eds.), *Mastering mobile learning: Tips and techniques for success*. Hoboken, NJ: Wiley.

Woodill, G., & Udell, C. (2015). Enterprise mobile learning: A primer. In C. Udell & G. Woodill (Eds.), *Mastering mobile learning: Tips and techniques for success*. Hoboken, NJ: Wiley.

Zhao, Y. (2012). *World class learners: Educating creative and entrepreneurial students*. Thousand Oaks, CA: Corwin.

MIX
Papier aus verantwortungsvollen Quellen
Paper from responsible sources
FSC® C105338

If you have any concerns about our products,
you can contact us on
ProductSafety@springernature.com

In case Publisher is established outside the EU,
the EU authorized representative is:
Springer Nature Customer Service Center GmbH
Europaplatz 3, 69115 Heidelberg, Germany

Printed by Libri Plureos GmbH
in Hamburg, Germany